Advances in Global Change Research

Volume 68

T0075450

Series Editors

Markus Stoffel, Institute of Geological Sciences, University of Geneva,
Geneva, Switzerland

Wolfgang Cramer, IMEP, Bâtiment Villemin, Europole de l'Arbois,
Aix-en-Provence, France

Urs Luterbacher, University of Geneva, Geneva, Switzerland

F. Toth, International Institute for Applied Systems Analysis (IIASA),
Laxanburg, Austria

This book series has been accepted for inclusion in SCOPUS.
Advances in Global Change Research
Aims and Scopes
This book series is aimed at addressing a range of environmental issues through state-of-the-art and/or interdisciplinary approaches. The books in the series can either be monographs or edited volumes based, for example, on the outcomes of conferences and workshops, or by invitation of experts. The topics that the series can consider publishing include, but are not limited to:

Physical and biological elements of earth system science, in particular

- Climate change
- Biodiversity
- Sea-level rise
- Paleo-climates and paleo-environments

Social aspects of global change

- Environmentally-triggered migrations
- Environmental change and health
- Food security and water availability
- Access to essential resources in a changing world

Economic and policy aspects of global change

- Economic impacts
- Cost-benefit analyses
- Environmental governance
- Energy transition

Methodologies for addressing environmental issues

- Planetary data analysis
- Earth observations from space
- Proxy data analyses
- Numerical modeling
- Statistical analyses

Solutions to global environmental problems

- Sustainability
- Ecosystem services
- Climate services
- Technological/engineering solutions

Books in the series should be at least 200 pages in length, and include a table of contents.
Images supplied in colour will be reproduced in both the print and electronic versions of the book at no cost to the author/editor.
Manuscripts should be provided as a Word document and will be professionally typeset at no cost to the authors/editors.
All contributions to an edited volume should undergo standard peer review to ensure high scientific quality, while monographs should also be reviewed by at least two experts in the field.
Manuscripts that have undergone successful review should then be prepared according to the Publisher's guidelines manuscripts: https://www.springer.com/gp/authors-editors/book-authors-editors/book-manuscript-guidelines

More information about this series at http://www.springer.com/series/5588

Anirudh Singh
Editor

Translating the Paris Agreement into Action in the Pacific

 Springer

Editor
Anirudh Singh
School of Science and Technology
The University of Fiji
Lautoka, Fiji

ISSN 1574-0919   ISSN 2215-1621   (electronic)
Advances in Global Change Research
ISBN 978-3-030-30213-9   ISBN 978-3-030-30211-5   (eBook)
https://doi.org/10.1007/978-3-030-30211-5

This Springer imprint is published by the registered company Springer Nature Switzerland AG.
The registered company address is: Gewerbestrasse 11, 6330 Cham, Switzerland

Associate Editor
Ravinesh C. Deo
Associate Professor
School of Agricultural, Computational and Environmental Sciences
University of Southern Queensland, Australia

To the people of the Pacific
– we can do it.

Foreword

Global warming and its consequences on climate change are generally considered to be one of the biggest threats to humanity and biodiversity and to the very planet we live on. Climate change impacts have created some of the greatest global challenges we are facing today, including rise in sea level, ocean acidification, extreme weather patterns, food security and the list goes on. We have reached the stage where the Pacific, including other Small Island Developing States (SIDS), is at the forefront of these disastrous impacts. This is akin to a full climate emergency and one that has forced us to work collaboratively in finding innovative and sustainable solutions that work best for the Pacific.

This fascinating book produced by the dedicated team of the University of Fiji, titled *Translating the Paris Agreement into Action in the Pacific*, focuses primarily on how Fiji is attempting to address the global issue of climate change mitigation through the various possible renewable energy projects on the ground that can support its Nationally Determined Contribution (NDC) Implementation Roadmap for 2017–2030. This Roadmap sets Fiji's pathway for its emissions reduction target under the Paris Agreement.

The book is divided into three separate sections. The first section (Introductory Concepts and Techniques) presents introductory concepts and techniques used in mitigation actions.

Despite the Pacific's negligible emission contributions, the Pacific SIDS have joined the rest of the world in the endeavour to reduce global GHG emissions. All the Pacific countries have rather ambitious Nationally Determined Contributions (NDCs), with Fiji and the Marshall Islands also developing Low Emission Development Strategies (LEDS) as recommended by the Paris Agreement. The central goal of Fiji's LEDS is to achieve net zero carbon emissions by 2050 across all sectors of its economy. These sectors include electricity and other energy use, land transport, domestic maritime transport, domestic air transport as well as agriculture, forestry and other land use (AFOLU) and waste.

To realise this goal, the Fiji LEDS specifies four possible low emissions approaches which include (1) business-as-usual (BAU) unconditional scenario, (2) BAU conditional scenario, (3) high ambition scenario and (4) very high ambition scenario. Some of the key actions for decarbonising Fiji's sector on 'electricity and other energy use' are economy-wide energy efficiency measures, capacity building and education, capacity building for renewable energy and smart grids and new solar, hydro, biomass, wind, waste-to-energy, biogas, geothermal and energy storage installations.

Part II (Mitigation Actions) of this publication investigates the potential contributions of several renewable energy projects towards greenhouse gas (GHG) reductions in Fiji. While this list is not exhaustive by any means, the discussion it initiates is important, for we do need to translate the aspirations, expressed in agreements, strategies and plans, into workable solutions. We need to continue this research, and we seek the best and most suitable solutions for our people. This is especially with regard to their energy needs, for we must keep in our minds that hundreds of communities in the Pacific have never enjoyed the benefits of electricity through conventional means in their lives.

Part III (Outcomes) of this publication summarises the main outcomes and appraises the significance of the mitigation action in the light of the objectives of Fiji's NDC Implementation Roadmap. This is a discussion that needs to continue as we continue to learn about their impact within our communities.

This begs a very important question of how these studies or renewable energy projects in Fiji can be emulated in other Pacific island countries (PICs). Two important concepts need to be taken into consideration. These are replicability and scalability. Project replicability focuses on the possibility of establishing similar project in another location or PIC. On the other hand, scalability relates to the potential to increase the size of project volume or installation in relation to location and geodemography amongst other factors. This is achievable through more collaborative actions and information-sharing between government, private sector and civil society in the PICs with academic and research institutions such as the University of Fiji as partners.

The Pacific Islands Development Forum and the University of Fiji, through the Centre for Renewable Energy (CORE), will continue the research into new and sustainable energy solutions that work for the Pacific people. The CORE will also continue providing valuable training and capacity development to make optimum use of these technologies for the benefit of the Pacific island countries and communities.

Secretary General   François Martel
Pacific Islands Development Forum
Suva, Fiji

Introduction

The diverse impacts of climate change, projected or already recorded, represent a key challenge for people across the globe, and particularly for the Pacific region. The Pacific island countries (PICs) are highly vulnerable to the impacts of climate variability and extreme weather events because of the geographic and socioeconomic factors specific to these countries. These include small populations, high dependence on natural resources and low elevation. The array of small low-lying Pacific islands and atolls such as Tuvalu, Kiribati and Marshall Islands is highly threatened by sea level rise and coastal erosion, making some parts of these islands uninhabitable.

All PICs, large or small, are confronted by extreme weather events such as cyclones and environmental impacts such as coral bleaching which have serious effects on their livelihoods and ecosystem services. Tropical Cyclone Pam (2015) and Cyclone Winston (2016), for instance, have led to major losses in Vanuatu and Fiji, respectively.

In addition to natural hazards, climate change imposes long-term risks, particularly coastal degradation, health impacts and agricultural losses. The projected reduction in average rainfall in Fiji is also a matter of concern, since 55–65% of the country's electricity supply is produced through hydropower. Natural disasters have a huge impact on the economic progress and environments of the PICs, posing a challenge to their ability to meet the objectives of their sustainable development goals.

In 1988, the UN General Assembly recognised the need to manage anthropogenic activities that affected global climate patterns. Subsequent international negotiations through the UNFCCC led to several countries ratifying UNFCCC's mission to limit the concentrations of GHG in the atmosphere to reasonable levels. Fiji duly signed the convention at its launch in 1992 and ratified it in 1993. Its Initial National Communication (INC) provided an overview of Fiji's national status, mostly relating to climate change issues. It presented a GHG inventory, a vulnerability assessment as well as climate change mitigation and adaptation strategies.

The Second National Communication (SNC) was submitted to the UNFCCC in 2013 as a follow-up to the INC. The SNC presented comprehensive information on national circumstances, a GHG inventory and reports on various climate change initiatives in Fiji. The highlights included the National Climate Change Policy which was developed in 2012. It also included Fiji's focus on mitigation options to reduce GHG emissions by reducing the use of fossil fuels in power generation and transportation through renewable energy (RE) and reduction in energy consumption via energy efficiency (EE) practices.

The SNC prepared the way for Fiji to submit its Intended Nationally Determined Contribution (INDC) to the UNFCCC in 2015 prior to COP21 in Paris. This focused on the energy sector and also took into consideration Fiji's forestry sector via the REDD+ programme.

The COP21 held in Paris in November 2015 finally provided the avenue for the 197 parties to the UNFCCC to take a strong stance on climate actions and investments considered necessary for a sustainable low-carbon future. The Paris Agreement on climate change undertook to hold global temperature rise to well below 2 °C above pre-industrial levels through reductions in GHG emissions. In its ambitious efforts to combat climate change and adapt to its impacts, the Paris Agreement pledged support to assist developing countries in fulfilling their obligations to the UNFCCC.

According to the agreement, each member nation is to prepare and maintain a Nationally Determined Contribution (NDC) towards global reductions in GHG emissions. In the case of the energy sector, such reductions can be effected either by replacing fossil fuel use by renewable energy (RE) or by adopting energy efficiency (EE) measures. The specific strategy to adopt is left to the individual members. It is noteworthy that Fiji was the first country to ratify the Paris Agreement.

The Fijian Government has demonstrated its strong commitment to global efforts in combating climate change on various platforms. Thus, it comes as no surprise that Fiji became the first Small Island Developing State (SIDS) to assume the Presidency of COP23, held in Bonn, Germany, in 2017. It is during the COP23 that Fiji launched its NDC Implementation Roadmap, which aims to reduce emissions from the energy sector by 30% by 2030. It also intends to reach close to 100% renewable energy power generation by 2030. Fiji's NDC provides a detailed temporal pathway towards achieving its target and calls for a reduction of 627,000 tCO$_2$ emissions per annum by the energy sector by 2030 through its power generation, demand-side efficiency and transport subsectors.

Suggested actions include the use of new renewable energy (RE) power generation, increased biomass plantation for fuel use and grid extensions. In addition, the effective implementation of the Roadmap is reliant on several key factors, which are important to ensure that the mitigation measures under the Roadmap can be aptly financed and executed. These factors include financing, governance and intuitional arrangements and monitoring and evaluation. Financial aspect is no doubt the major factor contributing to the successful implementation of the Roadmap. There is an absolute need for new financial strategies and mechanisms to reach the mitigation target.

The Roadmap is indeed laudable for its intended aims, but it raises several issues that need further attention. These include, amongst other things, the availability and identification of new RE sources and their method of utilisation. There are other requirements for RE production apart from the resources that need to be fully understood. There is also a need for a careful analysis of GHG emissions from the new RE resources and technologies to ensure that they do indeed contribute to a net reduction in emissions. Full life cycle assessments of the RE production will therefore have to be carried out to ascertain more realistic values of the actual reductions.

The projections of the Implementation Roadmap do not fully account for the possibility that the availability of RE resources may vary over the years. As most of these sources of energy, in particular wind, solar and hydro energy, are dependent on climatic conditions, this is a highly likely scenario, and a method is therefore required for forecasting the actual energy production in the future.

This book brings together the results of several renewable energy projects that shed important insights into many of these issues. It describes actions in the form of specific renewable energy projects that are verifiable and can be easily monitored for emissions reductions. It goes on to demonstrate how RE can be successfully used to reduce net GHG emissions in the Pacific and reveals how the 2 °C target of the Paris Agreement can be translated into verifiable and quantifiable actions that can be easily monitored.

School of Science and Technology   Priyatma Singh
The University of Fiji
Lautoka, Fiji

Contents

Part I
Introductory Concepts and Techniques

Chapter 1
Estimating Greenhouse Gas Emissions in the Pacific Island Countries

Francis S. Mani

Abstract A national Greenhouse Gas Inventory (GHGI) outlines estimates of emissions of greenhouse gases (GHGs) from various sectors of a country such as energy, agriculture, forestry and other land use (AFOLU), waste and industrial processes and product use (IPPU). The accuracy and consistency of the inventory is a basic requirement to ensure reliability of the estimates so that opportunities for potential reductions could be realized that would eventually lead to the development of low emission scenarios to achieve near zero emissions by 2050. An analysis of the second national communications of Pacific Island Countries (PICs) to UNFCCC shows that most of the emissions from PICs are from the energy sector and probably explains why Fiji's NDC Roadmap focuses on 30% emission reduction in the energy sector by 2030. This chapter discusses the IPCC 2006 guidelines to estimate emissions of CO_2 and other non-CO_2 greenhouse gases from different sectors. The uncertainties in emission estimates are discussed with more focus on data availability in the PICs. Research needed to derive country specific emission factors are also highlighted for certain sectors.

Keywords Greenhouse gas (GHG) · GHG inventory (GHGI) · Paris agreement · IPCC guidelines · Pacific Island countries (PICs) · Fiji NDC implementation roadmap

F. S. Mani (✉)
School of Biological and Chemical Sciences, Faculty of Science, Technology and Environment, The University of the South Pacific, Suva, Fiji
e-mail: francis.mani@usp.ac.fj

© Springer Nature Switzerland AG 2020   3
A. Singh (ed.), *Translating the Paris Agreement into Action in the Pacific*,
Advances in Global Change Research 68,
https://doi.org/10.1007/978-3-030-30211-5_1

1.1 Introduction

Estimating greenhouse gas emissions or developing a national greenhouse gas inventory is crucial in implementing mitigation policies and strategies to achieve climate goals. To execute an effective and feasible mitigation strategy it is critical to obtain robust emissions data before and after the implementation of the strategy to calculate the reduction of CO_2 equivalent achieved (Bi et al. 2011). The accuracy of the emission calculations depends on the consistency of the methodology applied, the robustness of the input data such as emission factors and other activity data required in the model (Kennedy et al. 2009).

The UNFCCC Article 4 – Commitments simply states "All Parties, taking into account their common but differentiated responsibilities and their specific national and regional development priorities, objectives and circumstances, shall:

- Develop, periodically update, publish and make available to the Conference of the Parties, in accordance with Article 12, national inventories of anthropogenic emissions by sources and removals by sinks of all greenhouse gases not controlled by the Montreal Protocol, using comparable methodologies to be agreed upon by the Conference of the Parties.

It is known that not all emissions can be measured. However, they can be estimated using credible methodologies that are generally accepted by the Conference of the Parties. To this end, the IPCC developed IPCC National Greenhouse Gas Inventory 1996 guidelines and good practice guidelines (GPG) 1996 which was then revised to IPCC National Greenhouse Gas Inventory 2006 guidelines and GPG2003. Currently a special Task Force is set up to refine the 2006 IPCC Guidelines for National Greenhouse Gas Inventories, and the final draft of this new methodology report titled "2019 Refinement to the 2006 IPCC Guidelines for National Greenhouse Gas Inventories" will be considered by the IPCC for adoption/ acceptance at its Plenary Session in May 2019. The development of such methodologies provide the much needed consistency in reporting emissions and providing comparable data to monitor progress in achieving reduction target set by international agreements.

The IPCC methodology is based on a simple inventory method highlighted in Fig. 1.1.

The complexity of the reporting methodologies depends on the extent of the availability of the activity data and the country specific emission factor. There are

Fig. 1.1 The basis of calculating greenhouse gas emissions from activity data and emission factor

three tiers in the methodology defined by IPCC 2006 namely, Tier 1, Tier 2 and Tier 3.

- **Tier 1** is a simple method that uses default emission factors and to some extent uses guided principles to estimate activity data and consequently has a large uncertainty associated with the estimates derived from such methodology.
- **Tier 2** is similar to tier 1 methodology except that country specific emission factors and other nationally available activity data are used in the calculation.
- **Tier 3** is a more complex method involving process based models with detailed, geographically specific data. Such methods provide robust and accurate estimate but requires thorough uncertainty assessments via means of detailed documentation procedures to ensure consistency and comparability between countries (Lokupitiya and Paustian 2006).

The greenhouse gases for which emissions are reported are those GHG that are regulated by the mandate of the Kyoto protocol and includes the following; Carbon dioxide (CO_2), methane (CH_4), nitrous oxide (N_2O), sulphur hexafluoride (SF_6), Hydrofluorocarbons (HFCs), perfluorocarbons (PFCs) and nitrogen trifluoride (NF_3) (Michael and Hsu 2008). The 2006 IPCC Guidelines for Greenhouse Gas Inventories, IPCC Good Practice Guidance (GPG) 2000 and 2003 takes a sectoral approach in estimating national greenhouse gas emissions and the following sectors (given below) are covered which are relevant for the PICs:

- Energy Sector
- Industrial Processes and Product Use (IPPU)
- Agriculture, Forestry and Other Land Use (AFOLU)
- Waste

The PICs have used the mentioned IPCC guidelines to prepare their national greenhouse gas inventories that were reported to the UNFCCC. Table 1.1 below gives sectoral emissions (CO_2eq in Gg/year) overview of the PICs as reported in their Second National Communication to UNFCCC.

It is apparent from the table that the energy sector is the major contributor in the PICs except in Tonga and Vanuatu. This is very obvious and is aligned with global scenario whereby cities in developed countries contribute over 67% to the global GHG emissions from fossil fuel use (Bi et al. 2011). In Vanuatu, the agriculture sector, particularly emissions from enteric fermentation in ruminants, was the dominant sector while land use changes in Tonga was the major emission source. The compiled data demonstrates the strength of each sector and clearly highlights which sector needs serious mitigation efforts in order to reduce the emissions in an effort to realize the objectives of the Paris agreement. Hence it is important that PICs develop capacity in generating valid and robust national greenhouse gas inventories using the IPCC guidelines for identification of opportunities for emission reduction. A classic example is Fiji's Nationally Determined Contributions (NDC) Roadmap which highlights 30% reduction in the energy sector by 2030 as the energy sector provides opportunities and feasible emission reductions.

Table 1.1 PICs emissions (Gg/year of CO_2 eq) from different sectors reported as Second National Communication to UNFCCC

	Cook Islands	Fiji	Kiribati	Marshalls	Nauru	Samoa	Solomons	Tonga	Tuvalu	Vanuatu
Energy	54.5	1570	64.0	85.0	13.3	174.4	350.6	94.9	11.2	122.4
AFOLU	0.08	977	0.84	NE	1.61	135.4	76.39	150.8	4.61	587.4
Waste	2.94	84	101.1	37.5	4.55	32.8	191.6	29.1	2.63	10.8
IPPU	0.59	NE	NE	NE	NE	9.51	NE	NE	NE	NE

Many developed cities have reported their carbon footprints using both GHG protocols and taking into account the Life Cycle Assessment (LCA) approach. The GHG protocol is based on the IPCC Guidelines whereas the LCA approach takes into account emissions from the initial stages to final stages of a product or service and is termed "cradle to grave" when considered from the resource extraction stage to the waste disposal stage (Reijnders 2012). In addition to GHG emissions, LCA offers the benefit of assessing other environmental impacts of products or services. It is becoming a widespread practice recently in greenhouse gas emission estimate that the IPCC guidelines are combined with the LCA approach to provide a more holistic estimate if a particular mitigation strategy is actually an emission reduction strategy (Ramaswami et al. 2008). The life cycle assessment of wind and solar energy, which are strong mitigating options in the energy sector, shows that they are not entirely emission free technologies. A comprehensive study involving published literature concludes that the LCA approach shows that the wind energy emits an average of 34.11 g CO_2eq/kWh and solar energy emits an average of 49.91 g CO_2eq/ kWh (Nugent and Sovacool 2014).

1.2 Energy Sector

The energy sector emits GHGs mainly through the combustion of fossil fuels and Volume 2 of IPCC 2006 guidelines states that the key source categories in the energy sector are fuel combustion activities (stationary combustion), fugitive emissions from the oil refineries and carbon dioxide transport and storage. For PICs, stationary combustion is the only relevant key source category (See Fig. 1.2) as the advanced technologies in exploration and exploitation of fossil fuel and injection and storage of CO_2 underground are not available. The major greenhouse gas emitted by the energy sector is CO_2 although there are very minimal emissions of CH_4 and N_2O as well.

1.2.1 Estimating Emissions from the Energy Sector

1.2.1.1 Stationary Combustion Sub-Sector

The emissions from the stationary combustion sub-sector mainly included the sum of emissions from main producers of electricity generation, on-site use of fuel to generate its own electricity such as in generators in commercial and institutional buildings, manufacturing and construction industries, mining and quarrying industries, agricultural industries and most importantly all emissions from fuel combustion in households such as use of kerosene or natural gas for cooking.

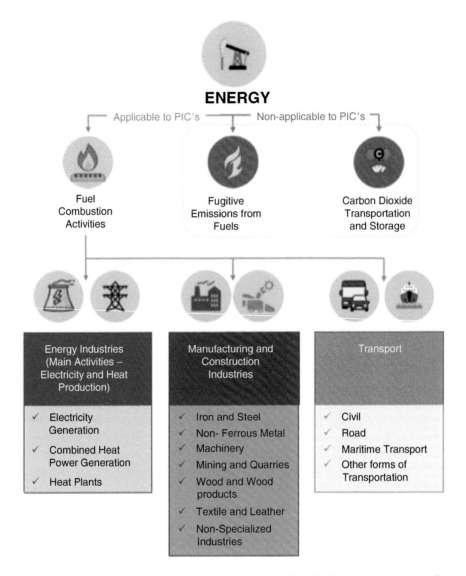

Fig. 1.2 The key source categories in the energy sector and the only relevant source category for PICs are fuel combustion activities

The Tier 1 approach requires data on the amount of fuel combusted in the source category and the default emission factor (See Table 1.2) as shown in the equation below:

$$\text{Emissions}_{GHG,fuel} = \text{Fuel Consumption}_{fuel} * \text{Emission Factor}_{GHG,fuel}$$

Table 1.2 Emission factors for different GHG and fuel type typically used in the Pacific Islands

Fuel Type		CO_2 (kg-CO_2/TJ)	CH_4 (kg-CO_2/TJ)	N_2O (kg-CO_2/TJ)
Crude Oil		73 300	3	0.6
Natural Gas Liquids		64 200	3	0.6
Gasoline	Motor Gasoline	69 300	3	0.6
	Aviation Gasoline	70 000	3	0.6
	Jet Gasoline	70 000	3	0.6
Jet Kerosene		71 500	3	0.6
Other Kerosene		71 900	3	0.6
Gas/Diesel Oil		74 100	3	0.6
Liquefied Petroleum Gases		63 100	1	0.1
Ethane		61 600	3	0.6
Coal		94 600	1	1.5
Coke		107 000	3	1.5
Residual Fuel Oil		77 400	3	0.6

Where;

- Emissions$_{GHG, fuel}$ = emissions of a given GHG by type of fuel (kg GHG)
- Fuel Consumption$_{fuel}$ = amount of fuel combusted (TJ)
- Emission factor$_{GHG, fuel}$ = default emission factor of a given GHG by type of fuel (kg gas/TJ). Refer to Table 1.2.

It should be noted that to use the above equation to calculate the emissions, the fuel consumption data in mass or volume units must be converted into the energy content of these fuels in terajoules (TJ). Tier 2 approach is used when a country specific emission factor for the source category and fuel for each gas is available. Tier 3 takes a more complex approach where different technologies to combust fuels are taken into account as some technologies maybe more fuel efficient and may influence emissions.

There have been some documented evidences on research into country specific emission factors to generate a more reliable estimate of GHG emissions from combustion sources. A study conducted on obtaining country specific emission factors for energy sector in Mauritius clearly demonstrated that emissions calculated with the experimentally derived country specific emission factors showed that the emission reported under the Third National Communication (TNC) was overestimated by 10% in the power subsector (Ramphull and Surroop 2017). Similarly in China the estimates from fossil fuel combustion were revised using two sets of comprehensive new measurements of emission factors as the fuel composition in China is known to vary widely from year to year especially for coal (Liu et al. 2015). The new revised emission estimate was lower than what was reported in the inventory earlier which was the consequence of the emission factor for Chinese coal being on average 40% lower than the IPCC recommended values (Liu et al. 2015).

1.2.1.2   Mobile Combustion Sub Sector

The greenhouse gas emissions from mobile combustion are due to major transport activity, i.e., road, off-road, air, railways, and maritime transport. The major greenhouse gas emitted is CO_2 with minute emissions of CH_4 (2%) and N_2O (1%). The following equation is used to calculate CO_2 emissions from the transport sector:

$$Emission = \Sigma \left(fuel\ type \times EF \right)$$

Where:

Emission = Emissions of CO_2 (kg)
Fuel = fuel sold (TJ)
EF = emission factor (kg/TJ)

The activity data needed to estimate emissions is the amount of fuel sold and to engage in higher tier approach than further classification is required in terms of amount of fuel sold to a particular type of vehicle with known emission control technologies such as catalytic converters. The aviation source category takes into account the emissions associated with the international bunkers and should be prepared as part of the national inventory but excluded from the national total and reported separately. It should be cautioned when considering emissions from biofuels used in vehicles. Only the CO_2 emissions from the fossil fuel component is accounted for whereas CO_2 emissions from the combustion of biogenic carbon is accounted for under the AFOLU sector and double counting should be avoided.

The national statistics for Fiji shows that a total of 707 million litres of fuel was imported of which 50% was re-exported to other PICs. Sectoral breakdown showed that 29% of the fuel stock was used for electricity generation, 64% was consumed by the transport sector; land transport (16%), air (26%), marine (22%) and the remainder was used for off grid electricity generation, household lighting and cooking (Holland et al. 2014). It was noteworthy that for PIC based scenario, emissions from the marine transport is sizeable portion and there have been some efforts in the emission reduction in maritime transport through sustainable sea transport research programme at the University of the South Pacific (Holland et al. 2014). The Maritime Technology Cooperation Center in the Pacific (MTCC-Pacific) provides initiatives for climate mitigation in the maritime industry and has contributed significantly to Kiribati's Nationally Determined Contributions (NDCs) and broader Sustainable Development Goals.

1.2.2   Uncertainties in Energy Sector Estimation

The uncertainties in estimating emissions in the energy sector arise from the fact that activity data is not available in the correct format to be used directly in the equation above. The fuel data is usually provided in litres and to convert it into mass, the density is used which is dependent on temperature. However the temperature effects on density are not considered and this could introduce a bias of ±3%. In compiling the national greenhouse gas inventory for TNC, it was noted that instead of fuel consumption data, the total amount of fuel imported was used to calculate the emissions for combustion activities. This may add bias to the emission estimate as the total fuel imported in the particular year is not what is consumed in that particular year. The quality of emission inventories for the most important greenhouse gas, CO_2, depends mainly on the accuracy of fuel use statistics. Ideally the fuel consumption data should be used for estimating emissions but the major challenge in the PICs is the unavailability of the sale data of different fuel types from the private oil companies. Although IPCC 2006 guidelines assign the uncertainty levels at 2% for emissions from fuels but for PICs it estimated to be approximately 10% due to unavailability of consumption data.

1.3 Agriculture, Forestry and Other Land Use (AFOLU) Sector

The AFOLU sector deals with the estimation of GHG emissions and removals from managed lands through biological and physical processes. Managed land is defined as land where anthropogenic activities influence the natural ecosystem for agricultural, ecological and social functions (IPCC 2006). This sector considers emissions from the agricultural subsectors such as CH_4 emissions from livestock enteric fermentation, CH_4 and N_2O emissions from manure management, CH_4 emissions from rice cultivation, direct N_2O emissions from N-based fertilizer application, indirect N_2O application from managed soils, CO_2 emissions from liming and urea application and non-CO_2 emissions from biomass burning. The Forestry and other land use sub-sectors take into account emissions and removals from forest land remaining and land converted from one category to another such as forest land converted to cropland or grassland or settlements. The forestry sector uses net changes in C stock over time to estimate CO_2 emissions and removals from dead organic matter, soil organic matter of organic and mineral soils and harvested woody products (HWP) for all managed lands.

In the PICs context, the key categories identified in the agricultural sector are CH_4 emissions from ruminant animals and CH_4 emissions from manure management whereas CH_4 emissions from rice cultivation and N_2O emission from fertilizer application are negligible. In Fiji the estimated emissions from the ruminants and manure management accounted for 37%, as reported in second national communication to UNFCCC, whereas Vanuatu recorded the highest emissions from the agricultural sector amounting to 86% in 2010.

1.3.1 Estimating Emissions from Forestry and Other Land Use

The key source category under this section is estimating emissions or removals from managed forest land. Chapter 4 of Volume 4 of IPCC 2006 guidelines describes three tiers in estimating changes in carbon stock from managed forests that have been under the forest land for over 20 years. The primary step in calculating emissions/removals from the forestry sector is to estimate the biomass gain or loss. There are basically two different methods for estimating biomass gains and losses: Gain-Loss method and a stock difference method. The Gain–Loss method is more appropriate for Tier 1 approach where country specific activity data are not available. There are seven basic steps outlined in Volume 4 of IPCC 2006 to estimate change in carbon stocks in biomass (ΔC_B) using the Gain-Loss method. The seven steps are as follows:

Step 1: Categorizing the area of forest land into appropriate forest types of different climatic or ecological zones.

Step 2: Estimate the annual biomass gain in forest land using equations 2.9 and 2.10 in chapter 2 of IPCC 2006 guidelines.

Step 3: Estimate annual carbon loss due to wood removals

Step 4: Estimate annual carbon loss due to fuelwood removals

Step 5: Estimate annual carbon loss due to disturbance

Step 6: from the estimated losses in steps 3–5 estimate the annual decrease in carbon stock due to biomass losses (ΔC_L) from equation 2.11 in chapter 2

Step 7: Estimate the annual change in carbon stocks biomass (ΔC_B) using equation 2.7 in chapter 2.

In addition to calculating the changes in biomass, changes in carbon stock from other carbon pools needs to be estimated such as dead organic matter (DOM) and soil organic matter (SOM) and emissions of CO_2 and non-CO_2 gases from forest burning. In Tier 1 approach it is assumed that changes in carbon stock due to DOM is zero. It is also noteworthy that in Tier 1 method, carbon stock changes for mineral soils it is assumed there is no change with forest management and it is assumed to be zero. However, for organic soils, carbon emissions due to drainage of forest organic soils are addressed. The C emissions can be calculated by multiplying the area of drained organic soil with the emission factor for annual losses of CO_2.

Basically similar approaches as above are applied to estimate emissions from croplands and grasslands. There is also guidance provided to report GHG emissions from managed wetlands particularly peat lands managed for peat extraction and lands flooded in reservoirs as flooded lands. The emissions due to croplands, grasslands, wetlands and settlements are detailed in chapter 5, 6, 7 and 8 of Vol 4, IPPC 2006 guidelines respectively.

1.3.2   Estimating Emissions from the Agricultural Activities

1.3.2.1   Enteric Fermentation

Methane emissions from enteric fermentation in ruminant animals and to a lesser extent, of non-ruminants are estimated as such:

$$\text{Emissions}\left(CH_4\right)_T = EF\left(\text{T}\right) \times \frac{N_T}{10^6}$$

Where;

Emissions $(CH_4)_T$ = methane emissions for animal category T, Gg CH_4 year^{-1}
$EF_{(T)}$ = emission factor for animal type, T, kg CH_4 head^{-1}
$N_{(T)}$ = number of heads for animal category T
T = animal category

1.3.2.2   Manure Management

This source category considers CH_4 emissions from anaerobic manure decomposition processes and direct and indirect N_2O emissions from animal excretion. Direct N_2O emission is related to the total amount of N in manure treated in different manure management systems (MMS). The indirect N_2O emission refer to N in manure that volatilizes as NH_3 and NO_x or lost through run-offs and leaching and when these N is deposited in some other place through the redeposition processes, it will be transformed by microbial activity into N_2O.

The methane emission from the manure management is estimated as such:

$$CH_{4(T)} = N_T{}^* EF$$

Where;

$CH_{4(T)}$ = CH_4 emissions in kg CH_4year^{-1} for animal category, T.
N_T = number of head f animal category T, heads year^{-1}
EF = default emission factors expressed in units of kg CH_4 head^{-1} year^{-1}.

To estimate the direct N_2O emissions from animal excretion in a particular MMS it is imperative to estimate the total N excreted from manure management systems for animal category as such:

Firstly, calculate the excretion rate per animal using the equation below:

$$Nex_{(T)} = N_{rate} \times TAM_{(T)} / 1000 \times 365$$

$Nex_{(T)}$ = Nexcreted in manure for animal category T, kg N animal^{-1} year^{-1}
$Nrate_{(T)}$ = default N excretion rate per mass, kg N (tonnes animal mass)$^{-1}$ day^{-1}
$TAM_{(T)}$ = typical mass for animal category T, kg animal^{-1}

Then calculate the manure N content in a particular MMS as such:

$$NE_{MS(T)} = \left(N_{(T)} \times Nex_{(T)} \times MS_{(S,T)} \right)$$

Where:

$NE_{MS(T)}$ = Total nitrogen excreted from MMS for animal category, T, heads year^{-1}.
$N_{(T)}$ = number of head f animal category T, heads year^{-1}
$Nex_{(T)}$ = annual N excretion for animal category T, kg N animal^{-1} year^{-1}
$MS_{(S,T)}$ = share of manure treated in each systems S for animal category T
T = animal category
S = manure management system

Hence direct N_2O emissions are calculated as follow:

$$\text{Direct emissions} \left(N_2O \right)_T = \Sigma \left[NE_{MS} \times EF_{3(s)} \right] \times 44 / 28 \times 10^{-6}$$

Where;

Direct emissions $(N_2O)_T$ = Direct N_2O emissions from MMS for animal category T, kg N year^{-1}
$EF_{3(S)}$ = Emission factor for direct N2O emissions from each MMS system, S, kgN20N/kg N.
T = Animal Category
S = manure management system (MMS)

The indirect emissions of N_2O is estimated as follows:

$$\text{Indirect emissions N2O} = NE_{MS(T)} \times \left[\left(\text{Frac}_{GASMS(S)} \times EF_4 \right) + \left(\text{Frac}_{LeachMS} \times EF_5 \right) \right] \times 44 / 28 \times 10^{6}$$

Where;

Indirect emissions $(N_2O)_T$ = Indirect N_2O emissions produced from the atmospheric deposition of N volatilized from manure management systems for animal category T, Gg N_2O year^{-1}
$NE_{MS(T)}$ = Total N excreted from manure management
$\text{Frac}_{GASMS(S)}$ = fraction of N from MMS that volatilizes as NH_3 and NO_x, kg N volatilized from each system S
EF_4 = emission factor indirect N_2O emissions from atmospheric deposition of N on soils.

$Frac_{Leach}$ = fraction of N leaches as NH_3 and NO_x
EF_5 = Emission factor indirect N_2O emissions from N leaching and run-off, kg N_2O/ kg N
T = animal category
S = manure management

1.3.2.3   Rice Cultivation

Methane emission from rice cultivation is the result of anaerobic decomposition of organic matter in paddy fields. The default IPCC seasonally integrated EF of 20 g CH_4 m^{-2} year^{-1} is used and this is further modified by the scaling factor for the water regime and application of organic amendments. For irrigated farms the scaling factor is one whereas the rainfed is 0.7 and 0 for upland and dry conditions. If organic manure or straw incorporation is applied in the rice paddies then a scaling factor of 1.4 is considered.

$$\text{Emissions}(CH_4) = \frac{EF \times SF_o \times \left(Ai + [Aj \times SFj]\right)}{10^5}$$

Where;

Emissions (CH_4) = Methane emissions per rice paddy, Gg CH_4 year^{-1}
EF = Seasonal methane emission factor, g m^{-2} year^{-1}
Ai,j = Rice paddy area harvested in the two water regimes, irrigated and rainfed, ha year^{-1}
SFo = 1.4 correction factor for organic amendments, for all countries
SFj = 0.7 scaling factor for Aj

In the PICs, Fiji has limited rice farming with total methane emissions of 0.09 Gg/year in 2004 and a recent assessment showed emissions to be 2.29 Gg/year in 2017 (Chand 2018). The increase in methane emission was observed due to the commitment from the Fiji Government to promote rice farming and to become self-sufficient by 2030 which led to an increase in land area of rice farming. The study also derived the emission factor for both rainfed and continuously flooded water regimes and it was noted that the country specific emission factors were within the range of the IPCC default emission factors (Chand 2018).

1.3.2.4   Synthetic Fertilizers

The application of nitrogen based synthetic fertilizers lead to direct emission of N_2O from the agricultural land due to microbial nitrification and denitrification processes. To estimate emissions of N_2O the following formula is used:

$$\text{Direct Emissions}(N_2O) = N \times 44 / 28 \times EF \times 10^{-6}$$

Where;

Direct emissions (N_2O) = Direct N_2O emissions from synthetic nitrogen additions
   to the managed soils, GgN_2O year^{-1}
N = Consumption in nutrients of nitrogen fertilizers, kg N input year^{-1}
EF = Emission factor for N_2O emissions from N inputs, kg N_2O–N/kg N$'$

The FAOSTAT database could be used to attain the main activity data on consumption of fertilizer in different years. The default emission factor for direct N_2O
emissions from the application of synthetic fertilizer is 1% of N-input (IPCC 2006).

1.3.3   Uncertainties in the Agriculture Sector

The major uncertainty usually comes from the Tier 1 (default) methodology of the
IPCC 2006 guidelines. The uncertainties in the agriculture sector emanate from the
reliability of the activity data and the emission factors. The Food and Agriculture
Organization (FAO) database is normally used to extract the activity data such as
ruminant animal population, however during the compilation of Fiji's SNC it was
noted that there were large discrepancies between the national statistics and the
FAO database on ruminant animal population. The FAO database for Fiji ruminant
population was projected from the animal surveys done in 1990s but in reality the
cattle and dairy industry suffered huge loss due to economic viability and the foot
and mouth disease that saw ruminant animal population dwindling in the recent two
decades. The emission factors in the agricultural sectors were derived from studies
in advanced western countries and have been used for tropical developing countries.
There is a dire need for research to derive country or region specific emission factors. A study carried out in Sub-Sharan African region taking into account the field
measurements of live weight, live weight change, milk production, dry matter intake
and local climatology showed that the country specific emission factors were
approximately 30–40% lower than IPCC default emission factors (Goopy et al.
2018). A study on measuring N_2O flux from the application of N-fertilizer in sugarcane plantations in Fiji suggested an emission factor of 5% as compared to the
default value of 1% (Nisbat 2018). A thorough assessment of published literature on
emission factors showed that the emission factor for direct N_2O emissions from
synthetic fertilizer ranges from 0.013% to 21% with very few studies done in the
tropical region (Nisbat 2018). Hence the default emission factor of 1% is a very
poor proxy and introduces large uncertainty in the estimates. It is highly recommended that country specific emission factor should be derived to enable N_2O emission estimates from synthetic fertilizer more robust.

1.4   Industrial Processes and Product Use (IPPU)

The IPPU sector is considered to be less significant in PICs compared to the energy and the AFOLU sector. This is primarily due to the absence of mineral industry, chemical industry and metal industry in PICs. The cement production process emits significant amounts of CO_2 during the clinker production. The default emission factor is 0.51 t CO_2/t clinker. In the Fiji's initial national communication to UNFCCC, it was reported that cement production emitted 45,000 tonnes of CO_2 but more recently the emission from cement production was reported to be zero in Fiji's second national communication to UNFCCC because there is no clinker production in Fiji. The clinker used in the cement production in Fiji is now sourced from outside. However, it should be cautioned that IPPU emissions are bound to increase in developing countries due to ODS substitutes in the refrigeration and air conditioning source category. The use of HFCs is on significant rise after the phasing out of HCFCs in 2013 under the amendments to the Montreal Protocol.

The common hydro-fluorocarbon (HFC) refrigerants used in commercial air conditioning systems include R-410A, R-407C and R-134a. Emissions of HFC-134a as approximated from atmospheric observations are 60% higher compared with the United Nations Framework Convention on Climate Change (UNFCCC) inventory from 2009 to 2012 (Xianga et al. 2014). The projected increases in recent years are the consequence of phasing out HCFC, which is an ozone depleting substance, leaving HFC as the most desirable replacement that could be used directly in equipment. Recent analysis shows that HFC emissions will significantly increase in developing countries provided there is no regulation on HFC consumption and emission. The projected emissions of HFC in developing countries will increase by as much as 800%, greater than in developed countries by 2050 (Velders et al. 2009).

It is a known fact that HFCs are not produced in PICs, however it is imported by Fiji and then re-exported to other PICs to be used as a charging gas in refrigeration and air conditioning (RAC) industry. Section 2F1 of Volume 3 of IPCC 2006 guidelines highlights the process of estimating emissions for HFC use in RAC application. The IPCC inventory software enables you to estimate actual emissions even if the historic data is not available. To enable estimation, the following data is required:

- Year of introduction of chemical
- Domestic production of chemical in the current year
- Imports of chemical in current year
- Exports of chemical in current year
- Growth rate of sales of equipment that uses the chemical

Tier 1 approach is simple and less data intensive than Tier 2 because emissions are carried out at the application level rather than for individual products or equipment types. There are two types of Tier 1methods; Tier 1a –Emission factor approach at the application level and Tier 1b – mass balance approach at the application level.

The emission factor approach takes into account the annual consumption multiplied by the composite EF for that specific application. The net consumption within Tier 1A is computed as:

$$Net \text{ consumption} = \text{Production} + \text{imports} - \text{exports} - \text{destruction}$$

The net consumption calculated is then used to calculate annual emissions as such:

$$\text{Annual Emissions} = Net \text{ Consumption} \times \text{Composite } EF$$

The mass balance approach estimates emission from assembly, operation and disposal of a pressurized system and does not rely on emission factor. The emissions can be computed as:

$$
\begin{aligned}
\text{Emissions} = {} & \text{Annual Sales of New Chemical} \\
& - \big(\text{total charge of New Equipment} \\
& - \text{Original Total Charge of Retiring Equipment}\big)
\end{aligned}
$$

For PICs, the emission factor approach is preferred as the emissions from different stages of assembly, operation and disposal of refrigerant in equipment is not documented. Although the emissions of HFCs are generally not estimated in the National Greenhouse Gas Inventory for PICs but there is a need for institutional arrangements between government agencies and private companies to look into proper record keeping of different types of HFCs imported and destroyed in the country. Since the GWP of HFCs are very high, a small emission of HFC can contribute significantly when expressed in CO_2 eq terms.

1.4.1 Uncertainties in the IPPU Sector

The major source of uncertainty in this sector is from the activity data on net consumption. During the compilation of the Fiji's SNC it was noted that there is very poor record of HFCs imported in the country due to mismatch of custom codes. Data availability on the amounts of HFCs imported in equipments and appliances is very limited and how much of these gases are released into atmosphere after the end life of the product is very scarce or non-existent. The emission estimate in this sector is highly uncertain but since it is not a key category source for PICs not much attention is given and is usually not estimated.

1.5 Waste Sector

The key gas emitted from the waste sector is CH_4, which is produced from the anaerobic degradation of organic matter. There are three sub-categories within this sector, namely: emissions from the landfill, waste water treatment and incineration of waste. The CO_2 emission from incineration of waste in the PICs is very small and almost negligible and has not been accounted for in the national greenhouse gas inventory. Hence the discussions are more focused on emissions from solid waste disposal sites (SWDS) and waste water treatment. The methane emissions from SWDS and wastewater treatment contribute 4% of the total global GHG emissions (Cai et al. 2014). However in the PICs the total emissions from the waste sector as highlighted in Table 1.1 tends to vary and can be as high as 30% in some countries although these estimations are subjected to approximately 30% uncertainty.

When organic matter (food waste, garden waste, paper, wood, textiles and diapers) decomposes in the absence of oxygen then CH_4 and CO_2 are produced. The CO_2 emissions is not accounted for in the greenhouse gas inventory because it is considered to be carbon neutral as this is the CO_2 released back into the atmosphere that were initially removed by these biomasses. Similar anecdote applies to emissions from the wastewater treatment, however in addition to CH_4 a negligible amount of N_2O is also produced in the wastewater sub-sector which is reported by many countries. Fiji's second national communication to UNFCCC clearly states that 3.12 Gg of methane was emitted by the SWDS and 1.10 Gg of methane was emitted by the wastewater treatment. Clearly, this demonstrates that the SWDS is a dominant sub sector within the waste sector and mitigation efforts in implementing integrated solid waste management (ISWM) policies could bring about fruitful reduction targets in the waste sector.

1.5.1 Estimating Methane Emissions from the Solid Waste Disposal Sites

There are numerous methods for estimating methane emissions from the landfill such as the mass balance method, flux chamber method, IPCC 2006 Waste Model and USEPA Landfill Gas Emission model (LandGEM) (Kamalan et al. 2011).

When waste is placed in the landfill, the degradable carbon content in the waste depletes or decays over a period of time and therefore First Order Decay (FOD) kinetics are used to explain the emissions from the landfill (Santalla et al. 2013). The two most commonly used FOD models are LandGEM and IPCC 2006 Waste Model. The major drawback for IPCC 2006 Waste Model is that it needs historical dataset back to 1950 for the amount of waste placed in the landfill. If such activity data exists then Tier 2 of the IPCC 2006 Guidelines will be used. However, in the case of missing data then amount of waste generated could be estimated from population and default waste generation rate. For reporting emissions from the waste sector to UNFCCC, it is strongly recommended to use the IPCC 2006 Guidelines.

1.5.1.1 IPCC 2006 Waste Model

The methane emissions from the SWD is calculated from the equation given below after taking into account the oxidation loss in the soil cover and if methane is recovered for flaring or utilisation for energy generation. Hence methane emissions are always lower than the methane generated and it is noteworthy that the methane recovered should be subtracted from the total methane generated and portion that is not recovered will only be subjected to oxidation.

$$CH_4 \text{ Emissions} = \sum_x \left[CH_4 generated_{x,T} - R_T \right] * \left(1 - OX_T \right)$$

Where;

CH_4 emissions = CH_4 emitted in year T, Gg
T = inventory year
R_T = recovered CH_4 in year T, Gg
OX_T = oxidation factor in year T, fraction

The CH_4 generation potential of a particular waste type will decrease gradually throughout the following decades. Hence the FOD model is built on the exponential factor that describes the fraction of decomposable degradable organic carbon (DDOC) that is degraded into CH_4 and CO_2 for each year. The DDOC is calculated as follows:

$$DDOC_m = W * DOC * DOC_f * MCF$$

Where;

$DDOC_m$ = mass of decomposable DOC deposited, Gg
W = mass of waste deposited, Gg
DOC = degradable organic matter in the year deposition, fraction, Gg C/Gg waste
DOC_f = fraction of DOC that can decompose
MCF = CH_4 Corrected factor for anaerobic decomposition in the year of deposition.

The default values for DOC for individual waste type are as follows: 40% for paper/cardboard, 24% textiles, 15% food waste, 43% wood, 20% garden and park waste and 24% diapers. It is also imperative for a highly reliable estimation that a country has undertaken waste characterization studies and in the absence of such data default values could be used for the Oceania region. DOC_F is fraction of degrad-

Table 1.3 MCF values for different types of SWDS

Type of site	Description	MCF values
Managed anaerobic	Controlled placement of waste, control of scavenging and fires and should meet one of the following criteria (i) cover material, (ii) mechanical compacting or (iii) levelling of waste	1
Managed semi-anaerobic	Controlled placement of waste and must include all of the following (i) permeable cover material, (ii) leachate drainage system, (iii) regulating pondage, (iv) gas ventilation system	0.5
Unmanaged deep	Not meeting the criteria of managed SWDS and has depths greater than or equal to 5 m and/or high water table at near ground level	0.8
Unmanaged shallow	Not meeting the criteria for managed SWDS and have depths <5 m	0.4
Uncategorized	Any SWDS that does not fall within any of the four categories above	0.6

able carbon content and usually a default value of 0.5 is used. The MCF values vary with the different types of SWDS and are given by IPCC 2006 guidelines as in Table 1.3.

Finally, the FOD kinetics is applied to calculate the methane generated, which is dependent on the total mass of decomposing material currently in the site. The FOD equation is mathematical expressed as:

$$CH_4 \text{generated}_T = (\text{DDOCmd}_T + \left(\text{DDOCma}_{T-1} * e^{-k}\right) * \left(1 - e^{-k}\right) * F * 16/12$$

Where;

DDOCmd$_T$ = DDOCm deposited into SWDS in year T, Gg
DDOCma$_{T-1}$ = DDOCmaccumulated in the SWDS at the end of the year (T-1), Gg
k = methane generation rate constant, k = ln2$*t_{1/2}$
$t_{1/2}$ = half-life time (y)
F = fraction of methane by volume in generated LFG
16/12 = the molecular weight ratio for CH$_4$/C

The methane generation rate constant is dependent on the local environmental conditions such as mean annual temperature(MAT) and pressure(MAP) and potential vapo-transpiration. Table 1.4 below summarizes the k values applied for different categories of waste depending on the environmental conditions for the tropical region.

A study conducted in a tropical landfill in Thailand highlighted that the methane generation rate constant, k, determined through field measurements was approximately 0.33 year^{-1}. This rate was much higher than the default value for bulk waste of 0.17 (Wangyao et al. 2010). This high rate was attributed to high moisture content of the waste in which the food waste was the main component that degrades rapidly. This degradation was further enhanced by tropical meteorology with high rainfall and high temperatures. This high methane generation rate would be applicable for

Table 1.4 Methane generation rate constant depending on type of waste and meteorological conditions for tropical regions

Type of waste		Tropical (MAT >20 °C)			
		Dry (MAP <1000 mm)		Moist and Wet (MAP>1000 mm)	
		Default	Range	Default	Range
Slowly degrading waste	Paper/textiles waste	0.045	0.04–0.06	0.07	0.06–0.085
	Wood/straw waste	0.025	0.02–0.04	0.035	0.03–0.05
Moderately degrading waste	Other (nonfood) organic putrescible/garden and park waste	0.065	0.05–0.08	0.17	0.15–0.2
Rapidly degrading waste	Food waste/sewage sludge	0.085	0.07–0.1	0.4	0.17–0.7
Bulk waste		0.065	0.05–0.07	0.17	0.15–0.2

tropical PIC countries as the waste generated are mostly food waste or organic waste with high moisture content. A feasibility study to capture and utilise methane from Naboro landfill in Fiji showed that methane generation is very fast given the high organic content of waste and tropical climatology and therefore such fast degradation rate posed real challenges in achieving high recovery rates using vertical wells as a mitigation option (Mani et al. 2016).

The above methodology is also used to estimate methane emissions if any emission reduction strategies are implemented by the countries. It is evident that the methane generated in the landfill is mostly from the DOC content of waste and therefore diverting organic waste from landfill through nationwide composting could lead to reduction in methane generated. In many developed countries landfill methane is recovered and utilised as either cooking gas or to run a gas turbine to generate electricity. There have been studies to demonstrate how to increase the efficiency of the recovery systems to enhance capture of landfill methane (Thompson et al. 2009; Spokas et al. 2006). This will increase the methane recovery factor in the equation above which will tend to reduce methane emission further. Such technology for methane recovery is expensive and is yet to be installed in the PICs landfills.

1.5.1.2 LandGEM Model

LandGEM is a Microsoft Excel based tool which incorporates the first order equation of decomposition rate to quantify methane emissions from the decomposition of urban wastes in landfill and has been developed by USEPA (Ghasemzade and Pazoki 2017; Chaudhary and Garg 2014). The LandGEM FOD model is based on the equation below:

$$Q_{CH_4} = \sum_{i=1}^{n} \sum_{j=0.1}^{1} kL_o \left(\frac{M_i}{10} \right) e^{-kt_{ij}}$$

Where;

Q_{CH4} = annual methane emission in the specified year of calculation (m^3/year)
I = 1 year time increment
n = (year of calculation) – (initial year of waste acceptance)
j = 0.1 year time increment
k = methane generation rate ($year^{-1}$)
L_o = potential methane generation rate (m^3/Mg)
M_i = mass of waste accepted in the i year (Mg)
t_{ij}= age of the j section of waste mass M_i accepted in the ith year

USEPA protocols state that the composition of waste used in the model reflects US waste composition of MSW, inert material and other non-hazardous wastes. For a landfill containing non-biodegradable waste (i.e., inert material), such as ash from waste combustion, this portion may be subtracted from the waste acceptance rates. LandGEM recommends subtracting inert materials only when documentation is provided and approved by a regulatory authority. LandGEM provides methane generation constant utilising both CAA (Clean Air Act) and AP42 standards. It is recommended to use AP42 default values for standard landfills. CAA default values have a high methane generation potential (L_0) of 180 m^3CH_4 Mg^{-1} waste (Scharff and Jacobs 2016).

LandGEM is based on the first-order decomposition rate equation and the inputs required in the model are similar to IPCC 2006 Waste Model such as design of the landfill, amount of waste placed, acceptance of hazardous waste, the methane generation rate constant (k), methane generation potential (L_o) and the years of waste acceptance. Default values for k and L_o can be used or site-specific values can be developed through field test measurement. Both FOD models were tested on Danish landfills and it was concluded that the LandGEM overestimated methane generation and the plausible explanation for such discrepancies was that the waste composition at Danish landfills was different to US waste composition of MSW. This highlights the point that waste composition is a critical factor in the model which then determines the methane generation potential and the methane generation rate constant.

1.5.2 Estimating Methane Emissions from the Wastewater Treatment

Methane emissions from the anaerobic wastewater treatment plants are estimated according to the procedure outlined in chapter 6 of Vol. 5 IPCC 2006 guidelines. The method for estimating CH_4 from wastewater handling requires three basic steps and the three worksheets provided by IPCC are used to calculate each of the following steps:

- Step 1: Estimation of Organically Degradable Material in Domestic Wastewater
- Step2: Estimation of CH_4 emission factor for Domestic Wastewater
- Step 3: Estimation of CH_4 emissions from Domestic Wastewater

To calculate the total organics in the wastewater in inventory year expressed as kg BOD/year, biologic oxygen demand (BOD) per capita in g/pers/year and the population (P) data is required. The equation used to calculate total organics is as follows:

$$TOW = P * BOD * 0.001 * 365$$

The emission factor for a particular treatment system (EF_j) is based on methane producing capacity (B_o) and the methane correction factor (MCF) and is calculated as follows:

$$EF_j = B_o \times MCF_j$$

The default value for B_o used is 0.6 kg CH_4/kg BOD and the MCF used is dependent on the treatment system as outlined in Table 6.3, Chapter 6 of Volume 5. The MCF for a well-managed aerobic plant is 0; for aerobic not well managed and overloaded is 0.3; septic system with anaerobic system is 0.5 and anerobic system is 0.8.

Methane emissions for this category are estimated as follows:

$$CH_4 Emissions = \left[\sum_{ij} \left(U_i * T_{ij} * EF_j \right) \right] * (TOW - S) - R$$

Where:

TOW = total organics in wastewater in inv year, kg BOD/year
S = organic component removed as sludge in inventory year, kg BOD/year
U_i = fraction of population in income group i in inventory year (See Table 6.5, IPCC 2006)
T_{ij} = degree of utilisation of treatment/discharge pathway or system, J, for each income group i in inventory year.
i = income group; rural, urban income and urban income low
j = each treatment/discharge pathway or system

There is N_2O emissions in kg N_2O-N/kg N to some extent in wastewater treatment and are calculated as follows:

$$N_2O\,emissions = N_{effluent} * EF_{effluent} \times 44 / 28$$

$EF_{effluent}$ is the emission factor for N_2O emissions from discharged wastewater and the default factor is 0.005 kg N_2O-N/kg N. The $N_{effluent}$ factor is nitrogen in the effluent discharge to aquatic environment, kg N/year and is calculated as:

$$N_{effluent} = \left(P * protein * F_{NON\,COM} * F_{IND\,COM} \right) - N_{sludge}$$

Where P is human population, protein is the annual per capita consumption, $F_{NON\ COM}$ is the factor for non-consumed protein added to the wastewater (default value of 1.1 is applied as per the IPCC 2006 guidelines), $F_{IND\ COM}$ is the factor of industrial and commercial protein co-discharged into sewer system (default value of 1.25 is applied) and N_{sludge} is the nitrogen removed in sludge, kg N/year (a default value of zero is applied). The N_2O emission from wastewater is not a key category in PICs and is not usually estimated in the national greenhouse gas inventory.

1.5.3 Uncertainties in the Waste Sector

As stated earlier if the default methodology is used for estimation then the uncertainty level is approximately 30%. In the PICs the uncertainty in estimation could be higher as the amount of waste generated is not very well recorded and the waste composition is not known. None of the SWD sites in Fiji apart from Naboro landfill have a weighbridge so the bulk weight is estimated by number of truckloads. To constraint the uncertainty in the waste sector it is strongly recommended to weigh the amount of waste deposited in the SWD sites and to carry out a thorough study on waste characterization of the waste generated in the PICs. If the studies revealed that the organic waste matter is higher than the default values, then certainly the emissions calculated previously were underestimated. In the wastewater sector the major challenge is the data limitations on the type of wastewater treatment systems such as the number of households that use septic tanks within the urban and peri-urban areas. It is very unclear as to how to assess emissions from such treatments.

1.6 Conclusion

It is obvious that a high degree of confidence in the activity data and country specific emission factors are mandatory for an accurate estimation of greenhouse gas emissions from different sectors. In the absence of activity data and emission factor, the default methodology or Tier 1 methodology can be used to enable calculation with highest uncertainty in the estimate. In the PICs, default methodology is used but there is a need for a strong institutional arrangement between government and private sectors to provide data in the correct format needed to estimate emissions. Once the country specific emission factors and activity data are present then a more robust inventory could be attained and consequently provide insights on potential sectors where emission reduction could be targeted effectively. A comprehensive review of the greenhouse gas inventory reported to UNFCCC indicates that more than 50% of emissions are from the energy sector in most of the PICs followed by agriculture and then the waste sector. To transform the Paris Agreement into reality, more focused attention is needed in the energy sector and removals of CO_2 of the AFOLU sector.

References

Bi, J., Zhang, R., Wang, H., Liu, M., & Wu, Y. (2011). The benchmarks of carbon emissions and policy implications for China's cities: Case of Nanjing. *Energy Policy, 39*, 4785–4794.

Cai, B.-F., Liu, J.-G., Gao, Q.-X., Nie, X.-Q., Cao, D., Liu, L.-C., Zhou, Y., & Zhang, Z.-S. (2014). Estimation of methane emissions from municipal solid waste landfills in China based on point emission sources. *Advanced Climate Change Resources, 5*(2). https://doi.org/10.3724/SP.J.1248.2014.08.

Chand, D. (2018). *Evaluation of methane emissions from the agriculture sector and high precision ambient methane measurements in Fiji*. Master of Science thesis, The University of the South Pacific, Fiji.

Chaudhary, R., & Garg, R. (2014). Comparisons of two methods for methane emission at proposed landfill Siteand their contribution to climate change: Indore City. *International Journal of Application or Innovation in Engineering & Management, 3*(5), 9–16.

Ghasemzade, R., & Pazoki, M. (2017). Estimation and modelling of gas emissions in municipal landfill (case study: Landfill of Jiroft City). *Pollution, 3*(4), 689–700.

Goopy, J., Onyango, A., Dickhoefer, U., & Butterbach-Bahl, K. (2018). A new approach for improving emission factors for enteric methane emissions of cattle in smallholder systems of East Africa – Results for Nyando, Western Kenya. *Agricultural Systems, 161*, 72–80.

Holland, E., Nuttal, P., Newall, A., Prasad, B., Veitayaki, J., Bola, A., & Kaitu'u, J. (2014). Connecting the dots: Policy connections between the policy and Pacific Island shipping and global CO_2 and pollutant emission reduction. *Carbon Management, 5*(1), 93–105.

IPCC. (2006). *2006 IPCC guidelines for National Greenhouse gas Inventories*, Prepared by the National Greenhouse Gas Inventories Programme, Eggleston, H. S., Buendia, L., Miwa, K., Ngara, T., & Tanabe K. (Eds). IGES, Japan.

Kamalan, H., Sabour, M., & Shariatmadari, N. (2011). A review on available landfill gas models. *Journal of Environmental Science and Technology, 4*, 79–92.

Kennedy, C., Steinberger, J., Gasson, B., Hansen, Y., Hillman, T., Havranek, M., Pataki, D., Phdungslip, A., Ramaswami, A., & Mendez, G. V. (2009). Greenhouse gas emissions from global cities. *Environmental Science and Technology, 43*, 7297–7302.

Liu, Z., Guan, D., Wei, W., Davis, S. T., Ciais, P., Bai, J., Peng, S., Zhang, Q., Hubeck, K., Marland, G., Andres, R. J., Crawford-Brown, D., Lin, J., Zhao, H., Hong, C., Boden, T. A., Feng, K., Peters, G. P., Xi, F., Liu, J., Li, Y., Zhao, N., & He, K. (2015). Reduced carbon emission estimates from fossil fuel combustion and cement production in China. *Nature, 524*, 335–338.

Lokupitiya, E., & Paustian, K. (2006). Agricultural soil greenhouse gas emissions: A review of national inventory methods. *Journal of Environmental Quality, 35*(4), 1413–1427.

Mani, F. S., Gronert, R., & Harvey, M. (2016). *Pre-feasibility study for methane recovery at Naboro Landfill*. Suva: Final Report on projects funded by PACE-Net Plus seed funding grants 2015.

Michael, J., & Hsu, J. (2008). NF_3, the greenhouse gas missing from Kyoto. *Geophysical Research Letters, 35*(12), 1–3.

Nisbat, Z. (2018). *Measurements of nitrous oxide in background air and assessment of nitrous oxide flux from sugarcane fields*. Fiji: The University of the South Pacific.

Nugent, D., & Sovacool, B. (2014). Assessing the lifecycle greenhouse gas emissions from solar PV and wind energy: A critical meta-survey. *Energy Policy, 65*, 229–244.

Ramaswami, A., Hillman, T., Janson, B., Reiner, M., & Thomas, G. (2008). A demand-centered, hybrid life cycle methodology for city-scale greenhouse gas inventories. *Environmental Science and Technology, 42*(17), 6455–6461.

Ramphull, M., & Surroop, D. (2017). Greenhouse gas emission factor for the energy sector in Mauritius. *Journal of Environmental Chemical Engineering, 5*, 5994–6000.

Reijnders, L. (2012). Life cycle assessment of greenhouse gas emissions. In W. Y. Chen, J. Seiner, T. Suzuki, & M. Lackner (Eds.), *Handbook of climate change mitigation*. New York: Springer.

Santalla, E., Córdoba, V., & Blanco, G. (2013). Greenhouse gas emissions from the waste sector in Argentina in business-as-usual and mitigation scenarios. *Journal of the Air & Waste Management Association, 63*(8), 909–917.

Scharff, H., & Jacobs, J. (2016). Applying guidance for methane emission estimation for landfills. *Waste Management, 26*(4), 417–429.

Spokas, K., Bogner, J., Chanton, J. P., Morcet, M., Aran, C., Graff, C., Moreau-Le Golvan, Y., & Hebe, I. (2006). Methane mass balance at three landfill sites: What is the efficiency of capture by gas collection systems? *Waste Management, 26*, 516–525.

Thompson, S., Sawyer, J., Bonam, R., & Valdivia, J. (2009). Building a better methane generation model: Validating models with methane recovery rates from 35 Canadian landfills. *Waste Management, 29*, 2085–2091.

Velders, G., Fahey, D., Daniel, J., McFarland, M., & Andersen, S. (2009). The large contribution of projected HFC emissions to future climate forcing. *Proceedings of the National Academy of Sciences, 106*(27), 10949–10954.

Wangyao, K., Yamada, M., Endo, K., Ishigaki, T., Naruoka, T., Towprayoon, S., Chiemchaisri, C., & Sutthasil, N. (2010). Methane generation rate constant in tropical landfill. *Journal of Sustainable Energy and Environment, 1*, 181–184.

Xianga, B., Patra, B., Montzka, S., Miller, S., Elkins, J., Moore, F., Atlas, E., Miller, B., Weiss, R., Prinn, R., & Wofsy, S. (2014). Global emissions of refrigerants HCFC-22 and HFC-134a: Unforeseen seasonal contributions. *Proceedings of the National Academy of Sciences, 111*(49), 17379–17384.

Chapter 2
Mitigating Through Renewable Energy: An Overview of the Requirements and Challenges

Anirudh Singh and Pritika Bijay

Abstract Renewable Energy (RE) provides one of the two main means of mitigating climate change. The requirements for the successful implementation of RE projects include RE and human resources, institutional capacity, science and technology infrastructure, policy and legislation, financing and the market-readiness of the renewable energy technology (RET).

This chapter considers each of these requirements in detail in the context of the Pacific Island Countries (PICs) with the help of illustrative examples from both the developed and developing countries, and in particular the Small Island Developing States (SIDS). Developing countries are forced to resort to the device of capacity building in an attempt to make up for their deficiencies.

The chapter then addresses several challenges to the implementation of RE projects in the SIDS. Perhaps chief amongst these is the better ability developed countries have in providing for the requirements for RE implementation than their developing counterparts. This differential ability is amplified in the case of the SIDS, and in particular the PICs. Factors that discriminate against them include their remoteness, lack of manufacturing infrastructure and human resources, finance, institutional frameworks, policy and their difficulty to attract investors. Barriers to RE implementation that are specific to the PICs include the small sizes of their economies, the lack of awareness amongst their decision-makers of the importance of supporting infrastructure development, lack of funding and appropriate institutional support.

RE does not always guarantee a reduction in emissions, and a full Life Cycle Analysis is needed to assess this ability, especially in the case of biofuels.

A. Singh (✉)
School of Science and Technology, The University of Fiji, Lautoka, Fiji
e-mail: anirudhs@unifiji.ac.fj

P. Bijay
Pacific Islands Forum Secretariat, Suva, Fiji

© Springer Nature Switzerland AG 2020
A. Singh (ed.), *Translating the Paris Agreement into Action in the Pacific*,
Advances in Global Change Research 68,
https://doi.org/10.1007/978-3-030-30211-5_2

Keywords Renewable energy (RE) · Climate change mitigation · Small Island Developing States (SIDS) · Pacific Island Countries (PICs) · Renewable energy resources · Institutional capacity · Policy · Science and technology infrastructure

2.1 Introduction: Why Renewable Energy?

The role of renewable energy (RE) in climate change mitigation can be best understood by first investigating the origins of greenhouse gases emitted to the atmosphere (also referred to simply as emissions) that cause global warming.

Man-made (or anthropogenic) greenhouse gas emissions are caused by several **sources** and **processes**, collectively called **activities**. In addition to sources of emissions, there may be entities that actually absorb or remove greenhouse gases from the atmosphere. These are called **sinks**, and are regarded as part of the activities.

To consider these activities further, they are divided into five groupings or **sectors,** consisting of **energy**, industrial process and product use (**IPPU**), Agriculture, Forestry and other Land Use (**AFOLU**), **Waste** and an **Others** sector (IPCC 2006). Of these sectors, the energy sector is by far the most significant, producing about 90% of the carbon dioxide and 75% of the total greenhouse gas emissions in developed countries (IPCC 2006).

The main reason for this dominant role of energy as an emitter of greenhouse gases is the high proportion of fossil fuels in the total global energy mix. As revealed in Fig. 2.1, fossil fuels make up some 78% of the global primary energy consumption today.

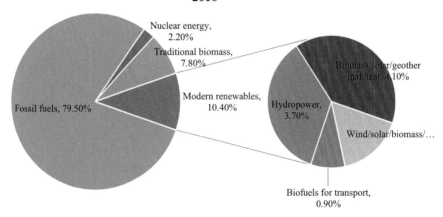

Fig. 2.1 Estimated renewable share of total final energy consumption, 2016. (Source: after REN21 2018)

Fossil fuels are used mainly in **stationary combustion** in power plants and refineries, and **mobile combustion** in the engines of transport vehicles. In both these processes, one of the final products is carbon dioxide, which contributes directly to greenhouse gas emissions. Reducing the percentage of fossil fuels in the global energy mix thus provides an obvious way of reducing the greenhouse gas emissions that cause global warming and the resulting climate change. One mechanism for achieving this is by replacing fossil fuels by renewable energy (RE), as the various types of RE normally produce little or insignificant amounts of carbon dioxide as compared to fossil fuels.

As Fig. 2.1 shows however, RE currently provides only 19% of the total primary energy requirements of the world. It is clear that to mitigate climate change effectively, there should be a concerted effort to increase this share of renewable energy. This chapter discusses how the Small Island Developing State of Fiji can contribute to this cause through the implementation of RE projects.

The other possible method for mitigating climate change involves the more efficient use of energy (thereby reducing the total energy consumed). Such **energy efficiency** measures can be achieved at both the supply (i.e. production) side and demand (or use) side of energy, and the latter has been suggested as a possible contender for the Fiji NDC Implementation Roadmap (Economy, 2017).

This chapter investigates the requirements (i.e. the pre-requisites) for the successful implementation of renewable energy projects that can bring about new reductions in greenhouse gas emissions, and considers the challenges and barriers encountered by the Least Developed Countries (LDCs) and Small Island Developing States (SIDS), including Fiji in particular, in meeting these requirements.

2.2   Requirements for Renewable Energy

2.2.1   Discovering the Requirements

There are certain obvious pre-requisites for the production of RE for human consumption. But the nature and extent of the full set of requirements often becomes clear only after one tries to actually implement the relevant technology. The case study in Box 2.1 is an attempt to illustrate how such a situation arises.

Box 2.1: Identifying the Requirements for Setting Up a Wind Turbine
Upon advice from her friend in the US, a cattle farmer in a small Pacific Island Country (PIC) decided to invest $200K in a wind turbine to provide for the 15 kWp power needed for her farm situated in a remote rural location. Her friend had a similar farm in Arizona USA, and he had made a similar investment very productively.

She tried to find local suppliers for the technology but was unable to do so. She was finally told by an engineering firm that she would have to import the technology as none was available on the market in their island nation. As there was no import precedence for such a commodity, it took a full 9 months for the local supplier and the farmer to have the project approved by the authorities and the import documentation completed. The supplier, a local engineering firm which had no prior experience in wind technology, finally managed to have the technology delivered to her farm where they proceeded to install the turbine and storage system.

They were confronted with some apparent teething problems with the system. The power generated, for instance, was much lower than the rated capacity stated in the documentation. The owner noticed that the blades were idle or hardly turning for long periods of time.

The local engineering firm could not help her. So she consulted the internet for possible leads into the problem and finally managed to locate a company abroad who was willing to advise her. They told her the problem was probably due to the low wind speed profile for the location of the turbine. The cut-in speed (i.e. the minimum wind speed required to turn the blades) of the turbine was 4 m/s. Evidently, this was too high for the average wind speed at the location.

As no detailed wind resource assessment had been carried out for the location, the value of this parameter was unknown. Meteorology Office data gave an average speed of 5.0 m/s for the entire region, but this said nothing about the actual average speed at the location of the turbine. A quick monitoring over a week using a hand-held anemometer (purchased on e-bay) gave an average wind speed of 4.2 m/s for the location.

After more internet research, the problem was finally identified as a poor selection of the exact siting of the turbine. Although the average wind speed for the region was ostensibly sufficient, the speed at the actual location was attenuated due to its low-lying position and shading effects caused by a cluster of rain-trees some 50 m away. The final advice that the owner received from her benevolent internet adviser was to do a thorough wind monitoring study of potential sites nearby and re-locate the turbine to a place with better potential. This was going to be a costly and time-consuming task, and still did not assure a reasonable power output.

The owner had learnt the hard way that there were many more requirements to the generation of wind power than those obvious at first site. She had also become grimly aware that living in a remote PIC had many disadvantages as compared to life in a developed country such as the USA.

This simple case study demonstrates how many essential requirements for the successful implementation of a RE project can remain hidden from view till one actually begins to go through the steps of implementing the project. It also reveals the significant handicap that developing countries have in venturing into RE projects as compared to the developed nations.

In the typical case, the requirements for the successful implementation of RE projects include the availability of the relevant **RE resources**, the required **human resources**, the **institutional capacity** and the **science and technology infrastructure**. They also include the **policy and legislation** as well as the **financing** needed to facilitate the project. It is also assumed that the technology being investigated is **market-ready**, i.e. that it is fully developed and has been in commercial use for some time.

The rest of this section considers each of these requirements in detail.

2.2.2 Renewable Energy Resource Requirements

Renewable energy comprises energy occurring naturally in various forms. Solar energy, for instance, is the energy arriving at the earth's surface from the sun in the form of solar radiation, while wind energy is the kinetic energy contained in the wind. In the utilisation of RE, such natural sources of energy are converted to other forms using the appropriate renewable energy technology (RET). In the case of wind energy for example, the kinetic energy of the wind is turned into electrical energy with the help of a wind turbine.

It is clear therefore that one of the most essential requirements for RE is the appropriate RE resource. Thus, to consider the successful implementation of hydro, biofuel and geothermal energy projects for instance, among the first requirements are the availability of flowing water at a height (called the **head**), edible or inedible vegetable oils and animal fats, and geothermal source of energy below ground respectively.

It must be noted that the availability of RE resources in any country depends on both the geography and geology of that country. One cannot, for instance have ample hydropower potential in a country that is non-mountainous and/or lacks water resources in the form of rivers and streams. Table 2.1 shows the relationship between the geography/geology and RE potentials for some of the Pacific Island Countries (PICs).

Note that the location of these nations within the tropics ensures that solar energy is available in abundance to all. Also, note the irregular distribution of geothermal energy (which depends on the geology) and hydro (which is only available in mountainous countries).

For the successful utilisation of a renewable energy (RE) resource, it is essential to obtain a quantitative assessment of its availability. As the resources usually vary with time, such measurements must be carried out over suitable time periods. RE resource monitoring projects of this nature form integral parts of the feasibility

Table 2.1 Availability of renewable energy resources in the PICs

Country	Geography/geology	Solar (kWh/m²/day)	Wind (m/s)	Hydro	Biomass/biofuel	Geothermal
Cook Islands	15 islands totalling 240 km² of land. Eight Southern islands are elevated and fertile, have approx. 90% of population. Northern islands are low-lying, coral atolls	5.5	6.1–7.5	No (implementation costs were too high)	No	Unknown
Fiji	Land area of 18,333 km² and includes 320 islands – one third inhabited, most population on the main islands of Viti Levu and Vanua Levu. Majority of the land mass is on continental-like volcanic islands that rise to well over 1000 m, many of the outer islands are low-lying raised coral or atoll-type islands	5.1	~7	Yes (Two grid-connected hydro stations (80, 40 MW) on Viti Levu, several micro-hydro on Viti Levu, Vanua Levu and Taveuni, untapped developable hydroelectric potential of 200 MW mostly on on Viti Levu	Yes (Biomass still provides a large fraction of cooking fuel in rural areas. Coconut oil could be used as a feedstock for bio-diesel. Sugar and timber industries produce industrial residues (bagasse and hog fuel) used for power generation	Yes (Assessments show potential at Labasa and Savusavu on Vanua Levu. Potential of 5–15 MW or more of power)
FSM	Land area of 702 km² with an exclusive economic zone exceeding 2.59 million km² distributed over four states: Chuuk, Kosrae, Pohnpei and Yap. Geology varies from high and mountainous to low coral atolls -majority of islands low-lying and resource poor	5.5	Limited assessment	Some potential (Pohnpei has ~ 4–5 MW of developable hydro potential on the Seniphen and Lehnmasi Rivers. Kosrae has a site on the Malem River, potential of ~ 35 kW)	Unlikely	Unknown

Kiribati	One raised coral island (Banaba) and 32 atolls in three island groups (Gilbert, Line and Phoenix) spread over an ocean area of 4200 km east to west and 2000 km north to south with a total land area of 811 km^2	6	No	No	Biomass, mostly in the form of coconut husks, shells and fronds. Biofuel in form of Coconut oil	No
Marshall Islands	Two groups of atolls and islands – Ralik in the west and Ratak to the east. Majuro is main island. 22 of the 29 atolls, and 4 of the 5 small raised coral islands are inhabited. The islands are typically several kilometres long but rarely over 200 m in width	Adequate solar resources	Potential-requires assessment especially Majuro and Ebeye	No	No, except coconut oil used as a biofuel to replace diesel fuel	No
Nauru	A single raised coral equatorial island with land of 21 km^2. Surface area is dominated by pinnacles and outcrops of limestone, after a century of mining of the high-grade tricalcic phosphate rock	6	4.2 (close to cut-off speed of 3.5 m/s)	No	No	No
Niue	Niue is a single raised coral island of 259 km^2	Yes (good)	Potential	No	No	No

(continued)

Table 2.1 (continued)

Country	Geography/geology	Solar (kWh/m²/day)	Wind (m/s)	Hydro	Biomass/biofuel	Geothermal
Palau	458 km² of land area spread over more than 200 islands – mostly on single reef structure that includes the heavily populated islands of Babeldaob, Peleliu and Koror. Babeldaob and Koror are mountainous and volcanic, but Palau also includes raised coral islands and atolls	5.5	No	No	No	No
PNG	Land area of 452,860 km². Mountainous with costal lowlands	Yes (large potential)	Yes (limited data)	Yes	Biomass available but largely inaccessible or unsuited. Possibility of Ethanol (alcohol) can from sugar-cane, molasses, sago palm, nipa palm and other crops	Yes
Samoa	2934 km² of land area, mostly on the islands of Savai'i (58% of land) and Upolu (38%). The climate is warm, humid and tropical with distinct wet and dry seasons	5	Potential	Yes	Limited practical energy potential of woody biomass waste until 2020. biofuels-coconut oil	Yes

Solomon Islands	Nearly 1000 islands – 350 populated – with 28,000 km² of land. The islands are mostly rugged and mountainous and the country is relatively rich in mineral, hydro and forest resources	5	Limited data	Yes	Not easy to assess the available biomass energy resource. biofuel-coconut oil	Yes
Tokelau	3 main atolls with a land area of 12 km² Each atoll consists of a lagoon enclosed by a curving reef with coral islets (motus), typically less than 200 m wide, separated by stretches of reef	5.5	No (only likely to be viable on the reef or in the lagoon)	No	Biofuel-coconut oil	No
Tonga	748 km² land area. Most islands have a limestone base formed from an uplifted coral formation, although some have limestone overlying a volcanic base	5.8	No (wind difficult to due to land issues and the prevalence of coconut trees – not enough fetch)	No	Biomass- available cooking and for drying crops, mainly fish and copra. biofuel-coconut oil	Potential presence due to volcanic activity in Tonga – no data
Tuvalu	Land area of 26 km² over 8 islands, the largest (Vaitupu), has an area of about 5.6 km², smallest (Niulakita), is only 0.42 km². Soil is low in fertility and only a narrow range of food plants can be supported	5.5	5.79 m/s	No	Biomass in the form of coconut trees	No

(continued)

Table 2.1 (continued)

Country	Geography/geology	Solar (kWh/m²/day)	Wind (m/s)	Hydro	Biomass/biofuel	Geothermal
Vanuatu	Land area of 12,200 km². 80 islands, 65 of which are populated. The islands are volcanic, mostly mountainous, with narrow coastal plains	6	Limited data on wind energy potential	Yes	Biomass-Yes as Vanuatu is heavily forested. biofuel-coconut oil	Yes (12 of Vanuatu's islands have thermal springs)

Data source: Singh (2012), Isaka et al. (2013), and similar sources
Note: *PNG* Papua New Guinea

studies of RE projects. To take the example of hydropower, the output power of an installation is approximated quite well by the expression

$$\text{Power}\left(\text{kilowatts}\right) = 10 \times \text{Q}\left(\text{cubic metres per second}\right) \times \text{H}\left(\text{metres}\right)$$

where Q is the volume flowrate of the water and H is the head (Twidell and Weir 2015). The Head is fixed for any location and thus needs to be determined once only. The volume flowrate Q of the water flowing through the turbine will, however, depend on the rainfall, and will therefore show a temporal (in particular seasonal) variation. This parameter must therefore be monitored throughout the four seasons for at least 1 year. As the seasons themselves vary in intensity over time either randomly or due to predictable effects such as the El Nino/Southern Oscillation, it becomes necessary to extend the monitoring period to several years.

The flow rate may be either measured directly using a flowrate meter, or inferred from other data such as the rainfall data which has been collected and maintained by the meteorological office over extended periods of time for the region. Measurement techniques for the flowrate include the simple (or bucket) method, the refined method, the sophisticated method or the Weir method (Twidell and Weir 2015). For spot checks of the flowrate, a flowrate meter is quite adequate.

2.2.3   Human Resource Requirements

No human endeavour can proceed without the input of human effort. In any initiative of national scale, human resources will be required at all stages of policy/decision-making, development and implementation. In the case of the implementation of RE projects, the human resources requirements may be conveniently partitioned into human input for the design and production of the technology, and that for its acquisition and use as an economic commodity.

Examples of human input for the technology development include project managers, scientists, engineers and factory floor workers for the design, implementation and evaluation/testing of the product. The trade and marketing of the technology considered as an economic commodity will require, amongst other things

- The services of customs agents, legal firms, specialised commercial bank staff for the import/export of the technology, as well as
- Human resources for the transportation and distribution of the commodity.

The nature of these human resources required by a country will clearly depend on whether the technology is developed/manufactured locally or imported from abroad. As much of the human capacity discussed above is available locally in developed countries, they are more likely to opt for manufacturing rather than importing the technology. The developing countries (in particular the LDCs and the SIDS) are largely reliant on the developed world for the supply of the technology, and require human resources mainly for importing, distribution, installation and maintenance of the renewable energy technology (RET). Their human resource requirements are therefore likely to vary significantly from those of developed nations.

2.2.4 Science and Technology Infrastructure

Most of modern technology, including RET is dependent on the results of scientific research and device fabrication, and their manufacture relies heavily on manufacturing technology. A science and technology infrastructure is therefore needed for the design, development and manufacture the RET. It typically consists of

- The physical infrastructure such as research laboratories, materials production and fabrication laboratories and the scientists needed to run these laboratories; and
- Governing institutions (an example being the Australian Research Council (ARC 2015)) that oversee the research and development in science and technology.

This requirement has been assessed to be the weakest link in the development chain for RE in developing countries (Singh 2012).

As an example of the role of science and technology in the RE sector, consider the Solar Photovoltaic (PV) system for energy generation. The components of a typical solar PV system consists of several components including an array of solar panels, an inverter, charge controller, battery storage system and the balance of plant. The solar array is constructed from individual solar panels connected together electrically. Each panel in turn is built from several solar cells that collect light energy for conversion to electricity.

The manufacture of these solar cells requires advanced research and development, including the production of ultra-high purity silicon and the fabrication of the solar cell device. Such research and development is only possible at dedicated research labs that only large established companies are capable of supporting. The same is true for the design and manufacture of the electronics used in the other components of the system. Because of these high technology demands, the manufacture of the basic components of solar PV systems normally remains confined to developed countries and other technologically advanced economies such as China and India.

2.2.5 Institutional Capacity

The production and use of RET requires support from several different types of institutions from within the public and private sectors. These institutions include

- Government institutions involved in facilitating and regulating the local production and use of the RET as well as its import;

- Policy institutions (usually affiliated with the government) needed to oversee and formulate policy;
- Financing institutions, including banks and other lending agencies, to provide for the financing of RE projects; and
- Training institutions (universities, technical and vocational institutes) needed for the training and development of the appropriate human resources.

Table 2.2 below lists some training institutions in Fiji that offer various levels of RE training.

2.2.6 Policy and Legislation

Policy and legislation are needed to guide and facilitate the development of the RE sector of a country or region. Quite often however, they also arise from the nation's commitments to regional or global goals.

The motivating factors for the development of renewable energy policy include

- Implementation of global agreements (e.g. UN's Climate Change agreements commencing from the Earth Summit and ending recently with the Paris Agreement; the Sustainable Development Goals (SDGs)
- Regional development goals (e.g. the Framework for Action on Energy Security in the Pacific (FAESP) (SPC 2011)
- National development goals (e.g. the Fiji National Energy Policy) (FDoE 2013)

Table 2.2 Some RE training institutions in Fiji and what they offer

Institution	Name of programme	Training/qualification offered
The University of the South Pacific (USP), Suva, Fiji	BSc (major in Physics); MSc in Physics PhD in Renewable Energy	Undergrad course PH301; Postgrad courses PH407, PH414, PH416; MSc thesis in renewable energy
University of Fiji, Lautoka, Fiji	PostGraduate Diploma in Energy and Environment(PGDEE), and Masters in Renewable Energy Management (MREM)	Postgrad courses REE400, REE401, REM400, REM401, REM402, REM403, REM404, REM406; minor thesis REM407
Pacific TAFE, The University of the South Pacific	Certificate IV in Resilience (Climate Change Adaptation & Disaster Risk Reduction). (*Certificate IV developed by European Union (EU) Pacific Technical and Vocational Education and Training in Sustainable Energy and Climate Change (EU PacTVET), which is being implemented by USP and Pacific Community (SPC)*)	CER41, CER42, CER43, and CER44

Usually the most immediate reason for developing RE policies is the need to provide adequately for the energy needs of a nation. Developing countries, especially the PICs, are heavily dependent on imported fossil fuels for their energy requirements, as a result of which a large component of their import bill are due to fuel imports (SPC 2012). The need to reduce this dependency provides a natural motivation for the introduction of the relevant energy policy.

RE policies are usually integrated into the country's National Energy Policy (NEP) framework. This is true for the policies of both developing and developed countries such as Fiji and Germany.

A National energy policy is usually stated in terms of a vision, mission and goals, and is enacted through an Act of Parliament. The basic structure and terminology of National Energy Policies include the following key concepts:

- Vision, mission, aims and goals;
- Energy sectors;
- Key priority areas;
- Energy targets;
- Energy strategies;
- Energy implementation plans;
- Monitoring and evaluation; and
- Energy Indicators.

Table 2.3 below highlights some of the salient features of Fiji's National Energy Policy.

Policy and legislation are powerful drivers in bringing about change in a country's RE sector. This is brought out well in Germany's Renewable Energy Sources Act 1991 (Singh 2012). Like many of the PICs, Germany has, in the past, been

Table 2.3 Fiji's National Energy Policy – overview of the Targets for Fiji's energy sector (FDoE 2013)

Indicator	Baseline	2015	2020	2030
Access to modern energy services				
Percentage of population with electricity access	89% (2007)	90%	100%	100%
Percentage of population with primary reliance on wood fuels for cooking	20% (2004)	18%	12%	<1%
Improving energy efficiency				
Energy intensity (consumption of imported fuel per unit of GDP in MJ/FJD)	2.89 (2011)	2.89 (−0%)	2.86 (−1%)	2.73 (−5.5%)
Energy intensity (power consumption per unit of GDP in kWh/FJD)	0.23 (2011)	0.219 (−4.7%)	0.215 (−6.5%)	0.209 (−9.1%)
Share of renewable energy				
Renewable energy share in electricity generation	56% (2011)	67%	81%	99%
Renewable energy share in total energy consumption	13% (2011)	15%	18%	25%

almost totally dependent on imported fossil fuels from its neighbours for its energy needs. Prior to 1991, the country was largely reliant on coal and nuclear energy for its power generation.

In 1991, motivated by concerns for nuclear safety after the1986 Chernobyl disaster and the country's commitment to climate change initiatives, the country passed the Electricity Feed-in Act to encourage the renewable energy to be fed into the grid (Leiren and Reimer 2018). Under this Act, grid operators were to give priority to renewable energy power providers, and a feed-in tariff was introduced for the first time to incentivise RE providers.

The 1991 Act had to face some legal challenges, resulting eventually in its replacement by the Renewable Energy Sources Act of 2000. Since its inception at the turn of the century, this Act has been highly successful in supporting the entry of renewables into the German grid system. As a result of this Act, the renewable energy share in total energy consumption tripled between 1998 and 2008 (Singh 2012). The Act has been revised several times, with the latest review occurring in 2017 (LSE 2019).

2.2.7 Finance

The availability of funding is a determining parameter in the success of a renewable energy project. Obtaining funding is rarely a straight-forward task of filling in a form with the necessary details. It is a complex process, and success usually depends on several key variables. They include the availability of opportunities, and knowledge and understanding of the mechanisms and instruments for obtaining financial assistance from the available sources.

Sources of funding consist of both global/regional financing agencies, such as the World Bank (WB), the Asian Development Bank (ADB), and more recently the Green Climate Fund, as well as national financial institutions including local development banks, commercial banks and other lending institutions. These institutions use specific financial mechanisms and instruments to evaluate and administer applications for funding. They also often provide mechanisms that are designed to facilitate loan approvals by other institutions. Making successful bids for funding depends on your understanding of how these institutions operate together, the nature and requirements of the financial mechanisms and instruments, as well as your skill in writing a successful funding proposal.

A financial mechanism for RE loans that was introduced by the Fiji Department of Energy in association with the Global Environment Fund (GEF) was the *Sustainable Energy Financing Project*. Box 2.2 below provides an account of how this scheme operated.

Box 2.2: The Sustainable Energy Financing Project (SEFP): An Example of an Enabling Financial Mechanism for RE Projects
Part of the requirement for obtaining funding for RE projects is a loan guarantee (or collateral) given to the loan provider. Any financial institution providing loans for RE projects will require assurance that someone (the guarantor) will pay back your loan in case you are unable to. A common difficulty with most Medium and Small to Medium Enterprises (MSMEs) is finding such guarantors, with the result that the application process ceases in mid-stream.

To alleviate this problem, the Global Environment Facility (GEF) in conjunction with the World Bank (WB) set up the SEFP (World Bank, 2019). This is a "Risk Sharing Facility (RSF) which is channelled through approved Participatory Financial Institutions (PFIs)" for making available 50% of guarantees to encourage financial institutions to provide loans for RE and EE investments.

The stated aim of the SEFP was

"to increase the adoption and use of RE technologies and more efficient use of energy through a package of incentives to encourage local financial institutions to participate in sustainable energy finance" (Sur 2017).

The project, which was initiated in June 2007 for Fiji, had the Fiji Development Bank (FDB) and the Australia and New Zealand (ANZ) bank as the starting PFIs. The technologies that were supported were

1. Solar Photovoltaic (Solar PV)
2. Pico-hydro
3. Coconut oil as replacement of diesel fuel in diesel gensets and transportation vehicles.

The Fiji Department of Energy (FDoE) was given the Executive Authority to implement and monitor the project (FDoE n.d.). ANZ was assigned the part of Fund Manager.

The FDB's role was to execute the primary function of the scheme, which was to issue loans for the prospective investors in Renewable Energy and Energy Efficiency (RE and EE) projects (FDoE n.d.).

Outcomes:

In 2011, the FDB began making loans of up to $FJ1.0m at an interest rate of 5% to MSME's. Successful projects funding applications included

- Eco-tourism retreats for Fiji's tourism industry, and PV power systems in the 100–300 kW range for

  - A Mariner
  - Poultry farm

(continued)

Box 2.2 (continued)

- – Garment factory
- – Supermarket
- – Tourism resort

By 2015, the FDB had made 26 loans totalling $FJ5.6 m. The overall outputs of the project included

- • US$21.5 M in loans for RE and EE equipment
- • 4.3 MW supply of RE power for over 100,000 people
- • More than 13,500 units of solar PV
- • 11 biofuel generators and 1 biofuel mill

In 2018, the SEFP was extended to include the Pacific Island Nation of Vanuatu, and the life-span of the project was increased to 31st December 2022.

The above example is a demonstration of the enabling effect that a carefully-designed financial mechanism can have in facilitating the implementation of renewable energy projects in the PICs.

2.3 Capacity Building

We saw above that developing countries lack several of the important requirements for facilitating the utilisation of their RE resources. These capacities therefore need to be developed for the specific needs through a process of capacity building.

This is conveniently performed in the PICs via capacity building projects, usually funded by Development Partners (i.e. countries and international organisations actively involved in the energy development of the PICs). Table 2.4 below provides a list of EU-funded energy capacity-building projects that were implemented in the Pacific over the past two decades.

2.3.1 Market-Readiness of Technology

Perhaps one of the most important requirements for the utilisation of RE is the renewable energy technology (RET) itself. This must meet certain minimum quality standards. In particular, the technology must be

- • Mature (i.e. tested through use and shown to be working and reliable); and
- • Market-tested (i.e. been in the market as a commercially-available technology for some time).

Table 2.4 EU-funded renewable energy capacity building projects in the Pacific (EDD 2014)

Project Name	Organisation	Dates	Description
European Union Pacific Technical Vocational Education and Training in Sustainable Energy and Climate Change Adaptation Project (EU PacTVET)	SPC and USP	August 2014 (53 months)	The EU PacTVET project is component three within the broader Adapting to Climate Change and Sustainable Energy (ACSE) programme. The project builds on the recognition that energy security and climate change are major issues that are currently hindering the social, environmental and economic development of Pacific African Caribbean and Pacific (P-ACP) countries
LifeLong Learning for Energy security, access and efficiency in African and Pacific SIDS (L³EAP)	Hamburg University of Applied Sciences, USP, University of Mauritius and Papua New Guinea University of Technology	11 October 2013–10 October 2016	The aim of the project was to strengthen the capacity of the partner HEIs in order to provide high level skills and training required for the energy labour market. In addition, the project also increased the academic and management capacity of university staff to modernise their educational and research programmes and activities, so as to develop labour market oriented lifelong learning concepts for the education of public and private sector staff in meeting the challenges of regional and national energy security and efficiency of supply
Small Developing Island Renewable Energy Knowledge and Technology Transfer Network (DIREKT)	Hamburg University of Applied Sciences, USP, University of Mauritius and University of West Indies	2009–2012	DIREKT was an international cooperation scheme involving universities from Germany, Fiji, Mauritius, Barbados and Trinidad & Tobago with the aim of strengthening the science and technology capacity in the field of renewable energy of a sample of ACP (Africa, Caribbean, Pacific) small island developing states, by means of technology transfer, information exchange and networking. Developing countries are especially vulnerable to problems associated with climate change and much can be gained by raising their capacity in the field of renewable energy, which is a key area of interest in climate change mitigation

(continued)

Table 2.4 (continued)

Project Name	Organisation	Dates	Description
Pacific Centre for Renewable Energy and Energy Efficiency (PCREEE) - A SE4ALL Centre of Excellence to Promote Sustainable Energy Markets, Industries and Innovation	SPC	September 2016 (48 months)	In line with the decisions of the Ministers of Energy of the Pacific Islands States and Territories (PICTs), the aim of the project was to establish and implement the first operational phase of the Pacific Centre for Renewable Energy and Energy Efficiency (PCREEE). The centre represents an innovative fusion of regional and international efforts and capabilities. Its design leverages a network of intra and extra-regional partnerships, serving as a "hub" for knowledge and technical expertise on matters related to sustainable energy projects' implementation. It will also serve as a facilitator for innovative partnerships with the private sector
SPC/GIZ Coping with Climate Change in the Pacific Islands Region (CCCPIR) Project	SPC and Deutsche Gesellschaft für Internationale Zusammenarbeit (GIZ)	2009–2018	The regional SPC/GIZ programme 'Coping with climate change in the Pacific Island Region' (CCCPIR) aims at strengthening the capacities of Pacific Island Countries (PICs) and regional organisations to cope with the anticipated effects of climate change that will affect communities across the region
			The CCCPIR is focusing on key economic sectors such as agriculture and livestock, forestry, fisheries, and tourism. Further focal areas are energy and education. Improving the sustainable supply of energy with a focus on enhancing renewable energy and energy efficiency is critical for PICs to increase the resilience of their economies. Integrating climate change considerations into primary and secondary education and technical and vocational training (TVET) is also vital to equipping young Pacific Islanders with the knowledge and skills required to cope with the effects of climate change

(continued)

Table 2.4 (continued)

Project Name	Organisation	Dates	Description
North Pacific ACP Renewable Energy and Energy Efficiency Project (North-REP)	SPC	2011–2016	North-REP was a multi-country programme where 3 of SPC island member countries (FSM, RMI & Palau) pooled their combined EURO 15.49 m of EDF 10 resources, which have been identified for the development of their energy sectors. SPC is entrusted with the responsibility of managing it. SPC's regional office for the North Pacific in Pohnpei, FSM houses the North-REP management office with Energy Specialists based in FSM, Palau and RMI
EU Adapting to Climate Change and Sustainable Energy (ACSE)	GIZ and SPC	2014 (ongoing)	The ACSE programme will help the 15 Pacific ACP countries (Cook Islands, East-Timor, Fiji, Kiribati, Federated States of Micronesia, Nauru, Niue, Palau, Papua New Guinea, Republic of the Marshall Islands, Samoa, Solomon Islands, Tonga, Tuvalu and Vanuatu) to address three main challenges common to all of them:
			1. Adapting to climate change;
			2. Reducing their reliance on fossil fuels; and
			3. Capacity building.
Support to the Energy Sector in 5 Pacific Island States (REP-5)	Pacific Islands Forum Secretariat (PIFS), IT Power (ITP), Transénergie, Ademe - French Environment and Energy Management Agency	2006–2010	The REP-5 Programme "Support to the Energy Sector in 5 Pacific Island States" was a regional initiative implemented by PIFS supported with funds from the ninth European Development Fund (EDF9). REP5 focused on renewable energy and demand-side energy efficiency activities. There are five countries participating in the REP-5 programme. These are the FSM, Nauru, Niue, Palau and RMI

Arguably one of the most mature RETs is hydro. This has been generating electricity successfully in various forms for more than 100 years (iha n.d.). During its early history (before electricity was discovered) hydropower was harnessed by waterwheels to perform mechanical tasks such as operating flour mills.

Photo-voltaic (PV) power generation has an encouraging history. The technology was developed several decades ago when solid state semiconductor devices first began appearing on the market. Up till a few years ago however, this RET was too expensive because of the price of the solar panels (also called modules) which provide the basic building block of these power systems. But the (relatively) recent entry into the energy market of China has helped to drive prices down, and as a result this technology is now widely used in all sizes, from roof-top solar systems to large solar farms, all over the world.

Among the least mature RETs are **wave energy** and Ocean Thermal Energy Conversion (**OTEC**). While several devices have been tested and show promise, many are still at the Pilot Project stage of their development. In the case of wave energy, the **Oscillating Water Column (OWC)** device has been tested at the 296 kW grid-connected Mutriku Power Plant in Spain since 2011 (Torre-Enciso et al. 2009), while the **Pelamis Wave Energy Converter** developed in Scotland (EMEC n.d.) has attracted considerable attention as a potentially viable wave energy device.

In the case of OTEC, an experimental power plant has been running on the Kona Coast in Hawaii for some time (Clancy 2016). Interest in the technology has been re-kindled in Korea more recently, and the Korea Research Institute of Ships and Engineering (KRISO) is in the process of developing a 200 kW plant to become operational in Kiribati from mid-2019 (Trauthwein 2016). While these developments are taking place however, both wave energy and OTEC fall short of the full status of market-readiness required of commercial technologies.

2.4 The Challenges

2.4.1 Differential Ability to Meet the Requirements

In the previous section, we have seen several examples where developing countries are unable to provide the requirements for the implementation of RE projects while their developed counterparts are much better equipped to do so. This differential ability between the rich and poor nations to meet the requirements is a key issue in the implementation of RE projects. And the situation is exacerbated by the fact that this disparity tends to depend on the nature of the developing country.

Human resources, infra-structure and finance, for instance, are generally lacking in all developing countries. These countries are not able to develop the science and technology infrastructure and are confronted with unique barriers in their efforts.

The Small Island Developing States (SIDS) however have additional issues due to their remote and/or isolated nature. This is especially important in the case of the PICs such as Fiji.

Remoteness leads to supply chain issues, seriously affecting the availability of resources and technology with consequent mounting costs. The lack of manufacturing infrastructure means that developing countries face enormous problems in producing the RETs themselves. This forces them to import the technology as the only option. This in turn requires them to establish a whole new import industry with its own infrastructure requirements.

A challenge that is non-discriminating between the rich nations and poor relates to the observation that not all RE resources will yield net emissions reductions in all situations. Some RE technologies, such as biofuels for transportation, may produce emissions during their production and use that are close to or may even exceed the emissions (from fossil fuels) they are trying to avoid. It thus becomes essential to obtain a careful assessment of these additional emissions.

To summarise, some of the major challenges and barriers encountered by the SIDS in implementing RE as clean energy to facilitate carbon emissions reductions include

- The lack of infra-structure, human resources, finance and institutional and policy frameworks;
- The difficulty in attracting investors; and
- Assessing the suitability of a particular renewable energy to produce net reductions in carbon emissions.

These issues are considered in more detail below.

2.4.2 Lack of Capacity

The lack of capacity in meeting the RE requirements noted above has a debilitating effect on developing nations in their pursuit for the adoption of the renewable energy option. What is of concern is their (sometimes) extreme difficulty in overcoming such deficiencies.

In the case of science and technology infrastructure, for instance, their attempts to redress the issue is often thwarted by the lack of a starting base, the relevant human resources as well as their physical remoteness from the developed countries. This is especially true of the PICs, which do not benefit from economies of scale due to their small populations.

The wind turbine example of Box 2.1 illustrates how the absence of a science and technology base can have a lasting effect on the implementation of a simple RE project. An issue that would plague that project through its entire lifetime would be service and maintenance of the turbine system. The only ostensible solution would be for the local engineering firm to acquire the relevant knowledge. This however would mean going through a steep learning curve for very marginal returns, a situation which provides little incentive to the prospective supplier to develop the service.

Among the specific barriers and constraints encountered by SIDS in their efforts to develop their RE sector are

- The small sizes of their economies which cannot grow or sustain the required research and technology structures and human capacity;
- The lack of appreciation by national decision-makers of the importance of such supporting infrastructure to development (Singh 2012);
- Lack of funding; and
- The lack of institutional and policy support that can facilitate such development of a science and technology base.

Table 2.5 shows how the SIDS, and in particular the PICs, compare with the developed nations in their ability to satisfy the requirements for the successful implementation of RE projects.

In the wake of the launching of the UN Secretary-General's Sustainable Energy for All (SE4ALL) initiative in 2011, the Fiji government carried out an analysis of its own energy access needs. Table 2.6 summarises some of the key gaps and barriers to ensuring energy access to all citizens of Fiji that were identified by the government in this study that concluded in 2014 (Government of Fiji 2014). It provides specific examples of the difficulties faced by the PICs as noted above, and also reveals the importance of relevant and complete data and information on energy needs as an additional requirement.

Table 2.5 Major differences between the rich and poor countries in their abilities to implement RE projects

	Requirement	Developed country competence	SIDS competence
1	Human resources (HR)	Full range of expertise available, training institutions available to train the HR	Generally lacking. Limited training institutional capacity means expertise has to be largely imported
2	Science and Technology infrastructure	Basic infrastructure available – depends on state of development of the country	Poor physical infrastructure such as laboratories, scarce scientific human resources; scientific governing institutions often totally lacking
3	Institutional capacity	Adequate governing, training and financing institutions	Poor or non-existent governing, training and financing institutions
4	Policy and legislation	Policies and legislations in place or can be speedily developed	Policies and legislations often lacking or at a developmental stage
5	Finance	Available via various types of lending agencies, schemes and mechanisms	Largely unavailable, lending agencies reluctant to lend due to collateral issues, financing largely foreign aid-dependent

Table 2.6 Typical gaps and barriers to ensuring energy access for all in Fiji (Government of Fiji 2014)

Requirement	Main gaps and barriers
Institutional and policy framework	1. Legislation for the relevant RE sector is not complete
	2. Data on energy does not exist or is unavailable to the public and potential investors
Programmes and financing	1. No rural electrification master plan exists
	2. Data on household cooking fuels is incomplete
	3. Detailed resource assessments are not available for all RE sectors
	4. Inadequate funding for rural electrification programmes by the government
Private sector investment	1. There is minimal private sector investment in power production, purportedly due to tariff and regulatory barriers
	2. Lack of net-metering legislation prevents individual households in participating in distributed electricity generation

2.4.3   Attracting New Players

Rising power demand and the need for rural electrification in the PICs has created an urgent demand for more power generation in the region. Such development however cannot proceed without additional funding. In the case of Fiji, almost all electrical power is produced by Energy Fiji Limited (EFL – previously FEA), the country's only national utility, which has limited capacity to cope with this extra burden. The only perceivable solution would seem to be sharing the additional generation burden with Independent Power Producers (IPPs) who are capable of sourcing their own funding.

While this solution might sound simple in principle, it has proven difficult to implement in Fiji's case, where the national utility has, over the years, only been able to acquire the services of two providers (the Fiji Sugar Corporation (FSC) and Tropic Wood) to share the grid load (FEA 2017). Part of the reason was the existence of several barriers to such an IPP strategy. In an effort to remove these barriers, the Fiji Department of Energy introduced the Fiji Renewable Energy Power Project (FREPP) (see Box 2.3 below) in 2011.

Box 2.3: The Fiji Renewable Energy Power Project

The Fiji Renewable Energy Power Project (FREPP) was a GEF-funded project implemented by the Fiji Department of Energy (FDoE) in partnership with the UNDP which aimed to remove barriers (including policy, regulatory, market, financial and technical barriers) to grid-connected power generation in Fiji. The project was initiated in 2011 and lasted till the end of 2014.

It consisted of four main components, addressing specific categories of barriers (EDD 2014):

(continued)

Box 2.3 (continued)

1. Energy Policy & Regulatory Frameworks (including a power development plan for Fiji);
2. RE Resource Assessments and RE-based Project Assessments (including feasibility studies of new power generation from various RE sources)
3. RE-based Power Generation Demonstrations, and
4. RE Institutional Strengthening (including a framework for IPP procurement for large-scale power generation, connecting small-scale RE technologies to the grid and transfer of regulatory functions for the power generation sector to government).

FREPP was expected to facilitate investments in RE-based power generation in Fiji, which would support the socio-economic development of the country, make use of its RE resources and reduce GHG emissions. By the end of the project in 2014 however, while some investors had signed MOUs with the grid-owner, no new power was being provided by IPPs apart from that produced by the two existing IPPs.

The situation had changed only slightly by 2018, with the addition of the 12MW rated capacity Nabou Biomass Power Plant operated by the Nabou Green Energy Ltd. in 2017 (Nabou Green Energy Ltd 2018). EFL has since noted the prospects of a 3–5 MW Solar farm in Western Viti Levu (FEA 2017), while discussions continue with other potential IPPs on new hydro generation in the interiors of Viti Levu and Vanua Levu. The totality of the power produced by all these new sources however, will fall very short of the country's power needs and the expectations of the country's NDC Implementation Roadmap (Fiji Ministry of Economy 2017).

As revealed in Box 2.3, the outcomes of FREPP have been far from satisfactory. Among the probable reasons for the lack of interest shown by investors, the most obvious is the lack of monetary incentive. The feed-in tariff rate of FJ$0.33 per kWh appears to be still too small. Overall, it is clear that there is a continuing need to improve the enabling framework that is needed to facilitate the development of the RE industry in Fiji.

2.4.4 Assessment of RE as a Mitigating Agent

Under what conditions does the utilisation of RE contribute to mitigation? To answer this question, one must first note that the production and use of RE itself produces emissions of greenhouse gases. RE only qualifies as a mitigating agent if it emits less greenhouse gases per unit energy produced than fossil fuel.

To assess the actual emission reduction potentials of various forms of renewable energy, one must note that RE projects invariably require fossil fuel use at some stage of their project development life cycle. The RE resources can be divided into two categories:

- Resources which do not lead to emissions except for certain phases of their life cycles, including those that have embedded emissions (i.e. emissions during their manufacture) only. The analysis of these resources and technologies can be easily dealt with using the concepts of carbon debts and their "payback periods".
- Others that produce emissions in a variable manner, depending on the exact production/use pathway chosen. The net emissions here can become quite significant, and may even become positive (Flugge et al. 2017; Hofstrand 2019). These need a comprehensive quantitative evaluation of the emissions (e.g. the carbon emission).

A Life Cycle Assessment (LCA) (Muralikrishna and Manickam 2017)is needed to carry out the assessment of the actual emission reduction potential of the RE resources. To facilitate this, it is convenient to picture the RE utilisation as a process in which a renewable energy fuel is converted using a renewable energy technology (RET) to produce the final useful form of energy. Note that the RET may involve either a single (one shot) energy conversion (e.g. solar PV cells) or several stages (e.g. biomass power generation, involving the burning of biomass in a furnace to produce heat, which is then converted sequentially to mechanical and electrical energy).

Emissions are possible from the following stages of this energy cycle:

(i) Fuel cultivation/preparation/transportation (e.g. biofuels);
(ii) Technology fabrication (e.g. embedded emission during the manufacture of solar PV cells); and
(iii) Technology installation (e.g. hydropower civil works, technology transportation and installation).

An LCA enables a quantitative evaluation of the emissions from RE relative to fossil fuels. A quick qualitative evaluation can, however, be obtained even without the use of the LCA technique. Table 2.7 shows how such an analysis can provide a useful assessment of the emissions from a RE source.

The table reveals that whereas the total emissions per unit output energy remains small in comparison with fossil fuel emissions for solar, wind and hydro, it can become significant, and indeed exceed that for fossil fuel in the cases of biomass power generation and biofuels for transportation. Extreme care must thus be exercised in the manner in which these latter RE fuels are produced and used.

Table 2.7 Comparison of life cycle emissions from various RE sources with fossil fuel emissions

	RE source	Emissions from fuel cultivation/transportation/manufacture	Embedded emissions in technology	Emissions during technology installation	Total emissions/MJ output	Comparison with diesel emission/MJ
1	Solar	NA	Yes	Yes	small	Much smaller/insignificant
2	Wind	NA	Yes	Yes	small	Smaller/insignificant
3	Hydro	NA	Yes	Yes	Can be significant during early lifespan	Small, and reduces quickly over the years to smaller
4	Biomass (power generation)	Can be large	Yes	Yes	Can be large	Can be larger
5	Biofuels (transportation)	Can be large	Yes	Yes	Can be large	Can be larger

2.5   Summary and Conclusions

- The energy sector is by far the most significant contributor to greenhouse gas emissions. Most of the emission is produced from the fossil fuel component of the global energy mix, of which renewable energy (RE) makes up a mere 19%;
- Climate change can be mitigated by increasing the share of RE in the global energy mix. The implementation of new RE projects thus plays a vital role in climate change mitigation;
- The key requirements for the implementation of RE are the RE resources, human resources, the science and technology infrastructure, institutional capacity, policy and legislation, finance, and the availability of market-ready technology;
- Developed countries have better capacity for the production, distribution and installation of renewable energy technologies (RETs) than developing countries. They face less challenges in meeting the requirements for the implementation of RETs than their developing counterparts, and are better-placed to cope with any challenge;
- Developing countries are faced with several challenges in meeting the requirements for the implementation of RE; and
- Developing countries are unlikely to have the resources for the production/manufacture of the technology, and usually have no alternative but to import from developed countries.

References

ARC. (2015). *About the Australian Research Council*. Retrieved January 8, 2018, from Australian Research Council: https://www.arc.gov.au/about-arc

Clancy, H. (2016, June 15). *Ocean energy: Will Hawaii take the plunge?* Retrieved January 15, 2019, from GreenBiz: https://www.greenbiz.com/article/ocean-energy-will-hawaii-take-plunge

EDD. (2014). Retrieved November 26, 2018, from Pacific Regional Data Repository (PRDP) for Sustainable Energy for All (SE4ALL): http://prdrse4all.spc.int

EMEC. (n.d.). *European marine energy centre*. Retrieved February 5, 2019, from PELAMIS WAVE POWER: http://www.emec.org.uk/about-us/wave-clients/pelamis-wave-power/

FDoE. (2013). *Fiji National Energy Policy 2013–2020*. Suva: Government of Fiji.

FDoE. (n.d.). *Sustainable Energy Financing Project (SEFP)*. Retrieved February 10, 2019, from Fiji Department of Energy: http://www.fdoe.gov.fj/index.php/power-sector/sustainable-energy-financing-project-sefp

FEA. (2017). *Fiji Electricity Authority Annual Report 2016*. Suva.

Fiji Ministry of Economy. (2017). *Fiji NDC Implementation Roadmap 2017–2030*. Suva: Fiji Ministry of Economy.

Flugge, M., Lewandrowski, J., Rosenfeld, J., Boland, C., Hendrickson, T., Jaglo, K., … Pape, D. (2017). *A life-cycle analysis of the greenhouse gas emissions of cornBased ethanol*. Report prepared by ICF under USDA Contract No. AG-3142-D-16-0243. January 30, 2017.

Government of Fiji. (2014). *Sustainable energy for all (SE4All): Rapid assessment and gap analysis*. Suva: Government of Fiji.

Hofstrand, D. (2019, February 10). *Greenhouse gas emissions of corn ethanol production*. Retrieved from Ag Marketing Resource Center Renewable Energy Newsletter: https://www.agmrc.org/renewable-energy/climate-change/greenhouse-gas-emissions-of-corn-ethanol-production

iha. (n.d.). *A brief history of hydropower*. Retrieved January 31, 2019, from International Hydropower Association: https://www.hydropower.org/a-brief-history-of-hydropower

IPCC. (2006). *2006 IPCC guidelines for National Greenhouse Gas Inventories, Prepared by the National Greenhouse Gas Inventories Programme* (H. S. Eggleston, L. Buendia, K. Miwa, T. Ngara, & K. Tanabe, Eds.). Hayama: Institute for Global Environmental Strategies

Isaka, M., Mofor, L., & Wade, H. (2013, November 13). *Pacific lighthouses: Renewable energy opportunities and challenges in the Pacific Islands region*. Retrieved November 13, 2018, from http://www.irena.org/publications/2013/Sep/Pacific-Lighthouses-Renewable-Energy-Roadmapping-for-Islands

Leiren, M. D., & Reimer, I. (2018). Historical institutionalist perspective on the shift from feed-in tariffs towards auctioning in German renewable energy policy. *Energy Research & Social Science, 43*, 33–40.

LSE (2019). *Renewable Energy Sources Act (EEG, latest version EEG 2017)*. Retrieved February 1, 2019, from London School of Economics and Political Science: http://www.lse.ac.uk/GranthamInstitute/law/renewable-energy-sources-act-eeg-latest-version-eeg-2017/

Muralikrishna, I. V., & Manickam, V. (2017). Chapter five – Life cycle assessment. In I. V. Muralikrishna & V. Manickam (Eds.), *Environmental management: Science and engineering for industry* (pp. 57–75). Oxford: Elsevier Inc.

Nabou Green Energy Ltd. (2018). *About us: Nobou green energy Ltd*. Retrieved February 8, 2019, from Nobou Green Energy: http://ngel.com.fj/

REN21. (2018). *Renewables 2018 global status report*. Paris: REN21 Secretariat.

Singh, A. (2012). Renewable energy in the Pacific Island countries: Resources, policies and issues. *Management of Environmental Quality: An International Journal, 23*(3), 254–256.

SPC. (2011). *Towards an energy secure Pacific: A framework for action on energy security in the Pacific*. Suva: SPC Suva Regional Office.

SPC. (2012). *Country energy security indicator profiles 2009*. Suva: Secretariat of the Pacific Community. Retrieved February 15, 2019 http://www.spc.int/edd/section-01/ energy-overview). http://edd.spc.int/section-01/energy-overview/179-country-energy-security-indicator-profiles-2009

Sur, M. (2017). *Pacific Islands – sustainable energy finance project: Restructuring (English)* Retrieved February 10, 2019, from The World Bank Group: http://documents.worldbank.org/curated/en/631501507081771057/Pacific-Islands-Sustainable-Energy-Finance-Project-restructuring

Torre-Enciso, Y., Ortubia, I., López de Aguileta, L. I., & Marqués, J. (2009). Mutriku wave power plant: From the thinking out to the reality. *Proceedings of the 8th European Wave and Tidal Energy Conference* (pp. 319–329). Uppsala: Uppsala universitet.

Trauthwein, G. (2016, January 29). *Energy from the ocean: The ocean thermal energy converter*. Retrieved February 10, 2019, from Marine Technology News: https://www.marinetechnology-news.com/news/energy-ocean-ocean-thermal-527332

Twidell, J., & Weir, T. (2015). *Renewable energy resources*. Taylor & Francis.

Chapter 3
Modeling and Forecasting Renewable Energy Resources for Sustainable Power Generation: Basic Concepts and Predictive Model Results

Ramendra Prasad, Lionel Joseph, and Ravinesh C. Deo

Abstract The utilization of renewable energy is an essential tool for the mitigation of the negative impacts of a changing climate on water resources, ecosystems and human lives. However, the intermittent nature of renewable energy poses a practical challenge for its wider applicability such as in electrical power grid utilization. Accurate modeling and forecasting of renewable energy resources, such as wind and solar energy are necessary to make the energy generation process easy and relatively reliable for utilization in grid energy systems. As the development of physics-based models can be economically costly and includes many assumptions and constraints, the emergence of data-driven and machine learning modeling approaches are becoming attractive and viable alternative tools. This chapter outlines the respective phases required for the development of machine learning models in the renewable energy generation sector, including the pertinent concepts and definitions that are used in time-series forecasting approaches. Such information provides a handy tool for engineers and renewable energy practitioners, as well as novice forecasters who wish to explore the practicality of methods for real-life forecasting of power. To provide insights into how machine learning could be used in the energy modeling sector, a 10-min wind speed forecasting case study for Rakiraki, Fiji is carried out using the widely adopted artificial neural network (ANN) model, and the results are compared with multiple linear regression (MLR) models. This serves as an example of how the different phases of model development need to be implemented in a power forecasting study. The present study reveals a better performance of the ANN model over the MLR model in a 10-min wind speed forecasting problem.

R. Prasad (✉) · L. Joseph
School of Science and Technology, Department of Science, The University of Fiji,
Lautoka, Fiji

R. C. Deo
School of Agricultural, Computational, and Environmental Sciences,
Centre for Sustainable Agricultural Systems & Centre for Applied Climate Science,
University of Southern Queensland, Springfield, QLD, Australia

© Springer Nature Switzerland AG 2020   59
A. Singh (ed.), *Translating the Paris Agreement into Action in the Pacific*,
Advances in Global Change Research 68,
https://doi.org/10.1007/978-3-030-30211-5_3

Keywords Forecasting · ANN · MLR · Wind energy · Renewable energy · Climate change

3.1 Introduction

The United Nations Framework Convention on Climate Change (UNFCCC) was a global treaty reached at the UN's Earth Summit in Rio de Janeiro in 1992 to combat and mitigate the drastic impacts of climate change. This action culminated in the Paris agreement of 2015, under which more than 190 nations of the world have committed themselves to reducing their greenhouse gas emissions.

The Parties to this agreement also include the (relatively low polluting) small Pacific Islands Nations such as Fiji which intends to reduce carbon emissions from its energy sector by about 30% by the year 2030 (Ministry of Economy 2017). Australia, the largest South Pacific Ocean country, also intends to reduce its greenhouse gas emissions by 26–28% below the 2005 levels by the year 2030 (Weiss et al. 2016). Amongst the other large players, the European Union has committed to reduce its domestic greenhouse gas emissions by at least 40% by the year 2030 in comparison to the base year of 1990 (Latvian Presidency of the Council of the European Union 2015).

With the commitment to reduce GHG emissions coming concurrently with the need to satisfy an ever-increasing demand for energy both in industrialized and developing nations, a global shift towards increasing the utilization of renewable energies for electricity generation has become apparent. This in turn means that more electricity will be generated from renewables, and will be fed into the respective national power grid in the future. For instance, the commitment of Fiji is to reach 100% electricity generation from renewables by 2030 (Ministry of Economy 2017). To support the growing renewable energy sector, there is a significant need for accurate and reliable resource planning and management (both on the supply-side as well as the demand-side). Such energy management can be enhanced by means of reliable modeling and forecasting approaches for renewable energy resources.

Modeling and forecasting provides a priori knowledge for human survival, including short and long-term prudent planning, efficient management of resources, and sustainability. The power generation from conventional non-renewable energy sources (*e.g.*, coal and gas) can easily be predicted and controlled on the basis of market demand and supply mechanisms. However, the power generation from clean and renewable resources are inherently non-programmable (Artipoli and Durante 2014). Renewable energy power generation particularly from solar (considering its diurnal variability) and wind resources are highly variable in nature that requires very short-term forecasts in order to generate and maintain it as a reliable resource. Despite the ability to modulate hydroelectric generation in a mechanistic manner, the knowledge of future rainfall trends, surface runoff, dam water level, and drought events are also important for sustained electricity generation from hydropower plants. Similarly, the production of energy from biomass and biofuels will require continuous biomass production (and supply of solar energy) to avoid brownouts for which effective future planning is required. A key issue with this energy resource is that production is largely contingent upon weather and climatic conditions that cannot be controlled. However, efficient forecasting techniques can be incorporated for better supply-side management and decision making *i.e.*, sustainable power generation using these cleaner, renewable energy resources.

Forecasting methods are generally of two types. Qualitative or judgemental forecasts are subjective in nature requiring one's experience and expertise. Quantitative forecasts are based on causality and employs probability theory in generating the forecasts. In order to implement a forecasting process, two classes of quantitative or regression models are used, *viz.* physical/dynamical (*e.g.*, Global Circulation Model – GCM) and statistical/data-driven models. A physical model is governed by the laws of physics, that incorporates the conservation of mass, energy, and momentum, enriched by meteorological information, that must be coupled with the atmospheric and oceanic dynamics (Artipoli and Durante 2014; CSIRO and Bureau of Meteorology 2015).

Physically-based energy modeling is usually performed utilizing the synoptic (large scale) climatic phenomenon by means of computational fluid dynamics knowledge (Artipoli and Durante 2014) that employ various assumptions and boundary layer conditions to be incorporated to force the predictive model. This process must take into account the large-scale climatic features such as the progression of high and low-pressure systems for wind generation, large-scale oceanic circulation, ocean currents, overturning and the global incident solar radiation. However, the physical processes at the finer spatial scales; such as localized solar radiation, precipitation processes, cloud formation, atmospheric and oceanic turbulence can be left out. Due to their coarse resolution and the challenges in emulation of physical processes, a physically-based model will have its own characteristic difficulties in simulating the accessibility of renewable energy resources at local scales.

Alternatively, the data-driven predictive model offers many advantages relative to the physically based model. Data-driven models 'learn' from the past trends within the observational dataset to make a future prediction (Deo and Şahin 2015b). These models have a number of advantages:

(i) they do not require initial forcings (Based on historical data)
(ii) they are easy to develop in comparison to the physically-based models
(iii) they are free from underlying boundary conditions and physical assumptions, and
(iv) they are cost-effective, fast and efficient.

In addition, a major benefit is that predictive models operate locally using point-based input data and modeling frameworks that can be easily used by renewable energy resource managers, researchers, investors, and other relevant stakeholders. Therefore, data-driven models can act as powerful localized forecasting tools for renewable energy generation and efficient smart-grid systems in particular for solar and wind energy.

For a developing nation like Fiji that has limited resources to develop and implement costly physically based modeling approaches, the data-intelligent models could serve as suitable and viable alternatives. Data intelligent models would not only be used as a forecasting tool, but can also be incorporated into smart grid systems to match the supply and demand enhancing the energy efficiency of large grid systems. Small-scale wind, photovoltaic and hybrid systems are common in small island developing states like Fiji and the efficiency of such systems could be enhanced via implementation of data intelligent machine learning approaches. In addition, these data-intelligent models can serve as an alert or awareness system, particularly for hydro and biomass energy production. The availability of these energy sources is contingent upon hydrological events. Drought also increases the number and severity of fires further exacerbating the biomass fuel supply in such events. In Fiji, during heavy rainfall events in the wet season, river flooding becomes imminent. In addition, six severe droughts events have also been noted between 1970 and 2016 (The Government of Fiji et al. 2017). Data intelligent models could forewarn the authorities and biomass energy managers of drought events and extreme hydrological events in their localized regions to promote prudent planning of hydro discharge rate and sustainable use of their biomass resources.

Despite the fact that the forecasting of renewable energy resources is a prerequisite for power generation, the forecasting studies in this area have been very limited in Fiji. So far only one study pertaining to a daily wind speed forecasting problem has been carried out by Kumar and Ali (2017) at Ellington Wharf in Rakiraki, Fiji from August 2012 to December 2016. Noting the significant gap in knowledge in this sector, this chapter outlines the basic and most important concepts relating to the forecasting of renewable energies using data-driven modeling approaches. The chapter examines the predictive model development and forecasting process, followed by a case study of 10-min wind speed forecasting using multiple linear regressions (MLR) model and the advanced machine learning models, and artificial neural networks (ANN).

3.2 Basic Important Definitions/Concepts

We begin by defining some basic concepts and principles used in modeling and forecasting.

Time Series
Regression-based forecasting involves time series data. A time series data is a sequential arrangement of observed data in a chronological order. An example of the wind speed time series data has been plotted in Fig. 3.1 which clearly illustrates the stochastic (*i.e.*, highly fluctuating) nature of the data series.

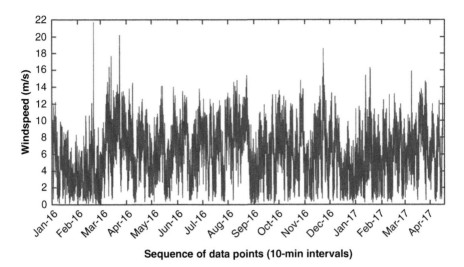

Sequence of data points (10-min intervals)

Fig. 3.1 Time series plot of 10-min wind speed data at the study site

Initially, a time series plot is constructed to observe the embedded patterns that may include:

(i) *trends* – depicted as a long-term increase or decrease in the plot.
(ii) *seasonality* – appears as recurring periodic patterns in a seasonal manner that could be weekly, monthly, annually, or at other regular intervals.
(iii) *cyclical behavior* – are the patterns that show a rise and a fall over a very long time, say several years and does not occur at regular intervals as in seasonal patterns.

It must be noted that usually, time series data have a combination of embedded patterns and care must be taken when modeling these series, especially when using statistical modeling approaches. Consequently, a time series can be classed as:

- *Stationary*–For a stationary time series, the probability laws governing the characteristics of the time series processes do not change over time. As a result, the mean is constant over time and there are no visible trends or periodicity in the time-series.
- *Non-stationary*–as opposed to stationarity, the non-stationary time series have varying governing probabilities induced by complex temporal behaviors such as trends, seasonal variations, periodicity and jumps that may affect the accuracy of data-driven models (Adamowski and Chan 2011; Adamowski et al. 2012).

Other pertinent terms include:

Forecasting horizon (*or lead time*): This describes the future periods for which forecasts are to be generated.
Forecast interval: This is the rate at which new forecasts are needed to be prepared.

Although at times these two terms are used interchangeably, they have distinctively different meanings. The lead time and forecast intervals largely depend on the nature of the problem. For example, we might want to conduct windspeed forecasting on a daily basis, for up to 4 weeks in the future. In this case, the forecast horizon or the lead time is 4 weeks, while the forecast interval is 1 day since new forecasts ought to be prepared on a daily basis.

The type of planning dictates the lead time. As a result, forecasting problems are further divided based on the lead times into:

- *Near-real-time forecasts*: this involves forecasting of events with lead times in minutes.
- *Very short-term forecasts*: involves hourly forecasting of events.
- *Short-term forecasts*: This involves forecasting events with lead times of days, weeks, and months, into the future.
- *Medium-term forecasts*: extends from 1 to 2 years into the future.
- *Long-term forecasting*: extends beyond that by many years such as 10-year projection or 30-year projections, etc.

Predictors are the input time series that are channeled into the model in emulating the target variable.

Target is the variable that needs to be forecasted.

3.3 General Predictive Model Development and Forecasting Process

The predictive model development requires strategic planning. Figure 3.2 shows a simplified schematic of the model development and forecasting process that is implemented once the specific forecasting problem and the forecast horizon are defined.

The model development commences with *data collation*. A lot of care needs to be taken in collating or acquiring the data since the outcomes solely depend on the features embedded within the historical data. The next step is the *data cleansing* stage whereby missing values are identified and appropriate imputation techniques need to be adopted. Also, it is important to identify and tag any outliers. Then a *preliminary statistical analysis* is carried out. In this step, the time series plot and the auto-correlation function (*ACF*) plots are designed to determine whether the time series plot exhibits non-stationarity features. The data distributions are also examined using skewness and kurtosis statistics. After this *feature selection* process is carried out. In this process, significant inputs/predictors are determined to forecast the target variable. Essentially, feature selection is critical in the development of parsimonious model as the irrelevant inputs can add unnecessary model complexity reducing model accuracy (Hejazi and Cai 2009; Maier et al. 2010), while the

Fig. 3.2 Simplified schematic of the model development and forecasting process

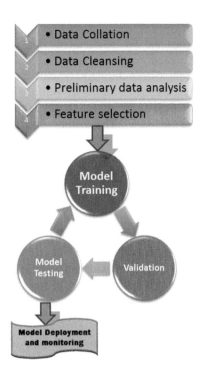

most relevant ones would ease the training process, and increase the interpretability and understanding of the dynamics of the system (Bowden et al. 2005).

For univariate forecasting method, where a memory of several (lagged) months in *windspeed* time-series could result from serial correlation in time-space (*i.e.*, persistence) (Chiew et al. 1998), the partial autocorrelation function (*PACF*) is utilized to determine the significant lags of one target that can be used as predictors. The *PACF* describes the supplementary information provided by the lagged time series after removing the effect of other variables. On the assumption that more than one explanatory lagged variable exists for a particular data, the *PACF* of the subsequent lag (τ) is determined by:

$$\varphi_\tau = \frac{\rho_\tau - \sum_{k=1}^{\tau-1}\varphi_{\tau-1,k} \bullet \rho_{\tau-k}}{1 - \sum_{k=1}^{\tau-1}\varphi_{\tau-1,k} \bullet \rho_{\tau-k}} \tag{3.1}$$

where ρ_τ is the autocorrelation coefficient of lag τ, while $\varphi_{\tau,k}$ is computed via the Durbin-Levinson recursion equation. Lagged series with statistically significant relationship (*i.e.*, at 95% confidence interval) are commonly screened as salient inputs (Ouyang et al. 2016; Ren et al. 2015; Seo and Kim 2016; Wang et al. 2013).

Following this, a three-step cyclic process is carried out *viz.*, *model training*, *validation,* and *testing*. The inputs and corresponding target are partitioned independently into training, validation, and testing sets to prevent the inclusion of future data which are not truly available at that particular time step, into the input time series and any unintentional introduction of bias in the forecast (Deo et al. 2016a; Kim and Valdes 2003). There is no set rule for data divisions, as researchers have used different training, validation and testing sets (Deo et al. 2016b), yet the common data partitioning is; Training-70%, validation-15%, and testing-15%. During *model training* apt modeling algorithm is utilized that can emulate the features of the time series. Then *model validation* using a separate data sub-set is carried out to determine the optimized modeling parameters while minimizing the chances of model overfitting. Then using another separate data subset, *model testing* is carried out to evaluate the performance of the model.

The model evaluation process can be carried out by means of many statistical metrics and diagnostic plots. Section 3.1 describes model evaluation metrics. These three phases are cyclic since if the developed model does not perform well on the testing data, then various optimization techniques are applied in the model training and subsequently tested until the satisfactory evaluation is achieved. Finally, the *model deployment and monitoring* phase is implemented as the predictive model is employed in the field for forecasting purposes.

3.3.1 Statistical Model Evaluation Metrics

Model evaluation is one of the critical phases in the model development process. An improper evaluation may lead to the low predictive performance of the model during deployment, culminating in inappropriate decision making. So far there is no solitary statistical metric that captures all the features of model performances. Hence, for apt model evaluations, a combination of statistical metrics are recommended (Chai and Draxler 2014). Given below is a listing of statistical measures with mathematical equations that are preferred for model evaluations. Note that in the following equations, *OBS* is observed values and *FOR* is forecasted values, i representing the occurrence time/place and N is the total number of data points.

The Pearson's correlation coefficient (r) provides the vital information on the strength and direction of a linear association between *OBS* and *FOR* values, and the square of this value (r^2) provide a measure of the covariance in *OBS* and *FOR* datasets. Without the correlation coefficient r, any assumption could be made regarding the associations of *OBS* and *FOR* values. An increase in the first variable associated with an increase in the second variable shows a positive correlation (+1 indicates a perfect increasing linear relationship) while a negative value show that an increase one variable corresponds to a decrease in the second variable (−1 indicates a decreasing linear relationship). 0 indicates no linear relationship at all between *OBS* and *FOR* values. However, a key weakness is that r is oversensitive to extreme values (outliers) (Legates and McCabe 1999; Willmott 1981).

(i) Correlation coefficient (r):

$$r = \frac{\sum_{i=1}^{N}\left(OBS_i - \overline{OBS}\right)\left(FOR_i - \overline{FOR}\right)}{\sqrt{\sum_{i=1}^{N}\left(OBS_i - \overline{OBS}\right)^2}\sqrt{\sum_{i=1}^{N}\left(FOR_i - \overline{FOR}\right)^2}} \tag{3.2}$$

The Willmott's Index (*WI*) or index of agreement has similar upper and lower bounds of 0 and 1 (perfect fit) that makes it useful. Yet, the meaning of zero in providing a convenient reference point is obscure (Dawson et al. 2007). An additional downside is that *WI* values greater than 0.65 could be recorded for poor model fits (Krause et al. 2005).

(ii) Willmott's Index (*WI*):

$$WI = 1 - \left[\frac{\sum_{i=1}^{N}\left(OBS_i - OBS_i\right)^2}{\sum_{i=1}^{N}\left(\left|FOR_i - \overline{OBS}\right| + \left|OBS_i - \overline{OBS}\right|\right)^2}\right], \; 0 \le WI \le 1 \tag{3.3}$$

The Nash–Sutcliffe Efficiency (E_{NS}), which is scaled version of mean squared error (Willems 2009) and is defined as one minus the sum of the absolute squared differences between the predicted and observed values normalized by the variance of the observed values during the period under investigation (Krause et al. 2005). The physical interpretation of the goodness-of-fit E_{NS} is very much clear with 1 representing a perfect model, 0 shows a model with no predictive advantage and negative values clearly depicts that the forecasted values diverge from the expected/observed values (Legates and McCabe 2013; Mehr et al. 2013). An interesting characteristic is that both *WI* and E_{NS} utilize squared values of residual terms making them oversensitive to the peak residual values (Legates and McCabe 1999; Willems 2009; Willmott 1981).

(iii) Nash–Sutcliffe Efficiency (E_{NS}):

$$E_{NS} = 1 - \left[\frac{\sum_{i=1}^{N}\left(OBS_i - FOR_i\right)^2}{\sum_{i=1}^{N}\left(OBS_i - \overline{OBS}\right)^2}\right], \left(-\infty < E_{NS} < 1\right) \tag{3.4}$$

The Legate-McCabe's index (*L*) does not utilize squared terms and gives errors and differences the appropriate weights (Legates and McCabe 1999). Since *L* takes absolute values into account, it is not overestimated. *L* is easy to interpret and is considered to yield a comparative model performance assessment (Legates and McCabe 1999).

(iv) Legates-McCabe's Index (*L*):

$$L = 1 - \left[\frac{\sum_{i=1}^{N} |FOR_i - OBS_i|}{\sum_{i=1}^{N} |OBS_i - \overline{OBS}|} \right], (0 < L \leq 1) \tag{3.5}$$

In terms of error measures, the absolute error measures, root mean square error (*RMSE*) and mean absolute error (*MAE*), extracts information on the average discrepancies between forecasted and observed values (Legates and McCabe 1999). The key difference is that the aggregation of residuals in *RMSE* is squared while the same aggregation in *MAE* is not squared. As a result, large errors have a relatively greater influence on the aggregates in *RMSE* than the smaller errors (Armstrong and Collopy 1992; Willmott and Matsuura 2005), while *MAE* equally evaluates all deviations from the observed values (Deo et al. 2016a). However, the key drawback is that both *RMSE* and *MAE* are scale dependent (in their absolute units) and cannot ideally be used to compare model performances in different measurement units or at different sites (Hora and Campos 2015).

 (v) Root mean square error (*RMSE*):

$$\text{RMSE} = \sqrt{\frac{1}{N} \sum_{i=1}^{N} (FOR_i - OBS_i)^2} \tag{3.6}$$

(vi) Mean absolute error (*MAE*):

$$MAE = \frac{1}{N} \sum_{i=1}^{N} |(FOR_i - OBS_i)| \tag{3.7}$$

For model comparisons at different sites, the scale-independent relative error measures: relative Root-Mean Square Error (*RRMSE*) and Mean Absolute Percentage Error (*MAPE*) are used. The equations of the *MAPE* and *RRMSE* are as follows:

 (vii) Relative root mean square error (*RRMSE*, %):

$$\text{RRMSE} = \frac{\sqrt{\frac{1}{N} \sum_{i=1}^{N} (FOR_i - OBS_i)^2}}{\frac{1}{N} \sum_{i=1}^{N} (OBS_i)} \times 100 \tag{3.8}$$

(viii) Mean absolute percentage error (*MAPE*; %):

$$\text{MAPE} = \frac{1}{N} \sum_{i=1}^{N} \left| \frac{(FOR_i - OBS_i)}{OBS_i} \right| \times 100 \tag{3.9}$$

A combination of evaluation metrics has been preferred by scholars such as coefficient of determination, r^2, and *RMSE* were accordingly used (Legates and McCabe 1999), while others used a combination of E_{NS} and *RMSE* (Humphrey et al. 2016; Mehr et al. 2014; Sajikumara and Thandaveswarab 1999). Yet, the best practice would be to use as many as possible to make a prudent judgment on the model performance and not to limit the assessments to only objective metrics that quantify the outcomes in a few numbers (Willems 2009). Additionally, subjective model performance assessments by means of various diagnostic plots such as scatter plots, boxplots, forecasting error histogram, time series graphs, and polar plots can also provide better a visual insight into the accuracy of the predictive model.

3.4   Brief Theory of Forecast Algorithms

3.4.1   *Artificial Neural Network (ANN)*

ANN is one of the widely used models developed by McCulloch and Pitts (1943) that mimics the functioning of the human brain. ANN is advantageous as it accepts non-Gaussian data and data with irregular seasonal variations. Additionally, ANN is very robust, performs well with limited available data and deals with outliers and noisy data (Jain et al. 1999).

The most commonly used type of ANN is the multilayer perceptron (MLP). The architecture of MLP consists of three or more layers that have an input layer, an output layer and one or more intermediate/hidden layers-which act as a collection of feature detectors as shown in Fig. 3.3.

The ANN is composed of simple non-linear elements called neurons that operate in parallel forming massive networks. The neurons are assigned with respective weights and they connect each of the nodes in preceding and following layers. During the training of the network, these weights are adjusted so that a particular input leads to a corresponding target output. The process of weight adjustments based on input/target comparison continues until the network output matches the target. As such many input/target pairs are needed during the training process. Principally, the receiving node sums the weights from the previous layer and adds a bias term, before processing the output via an activation function (Deo and Şahin 2015a). The common transfer functions include linear, tangent-sigmoid, and logarithmic-sigmoid.

From a mathematical viewpoint, an objective function, usually the mean squared error (MSE) in between the output and the target is minimized while determining the appropriate connecting weights and the added biases. A training algorithm such as back propagation (BP), the Levenberg–Marquardt (LM) or the Broyden–Fletcher–Goldfarb–Shanno (BFGS) quasi-Newton BP algorithms are employed in the minimization of the cost function, *MSE*.

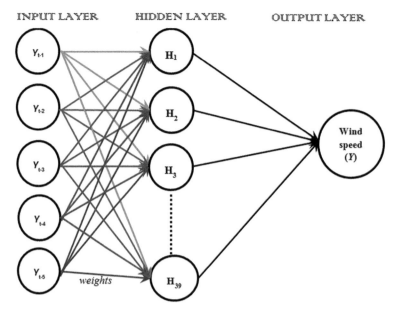

Fig. 3.3 Schematics of the Artificial Neural Network model

A key issue with all forecasting algorithm is overfitting or over-learning whereby the forecasting algorithm learns all the features within the training data and performs overly well, yet fails to perform well on unseen data. To prevent overfitting, an early stopping criterion is utilized in which case the *MSE* is monitored at each of the training and validation iterations and the training is stopped when the minimum MSE during the validation phase is achieved. The weights and biases are derived at this point in time. Then the final testing phase is carried out to assess the model performance on a new data set.

3.4.2 Multiple Linear Regressions (MLR)

A simple linear regression, with just one input and a target variable, explains the linear associations in between a predictor variable (x) and the target variable (y). Similarly, the MLR draws out the cause and effects between objective and predictor variables (Deo and Şahin 2017), however, MLR is an advancement of linear regressions whereby the target variable is a function of k predictors variables. The MLR model can mathematically be expressed as (Deo and Şahin 2017; Draper and Smith 1998; Montgomery et al. 2012; Rawlings et al. 1998):

$$Y = C + \beta_1 X_1 + \beta_1 X_1 + \ldots + \beta_k X_k \qquad (3.10)$$

where Y is the target/output variable, C is the intercept term, X is a vector of input/predictor variables and β are the regression coefficients.

The general linear regression equation estimated by the MLR model in this study is given as follows:

$$Y = 0.0098 + 0.8759Y_{t-1} - 0.0607Y_{t-2} + 0.0575Y_{t-3} + 0.0144Y_{t-4} + 0.0452Y_{t-5} \qquad (3.11)$$

where Y is the forecasted wind speed at 10-min intervals, and Y_{t-1}, $Y_{t-2}...Y_{t-5}$ are the lagged inputs of 10-min windspeed data were utilized as inputs. For more details regarding MLR readers can refer to Draper and Smith (1998), Rawlings et al. (1998) and Montgomery et al. (2012).

3.5 Wind Speed Forecasting Case Study

3.5.1 Study Site

In this chapter, the study site is situated in Rakiraki in the Western region of Viti Levu, which is one of the major islands in Fiji (latitude: 17.64°S and longitude: 178.15°E) as shown in Fig. 3.4. The site has an automatic weather station installed and monitored by the Fiji Meteorological Services (FMS) that does maintenance and data logging tasks. The height of the monitoring tower is 10 m above the ground, and the wind speed data used for this research have been sampled at a 10-min interval. The same monitoring tower is used to measure other meteorological parameters namely, wind direction, temperature, solar radiation, atmospheric pressure, relative humidity, and precipitation. These data are recorded on a data logger, which are directly transferred to the FMS server in Nadi via the internet. The data is retrieved on a daily basis and is checked for any irregularities.

3.5.2 Outcomes of the Case Study

This section presents the results obtained to evaluate the proposed artificial neural network (ANN) model against the comparative multi-linear regressions (MLR) model in forecasting 10-min wind speed data in Rakiraki, Fiji. These two models are chosen for this case study due to their ease of application and efficient nature of dealing with irregular seasonal variations and limited available datasets.

Initial data quality checking was performed and all missing data were statistically imputed using the hourly averages. Then preliminary data analysis was performed that revealed that the skewness was 0.012 and the kurtosis was −0.58. It shows the data is not skewed and probably is normally distributed. A maximum wind speed of 21.1 m/s was recorded while the least speed was of 0.1 m/s. The mean wind speed was 6.29 m/s while the median speed was 6.40 m/s.

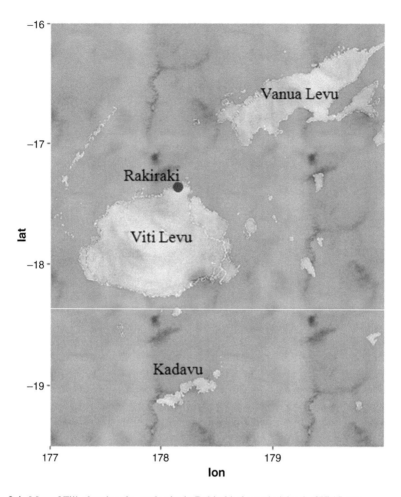

Fig. 3.4 Map of Fiji, showing the study site in Rakiraki, the main island of Viti Levu

For feature selection, partial autocorrelation function (*PACF*) was implemented. The input lags with *PACF* values outside of the 95% confidence level were selected to be important ones. Figure 3.5 shows the *PACF* analysis for the wind speed (m/s) time series up to the first 20 lags. In this study, the first five clear cut lags were used in the modeling process in order to develop a parsimonious model that could effi-ciently be applied to *near-real-time* (10-min forecast horizon) wind-speed forecast-ing purposes.

To forecast 10-min wind speed, two versatile predictive models were developed; the artificial neural network networks and the multiple linear regression model. Prior to model development, all data were normalized to be within the range of 0 and 1. The general linear regression equation estimated by the MLR model used in this study is represented by Eq. 3.11. For the input variable, antecedent 10-min wind speed data up to fifth 10-min interval was used as input. The multilayer perceptron

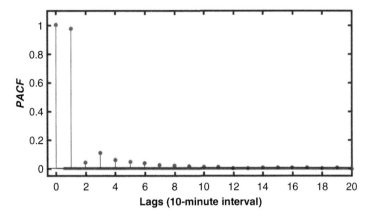

Fig. 3.5 Plot of the partial autocorrelation function (*PACF*) of wind speed time series. The blue lines show the 95% confidence intervals and the red stems shows the correlations of each lag

Table 3.1 Modelling parameters of ANN developed for 10-min wind speed forecasting

Number of neurons					
Input layer	Hidden layer	Output layer	Training algorithm	Hidden transfer function	Output transfer function
5	39	1	*trainlm*	*logsig*	*tansig*

(MLPs) is the simplest and the most commonly used neural network architecture and can be trained using many different learning algorithms (Adamowski and Karapataki 2010). In this research, MLPs were trained with a series of hidden neurons (hn) starting at hn = 1–40 in an incremental step of 1. The model architecture that performed the best in terms of the lowest mean square error (MSE) criterion was selected. To attain an accurate ANN model, various combinations of hidden transfer and output functions (*tansig, purelin, & logsig*) interchanged with the training algorithms (*trainlm & trainbfg*) were also trialed one by one resulting in 12 combinations. A total of 480 ANN models executable with unique hidden neuron (1–40) and hidden transfer and output functions (12) architectures were developed. The determination of the optimized network structure via a comprehensive trial and error is important since a small architecture can lack sufficient degrees of freedom to correctly learn the predictor data, while an unnecessarily large architecture may not converge in a reasonable modeling time, resulting in overfitting and memorization rather than generalization of the data (Karunanithi et al. 1994). After a comprehensive trial and error process, the 39 hidden neurons and 5 input layers with *trainlm* training algorithm as it had the optimum performance and the model developing parameters of ANN are furnished in Table 3.1 below.

A comprehensive statistical evaluation of ANN was carried out by means of r = correlation coefficient; WI = Willmott's Index; E_{NS} = Nash–Sutcliffe Efficiency, L = Legate-McCabes index, $RMSE$ = root mean square error and MAE = mean absolute error and was benchmarked against the MLR. The outcomes of these tests are

Table 3.2 Evaluation of ANN and MLR models, in terms of r = correlation coefficient; WI = Willmott's Index; E_{NS} = Nash–Sutcliffe Efficiency, L = Legate-McCabes index, $RMSE/MAE$ = root mean square/mean absolute error

Predictive model	r	WI	E_{NS}	L	$RMSE$ (m/s)	MAE (m/s)
ANN	0.978	0.977	0.956	0.816	0.648	0.469
MLR	0.977	0.976	0.955	0.814	0.653	0.473

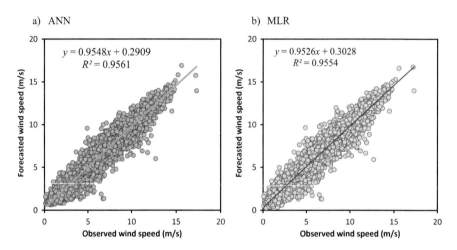

Fig. 3.6 Scatterplots comparing the observed and forecasted wind speed by the ANN and MLR models

summarized in Table 3.2. The selection criteria of each of the following metrics were used for assigning the best model. The tabulated values (Table 3.2) for model evaluation metrics evidently indicates that the ANN model is better at forecasting 10-min wind speed over the comparative MLR at the Rakiraki study site.

In addition, the diagnostic plots were utilized to alternatively evaluate the ANN model in forecasting 10-min wind speed data. Figure 3.6 plots the scatter-plot of observed and forecasted wind speed data with the coefficient of determination (R^2) during the testing phase. The optimal ANN achieved an R^2 value of 0.9561 which is greater in comparison to the MLR models that recorded the R^2 value of 0.9554. In addition to the use of scatter-plots, the model preciseness was also assessed using a histogram of forecasting errors (Fig. 3.7).

In accordance with Prasad et al. (2018), an ideal value of forecast error (FE) must be equivalent to 0, hence a better model is bound to have higher frequencies of the forecasting error values closer to a zero. Although both the models show FE of a closer range, but the FE values of ANN model display a higher percentage (57%) of the frequency distribution in the first error bracket (0–0.4 m/s), while the MLR model recorded 1% less (56%) in this error bracket. Additionally, MLR recorded a larger span of errors with 1% of errors in the largest error bracket (2.0–2.4 m/s),

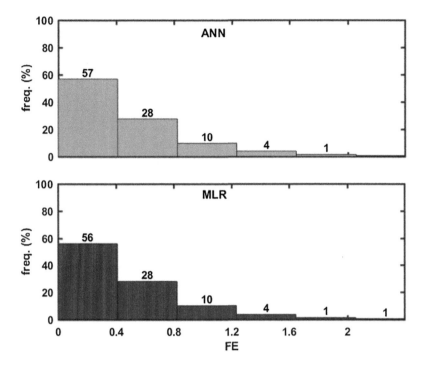

Fig 3.7 Histogram comparing the performance of ANN and MLR in terms of percentage frequency of forecasting errors

which was absent in the case of the ANN model. The error spread also substantiates better performance of ANN over the MLR model in wind speed forecasts.

Moreover, Table 3.3 displays further statistical analysis of observed and forecasted values during the testing phase. It can be seen that most of the forecasted values including maximum, upper quartile and lower quartile and the mean wind speeds are closer to the observed data with the ANN model, while only minimum and median forecasted values of the wind speeds are in close proximity to the observed values under the MLR model. The forecasted mean wind speed value is unchanged under the ANN model and slightly lower (-0.01 m/s) with the MLR model. The statistical analysis further reinforces the outcomes of the evaluation metrics (Table 3.2) and the diagnostic plots (Figs. 3.6 and 3.7) in exhibiting superior performance of the ANN model.

Finally, the relative error comparisons *RRMSE* and *MAPE* of ANN and MLR models are outlined in Table 3.4. These error indices are expressed in percentage and a lower percentage value indicates that there is a lesser error between the forecasted and the observed data. According to Li et al. (2013) the model performance under *RRMSE* are categorized as: excellent if *RRMSE* <10%, good if 10% $\leq$ *RRMSE* <20%, fair if 20% $\leq$ *RRMSE* <30% and is poor if *RRMSE* $\geq$30%. Accordingly, both models had excellent performances, yet both the relative error measures, *RRMSE* and *MAPE* validates that ANN was a better modeling choice over the MLR model.

Table 3.3 Statistical analysis of observed and forecasted values during the testing period from the respective ANN and MLR models in terms of maximum, minimum, lower quartile-LQ (Q_{25}), median (Q_{50}), upper quartile-UQ (Q_{75}), mean and range

Statistical property	OBS	ANN	MLR
Maximum (m/s)	17.3	**16.93**	16.76
Minimum (m/s)	0.1	0.62	**0.27**
Upper quartile (m/s)	8.8	**8.74**	8.71
Median (m/s)	6.8	6.86	**6.84**
Lower quartile (m/s)	4.3	**4.30**	4.33
Mean (m/s)	6.6	**6.60**	6.59
Range (m/s)	17.2	**16.30**	16.49

Table 3.4 Relative error comparisons of ANN and MLR models in wind speed forecasting using *RRMSE* and *MAPE*. Optimal models with the smallest forecasted error (%) are shown in **boldface**

	RRMSE (%)	*MAPE* (%)
ANN	**9.81**	**10.87**
MLR	9.88	10.93

3.6 Summary and Conclusions

The generation of energy from the traditional non-renewable sources can be easily modulated and predicted by energy market mechanisms, as the supply and demand can be matched. In contrast, the utilization of renewable energy resources are intermittent and therefore inherently not programmable by their nature. In the small island developing states such as Fiji, the use of these resources is compounded by the limited availability of capital resources. To counter these issues, efficient modeling and forecasting techniques become all important. Modeling and forecasting by means of data-intelligent machine learning algorithms are suitable alternatives due to their cost-effectiveness and ease of applications. Such model development, forecasting, model evaluations are described in this chapter. A case study of 10-min wind speed forecasting showed that ANN has better generalization capability in comparison to the MLR model. Machine learning models such as ANNs are powerful in terms of generalization capability and could be further explored in smart grid implementation for optimal energy generation from clean and renewable sources.

Acknowledgments The second author (LJ) is thankful to the Postgraduate Committee of the University of Fiji for providing the necessary funding to undertake the MREM studies. The authors also acknowledge the Fiji Meteorological Services for providing the wind speed data for this study.

References

Adamowski, J., & Chan, H. F. (2011). A wavelet neural network conjunction model for groundwater level forecasting. *Journal of Hydrology, 407*, 28–40. https://doi.org/10.1016/j.jhydrol.2011.06.013.

Adamowski, J., & Karapataki, C. (2010). Comparison of multivariate regression and artificial neural networks for peak urban water: Demand forecasting-evaluation of different ANN learning algorithms. *Journal of Hydrologic Engineering, 15*, 729–743. https://doi.org/10.1061//ASCE/HE.1943-5584.0000245.

Adamowski, J., Fung Chan, H., Prasher, S. O., Ozga-Zielinski, B., & Sliusarieva, A. (2012). Comparison of multiple linear and nonlinear regression, autoregressive integrated moving average, artificial neural network, and wavelet artificial neural network methods for urban water demand forecasting in Montreal, Canada. *Water Resources Research, 48*. https://doi.org/10.1029/2010wr009945

Armstrong, J. S., & Collopy, F. (1992). Error measures for generalizing about forecasting methods: Empirical comparisons. *International Journal of Forecasting, 8*, 69–80. https://doi.org/10.1016/0169-2070(92)90008-w.

Artipoli, G., & Durante, F. (2014). Physical modeling in wind energy forecasting. *DEWI GmbH Italy, 44*.

Bowden, G. J., Dandy, G. C., & Maier, H. R. (2005). Input determination for neural network models in water resources applications. Part 1 – background and methodology. *Journal of Hydrology, 301*, 75–92. https://doi.org/10.1016/j.jhydrol.2004.06.021.

Chai, T., & Draxler, R. R. (2014). Root mean square error (RMSE) or mean absolute error (MAE)? – Arguments against avoiding RMSE in the literature. *Geoscientific Model Development, 7*, 1247–1250. https://doi.org/10.5194/gmd-7-1247-2014.

Chiew, F. H., Piechota, T. C., Dracup, J. A., & McMahon, T. A. (1998). El Nino/southern oscillation and Australian rainfall, streamflow and drought: Links and potential for forecasting. *Journal of Hydrology, 204*, 138–149.

CSIRO and Bureau of Meteorology. (2015). *Climate change in Australia information for Australia's natural resource management regions: Technical report*. Canberra: CSIRO and Bureau of Meteorology.

Dawson, C. W., Abrahart, R. J., & See, L. M. (2007). HydroTest: A web-based toolbox of evaluation metrics for the standardised assessment of hydrological forecasts. *Environmental Modelling & Software, 22*, 1034–1052. https://doi.org/10.1016/j.envsoft.2006.06.008.

Deo, R. C., & Şahin, M. (2015a). Application of the Artificial Neural Network model for prediction of monthly Standardized Precipitation and Evapotranspiration Index using hydrometeorological parameters and climate indices in eastern Australia. *Atmospheric Research, 161-162*, 65–81. https://doi.org/10.1016/j.atmosres.2015.03.018.

Deo, R. C., & Şahin, M. (2015b). Application of the extreme learning machine algorithm for the prediction of monthly Effective Drought Index in eastern Australia. *Atmospheric Research, 153*, 512–525. https://doi.org/10.1016/j.atmosres.2014.10.016.

Deo, R. C., & Şahin, M. (2017). Forecasting long-term global solar radiation with an ANN algorithm coupled with satellite-derived (MODIS) land surface temperature (LST) for regional locations in Queensland. *Renewable and Sustainable Energy Reviews, 72*, 828–848. https://doi.org/10.1016/j.rser.2017.01.114.

Deo, R. C., Tiwari, M. K., Adamowski, J. F., & Quilty, J. M. (2016a). Forecasting effective drought index using a wavelet extreme learning machine (W-ELM) model. *Stochastic Environmental Research and Risk Assessment*. https://doi.org/10.1007/s00477-016-1265-z.

Deo, R. C., Wen, X., & Qi, F. (2016b). A wavelet-coupled support vector machine model for forecasting global incident solar radiation using limited meteorological dataset. *Applied Energy, 168*, 568–593. https://doi.org/10.1016/j.apenergy.2016.01.130.

Draper, N. R., & Smith, H. (1998). *Applied regression analysis*. New York: Wiley.

Hejazi, M. I., & Cai, X. (2009). Input variable selection for water resources systems using a modified minimum redundancy maximum relevance (mMRMR) algorithm. *Advances in Water Resources, 32*, 582–593. https://doi.org/10.1016/j.advwatres.2009.01.009.

Hora, J., & Campos, P. (2015). A review of performance criteria to validate simulation models. *Expert Systems, 32*, 578–595. https://doi.org/10.1111/exsy.12111.

Humphrey, G. B., Gibbs, M. S., Dandy, G. C., & Maier, H. R. (2016). A hybrid approach to monthly streamflow forecasting: Integrating hydrological model outputs into a Bayesian artificial neural network. *Journal of Hydrology, 540*, 623–640. https://doi.org/10.1016/j.jhydrol.2016.06.026.

Jain, S. K., Das, A., & Srivastava, D. K. (1999). Application of ANN for reservoir inflow prediction and operation. *Journal of Water Resources Planning and Management, 125*, 263–271.

Karunanithi, N., Grenney, W. J., Whitley, D., & Bovee, K. (1994). Neural networks for river flow prediction. *Journal of Computing in Civil Engineering, 8*, 201–220.

Kim, T.-W., & Valdes, J. B. (2003). Nonlinear model for drought forecasting based on a conjunction of wavelet transforms and neural networks. *Journal of Hydrological Engineering, 8*, 319–328. https://doi.org/10.1061//ASCE/1084-0699/2003/8:6/319.

Krause, P., Boyle, D. P., & Base, F. (2005). Comparison of different efficiency criteria for hydrological model assessment. *Advances in Geosciences, 5*, 89–97.

Kumar, A., & Ali, A. B. M. S. (2017). *Prospects of wind energy production in the western Fiji – An empirical study using machine learning forecasting algorithms.* In: 2017 Australasian Universities Power Engineering Conference (AUPEC), 19–22 November 2017. pp. 1–5. https://doi.org/10.1109/AUPEC.2017.8282443

Latvian Presidency of the Council of the European Union. (2015*). Intended nationally determined contribution of the EU and its member states.* UNFCCC.

Legates, D. R., & McCabe, G. J. (1999). Evaluating the use of "goodness-of-fit" measures in hydrologic and hydroclimatic model validation. *Water Resources Research, 35*, 233–241. https://doi.org/10.1029/1998wr900018.

Legates, D. R., & McCabe, G. J. (2013). A refined index of model performance: A rejoinder. *International Journal of Climatology, 33*, 1053–1056. https://doi.org/10.1002/joc.3487.

Li, M.-F., Tang, X.-P., Wu, W., & Liu, H.-B. (2013). General models for estimating daily global solar radiation for different solar radiation zones in mainland China. *Energy Conversion and Management, 70*, 139–148. https://doi.org/10.1016/j.enconman.2013.03.004.

Maier, H. R., Jain, A., Dandy, G. C., & Sudheer, K. P. (2010). Methods used for the development of neural networks for the prediction of water resource variables in river systems: Current status and future directions. *Environmental Modelling & Software, 25*, 891–909. https://doi.org/10.1016/j.envsoft.2010.02.003.

McCulloch, W. S., & Pitts, W. (1943). A logical calculus of the ideas immanent in nervous activity. *Bulletin of Mathematical Biophysics, 5*, 115–133.

Mehr, A. D., Kahya, E., & Olyaie, E. (2013). Streamflow prediction using linear genetic programming in comparison with a neuro-wavelet technique. *Journal of Hydrology, 505*, 240–249. https://doi.org/10.1016/j.jhydrol.2013.10.003.

Mehr, A. D., Kahya, E., Şahin, A., & Nazemosadat, M. J. (2014). Successive station monthly streamflow prediction using different artificial neural network algorithms. *International journal of Environmental Science and Technology, 12*, 2191–2200. https://doi.org/10.1007/s13762-014-0613-0.

Ministry of Economy. (2017). *Fiji NDC Implementation Roadmap 2017–2030.* Global Green Growth Framework, Ministry of Economy-Republic of Fiji.

Montgomery, D. C., Peck, E. A., & Vining, G. G. (2012). *Introduction to linear regression analysis* (5th ed.). Hoboken: Wiley.

Ouyang, Q., Lu, W., Xin, X., Zhang, Y., Cheng, W., & Yu, T. (2016). Monthly rainfall forecasting using EEMD-SVR based on phase-space reconstruction. *Water Resources Management, 30*, 2311–2325. https://doi.org/10.1007/s11269-016-1288-8.

Prasad, R., Deo, R. C., Li, Y., & Maraseni, T. (2018). Soil moisture forecasting by a hybrid machine learning technique: ELM integrated with ensemble empirical mode decomposition. *Geoderma, 330*, 136–161. https://doi.org/10.1016/j.geoderma.2018.05.035.

Rawlings, J. O., Pantula, S. G., & Dickey, D. A. (1998). *Applied regression analysis-a research tool* (2nd ed.). New York: Springer.

Ren, Y., Suganthan, P. N., & Srikanth, N. (2015). A comparative study of empirical mode decomposition-based short-term wind speed forecasting methods. *IEEE Transactions on Sustainable Energy, 6*, 236–244. https://doi.org/10.1109/TSTE.2014.2365580.

Sajikumara, N., & Thandaveswarab, B. S. (1999). A non-linear rainfall–runoff model using an artificial neural network. *Journal of Hydrology, 216*, 32–55.

Seo, Y., & Kim, S. (2016). Hydrological forecasting using hybrid data-driven approach. *American Journal of Applied Sciences, 13*, 891–899. https://doi.org/10.3844/ajassp.2016.891.899.

The Government of Fiji, World Bank, Global Facility for Disaster Reduction and Recover. (2017). *Fiji 2017: Climate vulnerability assessment-making Fiji climate resilient*. Washington, DC: World Bank.

Wang, W.-c., Xu, D.-m., Chau, K.-w., & Chen, S. (2013). Improved annual rainfall-runoff forecasting using PSO–SVM model based on EEMD. *Journal of Hydroinformatics, 15*, 1377–1390. https://doi.org/10.2166/hydro.2013.134.

Weiss, G., Fagan, E., & Holt, P. (2016). *Australias intended nationally determined contribution to a new climate change agreement*. Energetics Pty Ltd.

Willems, P. (2009). A time series tool to support the multi-criteria performance evaluation of rainfall-runoff models. *Environmental Modelling & Software, 24*, 311–321. https://doi.org/10.1016/j.envsoft.2008.09.005.

Willmott, C. J. (1981). On the validation of models. *Physical Geography, 2*, 184–194.

Willmott, C. J., & Matsuura, K. (2005). Advantages of the mean absolute error (MAE) over the root mean square error (RMSE) in assessing average model performance. *Climate Research, 30*, 79–82.

Chapter 4
Life Cycle Analysis as a Tool in Estimating Avoided Emissions

Anirudh Singh and Dhrishna Charan

Abstract Climate change mitigation requires assessment of both the total green-house gas (GHG) emissions of a country as well as the effectiveness of renewable energy (RE) resources and technologies in reducing (or avoiding) these emissions. GHG inventories can be calculated using the well-known IPCC 2006 guidelines. The technique of Life Cycle Assessment (LCA) provides a means of (amongst other things) carrying out detailed calculations of GHG emissions associated with the production and use of RE. This chapter provides a brief introduction to the technique of LCA and its use in the evaluation of RE-related GHG emissions. It begins with a simple case study that demonstrates the need for LCA in RE assessment. It next outlines the main features of the LCA technique, and ends with a brief description of LCA software available as freeware or commercially on the market.

Keywords Climate change mitigation · Life cycle assessment (LCA) · Renewable energy (RE) · Inventory analysis · Life cycle impact assessment · Functional unit · openLCA

4.1 Introduction

Since climate change is caused by the emission of greenhouse gases (GHGs), there is a need to estimate such emissions for any nation. These estimates are carried out according to the IPCC 2006 guidelines, and have been discussed in Chap. 1 and alluded to in Chap. 2.

The use of renewable energy (RE) in place of fossil fuels avoids the GHG emissions due to the latter, and provides one way of mitigating climate change. However, RE invariably requires some degree of fossil fuel use itself, with the consequent

A. Singh (✉) · D. Charan
School of Science and Technology, The University of Fiji, Lautoka, Fiji
e-mail: anirudhs@unifiji.ac.fj

© Springer Nature Switzerland AG 2020   81
A. Singh (ed.), *Translating the Paris Agreement into Action in the Pacific*,
Advances in Global Change Research 68,
https://doi.org/10.1007/978-3-030-30211-5_4

GHG emissions. Therefore, to ensure that the desired reductions in GHG have been attained, there is a need for a detailed assessment of the GHG emissions associated with the RE production and use. The process of Life Cycle Assessment (LCA) provides such a technique.

LCA provides an important tool in the assessment of the true ability of renewable energy sources and technology to reduce greenhouse gases by the burning of fossil fuels in the energy sector. It does so by estimating the totality of greenhouse gases emitted in the production and use of a unit quantity of energy by the renewable energy source or technology, and comparing it with the emissions from a similar amount of energy derived from the fossil fuel source that the renewable energy replaces. Box 4.1 below provides an example demonstrating the utility of this procedure in the case of a biofuel.

A **Life Cycle Assessment (LCA)** is a procedure, or methodology for solving complex problems that occur (for instance) in business and industry, and the energy, environment and development institutions of governments and private sectors. Broadly speaking, it considers the utility of a **product** or **process** for the purpose it was intended by assessing its true worth or quality as compared to other similar products or processes, and by considering its environmental impacts. What is unique about this problem-solving method is that it takes into account the **full life cycle** (dubbed "from cradle to grave") or at least part of it (e.g. "from cradle to gate") of the product or process being studied.

The LCA starts by defining the objective (i.e. the goal or purpose) of the exercise. This is called **Goal and Scope Definition**. It then collects all the information necessary to complete the investigation in the form of data tables called Inventories. This stage is called **Inventory Analysis** or **Life Cycle Inventory (LCI)**.

During its lifetime (**life cycle**) the product will go through several processes in which it is produced, used and disposed of. These processes are called **unit processes** and the system of processes representing the evolution of the product through its life cycle is called the **product system.** Figure 4.2 in Sect. 4.3 of this chapter provides an example of a product system where the product is a biofuel.

The investigation may require comparing the product with another product. This second product is called the **reference** and has a system of its own, called the **reference system**. If the objective of the investigation is to consider the environmental impact of the product, the next step or phase in the LCA is an impact assessment, called the **Life Cycle Impact Assessment** or **LCIA**.

Throughout the process of the investigation, the data/information collected is assessed and re-assessed in the light of new information – a process called **Interpretation**.

This chapter introduces these stages in more detail in Sect. 4.3 and the sections that follow, and provides a brief introduction to software available freely to assist in LCA calculations. The LCA procedure conforms to the ISO standards (ISO14040) that were first created in the 1997–2000 period and which were developed further in

Box 4.1: Producing Latoda Nut Biodiesel – The Need for a Full LCA

Latoda Methyl Ester (LME) is an alternative fuel for the petroleum diesel used in diesel engines. It can be produced from the oil derived from the nuts of the latoda tree which grows well in the tropics. An entrepreneur is interested in producing this biodiesel with a view to starting a business in supplying LME to the transportation industry.

He realises that he can also receive additional revenues in the form of carbon credits gained from the net reductions in carbon dioxide (CO_2) emissions this project will produce. For this to be possible however, he needs to ensure that the total CO_2 emissions produced in the production and use of the biodiesel as transportation fuel does not exceed (and is significantly lower than) that produced by the petroleum diesel it replaces. That is, he needs to ascertain that the total CO_2 emissions from a batch of biodiesel used for transportation in a year is less than the CO_2 produced by the petroleum diesel it replaces. The carbon credits are given only for the net emissions reductions.

When a certain volume of biodiesel is combusted in vehicle engines, the amount of CO_2 produced is the same as the amount the plant matter (in this case the latoda tree) absorbed from the atmosphere (via photosynthesis) in producing the corresponding latoda oil used in the LME production. Therefore, the net emissions from the combustion of the LME itself is nil. We say that the biodiesel is "carbon neutral".

However, the total production process of the oil and the LME biodiesel always involves some CO_2 emissions. This is because fossil fuels will inevitably be used at some stage of the biodiesel production process. To evaluate the net emissions, all these emissions will have to be taken into consideration. As they occur at various times and stages in the whole life cycle of the of LME production and use, one will have to do a full life cycle analysis of the emissions to ensure a complete accounting of all the emissions. This is done by carrying out an analysis of the full biodiesel product system, starting from the feedstock plantation stage and ending in the combustion of the final LME biodiesel product to produce the required energy.

The accounting can begin by calculating the emissions at each stage of the production process, starting from the emissions during the planting of the latoda tree (or even earlier). These individual calculations will then have to be added up and compared to the emissions from the petroleum diesel produced in its own life cycle to see whether the LME produces an overall net reduction in the CO_2 emissions to the atmosphere when it is used as an alternative fuel.

The individual processes we need to consider to describe the whole process are:

- Planting the latoda trees
- Harvesting the nuts
- Transporting to the oil extraction/purification mill

(continued)

Box 4.1 (continued)

- Extraction/purification of latoda oil from the nuts
- Transporting the purified oil to the LME-producing plant
- Producing the methanol used in the reaction
- The batch reaction for LME and separation of pure LME
- Transportation of the LME to the point of use
- The combustion of LME (in place of petroleum diesel)

Together, these processes form the LME biodiesel product system. Note that most (if not all) of the processes listed above involve

- Inputs of materials and energy, and
- Outputs of products, wastes and other materials.

The central process in the production of LME involves the production of the latoda oil and reaction of this oil with methanol in the presence of a catalyst such as sodium hydroxide.

$$\text{Latoda oil} + \text{methanol} + \text{sodium hydroxide (catalyst)} \rightarrow \text{latoda methyl ester} (LME) + \text{glycerol}$$

The inputs for this process include pure latoda oil, methanol, sodium hydroxide and energy (electricity/heat). The outputs are LME, glycerol, unreacted methanol/latoda oil and sodium hydroxide.

Our main interest is to see whether the whole process of LME production/ use will result in any reductions in CO_2 emissions when compared to petroleum diesel. To do this, we will have to compare like with like. Specifically, we will need to specify how much of the petroleum diesel we are considering (and choose the same amount of LME). We could measure this in kilograms, and compare the amounts of CO_2 produced by 1 kg of LME with that of 1 kg of petroleum diesel.

Now suppose we started with a plantation of latoda trees that produced 200 tonnes (i.e. 200,000 kg) of nuts, which eventually resulted in the extraction of 90,000 kg of pure oil and the production of 50,000 kg of LME. To determine the emissions from 1 kg of LME, we will need to (notionally) normalize the production by dividing all inputs and outputs by 50,000. We see that we need to scale the inputs and outputs so that the final product is 1 kg of LME.

We will be able to choose a better basic unit (instead of the kilogram) if we first considered what purpose we will be using these fuels for, and then decided what the basic unit for comparison should be. The function of the fuels is to provide energy to the diesel vehicle, so a more appropriate functional

(continued)

Box 4.1 (continued)

unit for comparison would be one unit of energy, such as 1 Mega-Joule (MJ). Whatever unit we choose; all the results will have to be normalized to it.

It will assist the process if we constructed a large table that showed the inputs and outputs of all the processes of the LME product system arranged sequentially. That is, we need to create an inventory of the whole process, and this inventory will play a central part in our assessment.

Once we have carried out this evaluation of CO_2 emissions for LME, we will have to repeat the whole procedure to evaluate the emission for each functional unit of petroleum diesel to be able to determine whether the LME really saves any CO_2 emissions when it is used instead of petroleum diesel. This will (ideally) require similar calculations on the reference system provided by the diesel product system. This however may not always be necessary as

- The amount of CO_2 released during the combustion of one functional unit of petroleum diesel usually far outweighs the other emissions during its production cycle (so we can ignore emissions during the other processes), and
- The amount of emission from the combustion of a unit of petroleum diesel has already been calculated and tabulated for us by the experts.

So to complete the LCA exercise, we simply calculate the emissions from the LME process and subtract it from the known value for petroleum diesel that we can easily calculate using information from the literature.

2006 into the current ISO14044 standards (International Organization for Standardization 2006). Before proceeding further with the formal study of LCA however, it will be useful to consider a simple case study that illustrates why an LCA is needed in a practical situation.

4.2 A Simple Case Study

Box 4.1 considers the production of a fictitious biofuel (LME) and demonstrates the need for a systematic approach such as an LCA to assess whether the use of the biofuel will really bring about any net savings in CO_2 emissions as compared to fossil fuels.

The above example emphasizes the need for a detailed and systematic analysis of all aspects relating to the production/use of biofuels in order to assess whether they bring about any savings in GHG emissions. The steps described there follow closely those of a formal LCA. These are described below in more detail.

4.3   Introduction to the Stages of LCA

An LCA is carried out in four steps or phases (GaBi Solutions 2018)

1. Goal and Scope Definition – during which the application and aims of the study are defined, and the all aspects of the product system to be studied are described
2. Inventory analysis – in which an **inventory** of all processes, materials and energy used, and the products produced and wastes generated are documented in detail
3. Impact assessment – where the particular environmental effect of the product that is of interest is evaluated in measurable terms
4. Interpretation – this is an ongoing process during the LCA where the results of the data collection and analysis exercise are reconciled against the goals and scope of the study.

Figure 4.1 shows diagrammatically how the four phases of an LCA investigation of a product or process are related to each other. Note the reversible arrows between the phases. They emphasize that the LCA investigation is an interactive process where each phase is tested by returning to the original to ensure, for instance, that the goal and scope of the investigation has been adhered to.

The aim of the LCA exercise is to obtain a holistic and comprehensive understanding of the product over its entire lifetime (which is represented by the **product system**), and to use this knowledge to make informed decisions. This is done by collecting data on each of the process (called a **unit process**) comprising the product system and compiling the data in a table called the **Life Cycle Inventory (LCI) table**. This table contains all the inputs and outputs (also called **the input and output flows**) for the processes that provide the building blocks of the product system, and can then be used for whatever specific needs the user specifies.

In the case where the interest is in the environmental impact of the product, a **Life Cycle Impact Analysis (LCIA)** is next carried out to assess the extent of this

Fig. 4.1 The four phases of the Life Cycle Assessment process

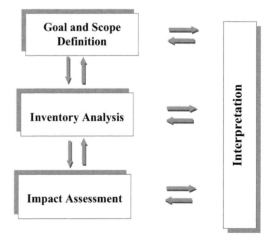

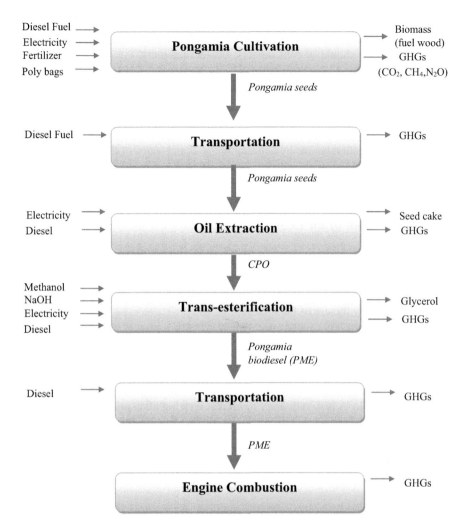

Fig. 4.2 The product system for the biodiesel pongamia methyl ester (PME), showing the important inputs, outputs and intermediate products for each unit process. (Data for the inputs and outputs are compiled in the LCI table. After Chandrashekar et al. 2012)

impact. Figure 4.2 gives an example of a product system for the production of the biodiesel pongamia methyl ester (PME), and indicates the essential input and output flows as well as the intermediate products for each of its unit processes.

Another use for LCA commonly encountered in the study of renewable energy resources is the determination of the energy balance of a fuel. In this application, the **Net Energy Balance (NEB)** of the fuel is determined by comparing the amount of energy a certain quantity of the fuel produces when it is burnt (the energy output) to the amount of energy used, over the entire life cycle, to produce this amount of fuel (total energy input). The ratio of the output energy to the input energy is called the **Net Energy Ratio (NER)**. For a fuel to be useful, its NER should be greater than unity and its NEB should be considerably greater than unity.

4.4 Goal and Scope Definition

The LCA investigation begins by defining the goal and scope of the exercise that is to be carried out.

4.4.1 The Goal of an LCA

The **goal of the LCA study** is first defined by the intended application, or purpose of the study. In the above example, this was to compare the GHG emissions of a biodiesel (LME) to that of petroleum diesel to ascertain the amount of emissions that are avoided by the use of the former.

4.4.2 The Scope of an LCA

The scope of the LCA is next defined by **characterizing all aspects of the product system** that is to be considered. In particular, this is done by defining

- The function of the product
- The functional unit of the product system and the reference flow
- The product system boundaries

4.4.2.1 The Function of the Product

The function of the product

 (i) Defines the purpose or function of the product, i.e. what it is for (e.g. providing fuel for diesel engines)
(ii) Documents the functionality of other products that this product will be compared to (e.g. petroleum diesel in the case of biofuels)

4.4.2.2 The Functional Unit

The functional unit (FU) of the product system is a quantitative measure of the function of the product. It is a basic unit or quantity to which all inputs and outputs to the product system can be associated with. In the case of a biofuel, it could be the unit of energy to be used for measuring the energy content of the fuel. The FU provides a means by which two different product systems can be compared for any reason (e.g. comparing the environmental impacts of using biodiesel and petroleum diesel).

4.4.2.3 Reference Flow

The reference flow is related to the functional unit, and provides a measure of the product components and materials needed to fulfil the function, as defined by the functional unit. It is also the output flow that leads from one process to another in a product system diagram (see later), and serves as a quantitative reference of each process.

4.4.2.4 System Boundary

The system boundary defines the processes that will be included in the LCA study. The system boundaries are defined by the **Cut-off Criteria**. These define the materials and energy that are to be included in the product system.

The Cut-off criteria can also be used to define the number of processes from the full process chain to be included in the LCA. These can be based on several considerations, including, for instance, the environmental impacts of particular processes (which would need to be included). Four major options to define the system boundaries are:

- **Cradle to Grave**
- This includes the materials and energy production chain and all process from raw material extraction to production, transportation, use and end-of-life treatment of the product
- **Cradle to Gate**
- Includes all materials, energy and processes from the raw materials extraction to the production (gate of factory) stage, and is useful for assessing the environmental impact of the product production process
- **Gate to Grave**
- Includes all processes from the use to the end-of-life stage, and is used to determine the environmental impact of the product use.
- **Gate to Gate**
- Includes the processes of the production phase only, and is used to determine the environmental impact of a single production process.

4.5   Inventory Analysis and Life Cycle Inventory (LCI)

Once the goal and scope of the study has been determined, one is in a position to determine which processes to include in the study and what type of data to collect.

The next stage is to collect quantitative data on the inputs, outputs, products and wastes for each of the process, and to compile them in a **Life Cycle Inventory (LCI) table** after they have been properly normalized with respect to the functional unit. This is the most detailed and laborious part of the LCA process, and the data has to be refined iteratively as more is learnt about the system with increasing data.

In the case of pongamia biodiesel production, the chain of processes in a cradle to grave analysis begins with the preparation of the pongamia seedlings, their planting, and the harvesting of the pongamia pods. These are transported to the oil mill where the crude pongamia oil (CPO) is extracted using an oil press and filter press, with the production of pongamia seedcake as the main by-product/waste. The CPO is then transferred to a trans-esterification plant where it is reacted with methanol in the presence of sodium hydroxide as a catalyst. This produces pongamia methyl ester (PME) as the main product and glycerol as a by-product.

As a pre-requisite to the collection of data on these processes, it is important to prepare a complete list of all the types of data that need to be collected. Table 4.1 attempts to categorize these data requirements for each unit process of the PME product system.

Table 4.1 Data requirements for inputs, outputs and emissions for each unit process of the PME product system

| Unit process | Data requirements | | |
	Material requirements	Products/intermediate products/by-products	Emissions
Cultivation			
Seedling preparation	Number/amount of polybags, fertilizer, machinery, fuel, labour		Amount of CO_2/ CH_4/N_2O released
Land preparation	Land area, plants/ ha., fertilizer/ ha, diesel and equipment used		Amount of CO_2/ CH_4/N_2O relesased
Harvest and transportation	Number of pongamia pods/tree/ year, kg of seeds/pod, distance to oil mill, diesel use		Amount of CO_2/ CH_4/N_2O released
Oil extraction			
Oil press and filter press	Kg of oil (CPO) per kg of seeds; amount of diesel fuel use, electricity use	Kg of seedcake/kg of seed produced	Amount of CO_2/ CH_4/N_2O released
Trans-esterification			
Esterification reaction	Volume of batch, amount of CPO, NaOH used	Amount of PME, glycerol produced	Amount of CO_2/ CH_4/N_2O released

After Chandrashekar et al. (2012)

Table 4.2 A simple life cycle inventory of energy inputs and outputs for the PME system

Unit process	Quantity ha^{-1} year^{-1}	Energy input (MJ)	Energy output (MJ)
Cultivation			
Diesel (L)	22.0	825.0	
Polybag use	330	229.3	
Oil extraction			
Diesel (L)	11.0	412.5	
Electricity (kWh)	60.0	216.0	
Trans-esterification			
Electricity (kWh)	34.0	122.4	
NaOH (kg)	0.0017	39.6	
Methanol (L)	42.5	562.3	
PME biodiesel(kg)	120.0		4378
Total (MJ)		2407	4378

After Chandrashekar et al. (2012)

An appropriate functional unit (FU) for this product system is 1 MJ of energy in the final product (PME). This choice allows a comparison with the petroleum diesel reference system to be carried out conveniently.

The next task is to obtain a complete inventory of the material and energy inflows and outflows for the entire system by quantifying the inputs and outputs. This life cycle inventory (LCI) is often based on a convenient unit of output or input resource. In the case of biofuels, it is not infrequent to carry out this assessment for 1 ha of land used for the plantation of the feedstock.

The following basic information for the preparation of the LCI for the pongamia system has been gathered by (Chandrashekar et al. 2012). One hectare of land can accommodate 300–330 pongamia trees, which start bearing in 5 years, reaching peak production at about 10 years and continuing to produce till the tree is 80 years old. The seed yield is 3–5 tons ha^{-1} from 10 year old trees. Oil (CPO) yield is 270–300 kg oil from 1000 kg seeds. In the esterification process, 100 L (i.e. 85 kg) CPO requires 20 kg methanol and 0.8 kg NaOH to produce 85–90 kg PME and 15–16 kg glycerol. The energy requirements for the entire pongamia system are 45 kJ per FU of PME produced. One FU require 28 g of PME biodiesel, which is made from 33 g of CPO which is extracted from 100 g of seeds or 200 g of pongamia pods.

Table 4.2 shows a simple life cycle inventory for the energy inputs and outputs of the pongamia biodiesel system, including the essential inputs and outputs. Note that this table does not include any information on the by-products, which can have a significant impact on the total output energy.

The information in the above inventory may be used to estimate the Net Energy Balance (NEB) and Net Energy Ratio (NER) for the PME production relevant to the specific conditions to yield a NEB of 1971 MJ ha^{-1} year^{-1} and a NER of 1.82. Note that the inclusion of by-products such as seek-cake or glycerol would change the results significantly.

4.6   Life Cycle Impact Assessment (LCIA)

If the aim of the investigation is to assess the environmental impact of the production/use of the product being studied, then a Life Cycle Impact Assessment is carried out. The environmental impacts may vary in nature from ecological effects, to the toxic and other effects of environmental pollutants, the impact of land use as well as other impact categories. In a complete LCIA, the potential environmental impacts due to all these activities and processes are evaluated quantitatively by first arranging these impacts in terms of impact categories (a process called **classification**).

Some of the important impact categories include **Acidification, Eutrophication, Fresh Water Toxicity, Marine Water Toxicity, Terrestrial Toxicity, Human Toxicity, Global Warming, and Photochemical ozone creation.** Table 4.3 provides a brief description of some of these categories.

After the emissions have been classified into impact categories, they must be characterized to take into account their relative strengths to produce their respective environmental impacts. This is done by multiplying the emission unit with the appropriate characterization factor.

To take the example of global warming, suppose we wish to evaluate the total global warming impact potential of 150 kg of CO_2, 5 kg of CO and 6 kg of CH_4. Each of these gases has a different ability to contribute to global warming. This is reflected in its **characterization factor** (also called global warming potential in some literature), which is evaluated relative to CO_2 as the reference substance. The

Table 4.3   Some of the main environmental impact categories considered in most LCIA

Impact category	Description
Abiotic resource depletion	Extraction of minerals, fossil fuels and other resources needed for the production of the product
Climate change (global warming)	Global warming due to the emission of greenhouse gases (GHGs) into the atmosphere
Stratospheric ozone depletion	Reductions is stratospheric (high altitude atmosphere) ozone levels due to emission of ozone-destroying gases
Human toxicity	Emission of toxic substances (having harmful effects on human health) into the biosphere
Eco-toxicity (terrestrial, marine, fresh-water)	Impact of toxic substances on terrestrial, marine and freshwater ecosystems
Photo-oxidant formation	Formation of reactive substances (e.g. photo-chemical ozone creation) that are harmful to both human health and ecosystems
Acidification	Emission of acidifying agents such as SO2, NOx, which impact adversely on soil, ground water and ecosystems
Eutrophication (Nutrification)	Emission of macro-nutrients (including N and P from fertilizers and sewage streams) into the environment, leading to abnormal productivity

Stranddorf et al. (2005)

Table 4.4 Converting emissions into global warming impact potential for some common greenhouse gases

Emission	Impact category	Characterizing factor	Impact potential
150 kg CO_2	GWP	1	$150 \times 1 = 150$ kg CO_2 eq
5 kg CO	GWP	3	$5 \times 3 = 15$ kg CO_2 eq
6 kg CH_4	GWP	25	$6 \times 25 = 150$ kg CO_2 eq
Total impact potential			315 kg CO_2 eq

factors are 1, 3 and 25 for CO_2, CO and CH_4 respectively, and the impact potential for each gas is calculated by multiplying its mass with its characterization factor. The final result is obtained by summing over all the three gases. The result of these calculations are shown in Table 4.4.

4.7 LCA Analysis Software

Life cycle assessments can be done using software that has been specifically designed for carrying out LCAs. The basic function of all LCA software packages is to determine the energy and mass balances of a product or process and to assign emissions and energy usage to them (Ormazabal et al. 2014). Essentially, the software will carry out the LCA analysis following the steps described in the previous sections of this chapter. However due to the logic/functions built into the program, getting the results of an LCA will be easier and less time consuming. Uncertainty calculations, impacts evaluation and several other analyses is possible with LCA software which comes with in-built functions to explore all aspects of an LCA (Leyens 2010).

Currently, there are several LCA software tools available in the market. These can be either purchased or obtained for free. OpenLCA, SimaPro and GaBi are some commonly used LCA software. The majority of the LCA software available are **closed sourced** (i.e. not free downloads) and have high licence pricing. There are some free software and tools available, the best of these being OpenLCA, which is open-source software available for use for free. The features of openLCA are described below in more detail.

4.7.1 OpenLCA

OpenLCA has been developed by GreenDelta, with the support of PE International (makers of GaBi), PRé Consultants (creators of SimaPro) and United Nations Environment Programme (Ormazabal et al. 2014). OpenLCA can be downloaded free of cost from *openLCA downloads* website: https://www.openlca.org

4.7.1.1 Modelling in openLCA

OpenLCA works by modelling **flows, processes, product systems** and **projects** with a view to (amongst other things) quantifying the environmental impacts of product systems (GreenDelta GmbH 2019). In the case where the interest is in quantifying the environmental impacts of a product, a Life Cycle Impact Assessment (LCIA) method has to be **imported** (i.e. brought from another source and added) to openLCA. The openLCA LCIA method is available online for free and can be used for all data sets provided through the **openLCA Nexus platform**.

Modelling in openLCA begins with creating or importing a database which contains the life cycle data. Creating a database requires starting from scratch and modelling all the database elements which are needed for the LCA exercise. On the other hand, an imported database from the openLCA nexus platform includes all the reference data for conducting LCA. OpenLCA is thus instrumental in making work easier by providing quick access to a large collection of datasets and databases which contain well documented process data for various product systems. Access to such databases provides relevant process and flow information which helps to reduce the time needed to carry out the LCA (Haselbach and Langfitt 2015).

The database needs to be populated with flows, processes, product systems and projects. Let's work through these elements with an example. Figure 4.3, gives an example of the various modelling elements in an openLCA for two product systems. The first is the biodiesel pongamia methyl ester product system that has been described in Sect. 4.3 and Fig. 4.2 and the second product system shown is the petroleum diesel system, to which the first needs to be compared while assessing avoided greenhouse gas emissions. All the modelling elements, including the flows, processes, product systems and projects, are shown in Fig. 4.3. A screenshot of the openLCA software is shown in Fig. 4.4 to show the location of each of the **tabs** in the software required in the modelling of these systems.

Flows in OpenLCA are all product, material or energy inputs and outputs of unit processes of a product system, characterized by the name, type and reference flow property (GreenDelta GmbH 2019). For the case of the PME product system, the input flows for the process of pongamia cultivation are poly bags, diesel fuel, electricity, fertilizer, while the output flows are biomass, greenhouse gases and Pongamia seeds. All flows apart from the reference flow are modelled in the *processes* tab, which is filled after filling in the *flows* tab in openLCA.

The initial steps when modelling the flows in openLCA is to add the output flow that serves as a quantitative reference for each process. Any flow can be chosen as the reference flow in openLCA. Usually a quantifiable output product, one which is necessary for a product system to deliver is chosen as the reference flow (Weidema et al. 2000). The arrows in Fig. 4.3 show the reference flow/output for each process. For instance, the reference flow for the process of pongamia cultivation is pongamia seeds. This is a product flow which is used by the next process (transportation) as an input. The location of the respective tabs in the OpenLCA software is shown in Fig. 4.4.

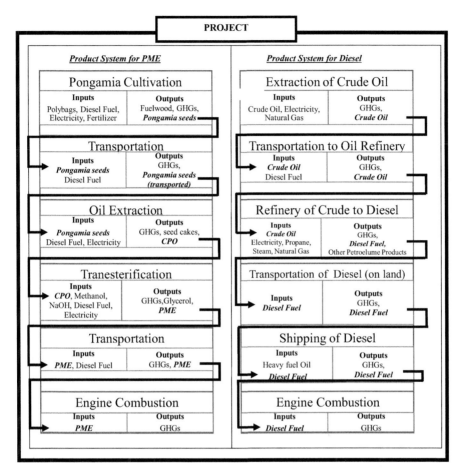

Fig. 4.3 Example of a project with two product systems, showing the quantitative reference flows (indicated by arrows), the processes and the input and output flows for each

The flow type and flow property of the reference flow will need to be defined in the flow tab. Other quantitative reference flows from Fig. 4.3 for the PME product system has been identified with use of arrows and include pongamia seeds (transported), CPO, PME and PME (transported) for the processes of pongamia seeds transportation, oil extraction, trans-esterification, PME transportation and engine combustion respectively. The reference flows for the diesel system can be modelled likewise. These will need to be modelled in the flow section before proceeding with processes. An example of the reference flows is shown in Fig. 4.4 under the flow tab.

Processes transform the inputs into outputs. In the case of the pongamia cultivation process, polybags, diesel fuel, electricity and fertilizer are inputs required for the process. The outputs formed are fuelwood and GHG emissions together

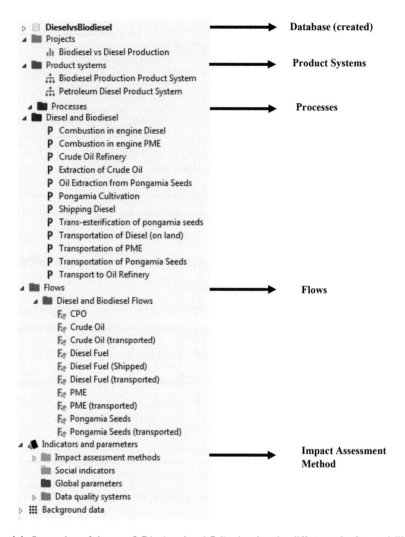

Fig. 4.4 Screenshot of the openLCA, (version 1.7.4), showing the different tabs for modelling flows, processes, product systems, impact assessment method and projects in an openLCA software

with the reference output of Pongamia seeds. When creating Processes in open-LCA, all the remaining inputs and outputs of the process will need to be modelled in. The reference created in the initial steps has to be set as the quantitative reference for the cultivation process. This will then automatically appear as an output of the process. All other processes for the PME and diesel system can be defined in the same way.

The **product system** contains all the processes under study. For instance, the processes of the PME product system include the pongamia cultivation, transportation of pongamia seeds, oil extraction, trans-esterification, transportation of PME and engine combustion of PME. Likewise, the product system for the diesel which is the reference system consists of the processes of crude oil extraction, transportation of crude oil to refinery, production of diesel from crude oil in a refinery, transportation of diesel in land, shipping of diesel to destination and combustion of diesel in an engine.

In a manner analogous to defining a reference flow/output in the flow tab, a quantitative reference for a process known as a **reference process** needs to be defined for a product system in openLCA. The last process of a product system is usually chosen as the reference process. For instance, when modelling in the PME product system, the reference process will be the combustion of PME in the engine. Similarly, the reference process for the diesel product system will be the combustion of diesel in the engine. The other processes of a product system should be linked to the reference process. The impacts of a product system are calculated at this stage using the impact assessment method imported. Impacts such as depletion of fossil fuel, terrestrial exotocity, acidification potential, climate change global warming potential for 100 years, levels of eutrophication, ozone layer depletion, human toxicity together with numerous other impacts can be calculated in openLCA.

Projects in openLCA are used for comparison purposes between two or more product systems. For instance in order to compare the environmental impacts of the PME and the petroleum diesel product systems, a project of *PME vs Diesel Production* can be created under the Project tab. Figure 4.4 shows the all elements that need to be modelled under the project.

4.7.2 Simapro

SimaPro developed by PRé Sustainability, is an example of closed source software. Closed source software is distributed under a licensing agreement to users who purchase the software (Singh et al. 2015). In other words, only the original authors of the software can access and modify the source code. Anyone interested in using Simapro, has to pay to get access to the software and to use it. Paying for licensing also means that the level of support that a user receives when using Simapro is superior since the payment ensures that full technical service is provided to users of the software.

Most of the LCA software in use currently have similar features with the main difference seeming to lie in their user interfaces (Haselbach and Langfitt 2015). However, Simapro is integrated with more databases compared to openLCA, thus simplifying the LCA process further. With many useful graphical features, the process of displaying and presenting the results of the LCA become easier.

References

Chandrashekar, L. A., Mahesh, N. S., Gowda, B., & Hall, W. (2012). Life cycle assessment of biodiesel production from pongamia oil in rural Karnataka. *Agricultural Engineering International: CIGR Journal., 14*(3), 67–77.

GaBi Solutions. (2018). Retrieved April 2019, from LCA and introduction to GaBi: http://www.gabi-software.com/international/support/gabi-learning-center/gabi-learning-center/part-1-lca-and-introduction-to-gabi/

GreenDelta GmbH. (2019). *OpenLCA 1.8: Basic modelling in openLCA*. Berlin: GreenDelta.

Haselbach, L., & Langfitt, Q. (2015). *Welcome to the life cycle assessment (LCA) learning moduel series*. Retrieved March 20, 2019, from University of aAlaska fairbanks: http://cem.uaf.edu/media/138876/module-g1-general-paid-lca-tools-25-feb-2015.pdf

International Organization for Standardization. (2006). *ISO 14040:2006*. Retrieved from International Organization for Standardization: https://www.iso.org/standard/37456.html

Leyens, T. (2010). *Life cycle analysis of biofuels*. Gainesville: HELMo University College.

Ormazabal, M., Jaca, C., & Puga-Leal, R. (2014). Analysis and comparison of life cycle assessment and carbon footprint software. In *Eight international conference on management science and engineering management* (pp. 1521–1530). Berlin: Springer.

Singh, A., Bansal, R. K., & Jha, N. (2015). Open source software vs proprietary software. *International Journal of Computer Applications, 114*, 26–31.

Stranddorf, H. K., Hoffmann, L., & Schmidt, A. (2005). *Environmental news*. Retrieved April 2019, from Impact categories, normalisation and weighting in LCA: https://www2.mst.dk/udgiv/publications/2005/87-7614-574-3/pdf/87-7614-575-1.pdf

Weidema, B., Wenzel, H., Petersen, C., & Hansen, K. (2000). Retrieved April 16, 2019, from The Danish Environmental Protection Agency: https://www2.mst.dk

Part II
Mitigation Actions

Chapter 5
An Assessment of the Hydro Potential for Viti Levu, Fiji Using GIS Techniques

Sanjay Raj Singh

Abstract Hydropower helps countries meet their energy needs in an economically, environmentally, and socially sustainable way while saving money and increasing energy security and self-reliance. The increased urbanisation, industrialisation and grid extensions has led to an increased demand for electricity in Fiji. Energy Fiji Limited (EFL) along with the Fiji Department of Energy (FDoE) is working actively with consultants in order to determine new potential hydropower sites in Fiji. This work is directed towards achieving the targets of 90% renewables by 2030 as stipulated in Fiji's Nationally Determined Contribution (NDC) Implementation Roadmap. The upper stream of Qaliwana river and the lower stream of Ba river has already been proposed as new potential sites for hydropower generation in Fiji.

This study provides the current status of hydropower production in Fiji and provides a detailed account of the methodology used to perform resource assessments at various sites in Vitilevu to determine the new hydropower capacity. This technique entails the in corporation of spatial statistical modeling tool in the Geographic Information System (GIS) platform. The input data (including topographic characteristics and precipitation) was compiled and analysed using GIS data layers.

The preliminary results presented were based on a suitability map which showed that approximately 40% of land mass area on the Island of Vitilevu had potential for hydropower in consideration to factors such as protected forest covers, fault lines, slope and water retention properties of soil. The suitability map incorporates both existing hydropower generation sites as well as new potential sites identified by using the GIS technique. Thus this study reveals the existence of other potential hydropower sites not identified by traditional methods.

Keywords Hydropower · Sustainable · Geographic information system (GIS) · Hydro power capacity

S. R. Singh (✉)
School of Science and Technology, The University of Fiji, Lautoka, Fiji
e-mail: sanjays@unifiji.ac.fj

© Springer Nature Switzerland AG 2020   101
A. Singh (ed.), *Translating the Paris Agreement into Action in the Pacific*,
Advances in Global Change Research 68,
https://doi.org/10.1007/978-3-030-30211-5_5

5.1 Introduction

5.1.1 The Power Scenario in Fiji

Fiji being part of small island developing nation in the Pacific highly depends on imported fuels. The energy sector of Fiji has a significant impact on Fiji's macro-economy through importation of fuels at relatively high and volatile fuel prices. This has led to high import payments thus reducing gross domestic product (GDP) of the nation. In order to safeguard the foreign reserves and reduce reliance on imported fuels ensuring macro-economic stability of Fiji, there is an urgent need to determine alternative fuels in the form of renewable and indigenous fuels (Sustainable Energy for All (SE4All): Rapid Assessment and Gap Analysis Fiji).

The major renewable energy source in Fiji is hydropower. More importantly hydropower provides a a clean, renewable and reliable energy source that serves the national environmental and energy policy objectives quite well (Kurse et al. 2010). Hydropower being highly reliable and efficient energy source is also dispatchable in nature. However, hydropower potential varies seasonally with the rainfall patterns unlike wind which has far greater variability within shorter time periods. Most importantly the greenhouse gas (GHG) emissions of hydropower are the least among all power sources. The available energy can be harnessed by utilizing the natural water flows at possible river locations. It has to be noted however that while many of the rivers of the world have abundant hydropower potential, these potentials have been optimally utilized only in countries which are technologically advanced (Kurse et al. 2010).

Energy Fiji Limited (EFL) is the only power utility company in Fiji which was established in 1966 and has been responsible for generation, transmission and retail of grid electricity on Vitilevu, Vanualevu and Ovalau ever since. Diesel generator sets, micro hydro systems or generators running on biofuels are used to electrify the rest of the islands. The Fiji Department of Energy (FDoE) is also responsible for electrification using off-grid extensions for populations in remote areas (Prasad et al. 2017). Currently, hydropower is the dominant renewable energy source in Fiji and contributes 53.05% of the islands energy needs (EFL annual report 2016). However, there are some variations in the yearly electricity generated using hydropower. These variations are due to difference in the amount of precipitation per year, natural disasters and technical and mechanical problems encountered at the hydropower stations and sub-stations. In 2016, the electricity generation mix was as follows: 53.05% hydro, 45.45% diesel generators, 0.39% from Butoni wind farm and 1.11% from the IPP's (Independent Power Producers) using biomass energy, (see Table 5.1). However, there are variations in the total electricity generation mix over the years. Table 5.2 summarises the current hydropower generation sites on the Island of Viti Levu with their installed capacity and year of commissioning.

Table 5.1 Percentage
electricity generated in 2016

Power generation mix	Percentage
Hydro	53.05%
Wind	0.39%
IPP's	1.11%
Industrial diesel oil	45.45%

Source: EFL annual report 2016

Table 5.2 Hydropower
potentials in Fiji at various
sites with total installed
generation capacity of
258.9 MW

Location/site	Installed capacity (MW)	Year of commission
Monasavu Wailoa	83	1983
Nadarivatu	42	2012
Wainikasou	6.6	2004
Nagado/Vaturu	2.3	2006

Source: EFL Power Development Plan, EFL
Presentation Energy Forum 2013

5.1.2   Grid-Electricity Demand Trend

The hydro electricity power generation in Fiji from 1998 to 2014 ranged from 46% to 85% (Prasad et al. 2017). The major driving variables for the varied percentages of hydropower generation among different years are the rainfall based on seasonality pattern and the peak power demand. As per Fiji Bureau of Statistics data from 1976 to 2014, there has been an increasing electricity demand with an overall 368% increase in the last 38 years, (see Fig. 5.1). There is an average annual increase of 4.3% in grid electricity demand (Prasad et al. 2017). As revealed in Table 5.3, it is obvious that there are distinct "3 steps" in the grid-electricity demand: (i) prior to 1983 (ii) between 1984–2000 and (iii) 2001–2014, (Prasad et al. 2017). The electricity demand was 24% from 1983 to 1984, which increased by 25% from 2000 to 2001. Interestingly, an escalated demand for electricity during the period 1983 to 1984 was catered by installation of the new 80 MW Monasavu hydro power scheme. However, there was an escalation for power demand between the periods 2000 and 2001 due to the increased number of people connected to the grid as a result of an increase in the economic activity despite the political unrest in 2000. Dips in electricity demand trend were observed during 1987, 2000, 2006 and 2010 and probably these dips are due to political unrest in Fiji. Year 2010 experienced a dip due to increased tariff rate and this has certainly contributed to a fall in electricity consumption (Prasad et al. 2017).

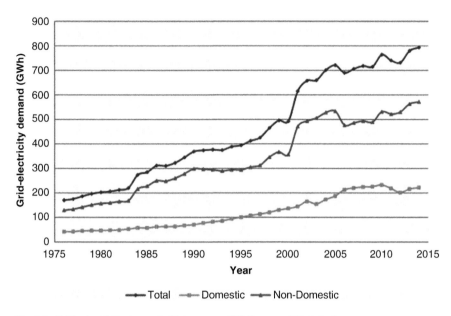

Fig. 5.1 Grid- electricity demand. (Data source: Fiji Bureau of Statistics)

Table 5.3 Structural change in total grid-electricity demand

Period	Total % increase	Average annual % increase	Structure % increase
1976–1983	68	3.82	
1984–2000	58	3.79	1983–1984 = 24
2001–2014	29	2.04	2000–2001 = 25

Source: Prasad et al. 2017, Energy p. 863

5.1.3 Current Energy Situation for Fiji

Fiji has progressed significantly well over the last 10 years in providing access to modern energy and increasing the share of percentage renewable energy sources in the electricity generation. Through Fiji Department of Energy's initiative, new bio-fuel, wind, solar and hydropower plants have been installed. Moreover, rural electri-fication has also advanced and the energy efficiency promotion such as energy labeling and promotion of energy efficient vehicles to replace inefficient vehicles and appliances are extensively practised (Sustainable Energy for All (SE4All): Rapid Assessment and Gap Analysis Fiji). Fiji is progressing well in providing rural electrification to its people. In 2003, the rural electrification was approximately 69% followed by marked improvement in 2007 with a rise to approximately 82%. Moreover, EFL has been performing relatively well in comparison to other Pacific Island utilities in the region in order to achieve its renewable energy target of 90% for grid energy supply as stipulated in Fiji's National Energy Policy (Table 5.4).

Table 5.4 Targets for Fiji's energy sector to 2030

Indicator	Baseline	Targets 2015	2020	2030
Access to modern energy services				
Percentage of population with electricity access	89% (2007)	90%	100%	100%
Percentage of population with primary reliance on wood fuels for cooking	20% (2004)	18%	12%	<1%
Improving energy efficiency				
Energy intensity (consumption of imported fuel per unit of GDP in MJ/FJD)	2.89 (2011)	2.89 (0%)	2.86 (−1%)	2.73 (−5.5%)
Energy intensity (power consumption per unit of GDP in kWh/FJD)	0.23 (2011)	0.219 (−4.7%)	0.215 (−6.15%)	0.209 (−9.1%)
Share of renewable energy				
Renewable energy share in electricity generation	60% (2011)	67%	81%	100%
Renewable energy share in total energy consumption	13% (2011)	15%	18%	25%

Source: Sustainable Energy for All (SE4All): Rapid Assessment and Gap Analysis Fiji p. 3

With the utilisation of considerable local renewable energy resources such as untapped solar potential and geothermal energy, the power sector in Fiji can save fossil fuel imports worth approximately 100 million FJD per annum (Sustainable Energy for All (SE4All): Rapid Assessment and Gap Analysis Fiji sited in EFL annual report, 2011). Currently, some 60% of grid electricity is generated using renewable sources (Sustainable Energy for All (SE4All): Rapid Assessment and Gap Analysis Fiji cited in EFL annual report, 2011).

There is an immediate need to promote renewable energy in the transport sector for both land and maritime to replace fossil fuels. Particularly, it is the transport sector that depends heavily on fossil fuel. This issue can be solved by using alternative sources such as bio-fuels, hybrid and electric cars, and improving the efficiency of the vessels and engines.

Currently Fiji has a total installed generation capacity of 258.9 MW at its four major hydroelectric schemes located on the main island of Vitilevu, out of which 0.18 MW is installed by FDoE for off-grid power. Moreover, there are three co-generators in biomass production in Fiji who intend to increase the percentage share of renewable energy, increasing energy security and energy efficiency of the nation. These independent power producers (IPPs) are contracted through EFL and feed power generated to the EFL's main grid. Fiji sugar co-operation Limited (FSC) uses bagasse, a by-product of sugarcane to generate power for sugar manufacturing and surplus generated power is fed to EFL grid. Similarly, Tropik Woods Fiji Limited and Nabou Biomass Power Plant use wood biomass to produce electricity which is fed to the EFL grid system (IRENA 2015). The respective IPPs have the following installed capacities: FSC (Lautoka mill) –12 MW and 5 MW co-generation plants, FSC (Labasa mill) – a 10 MW and a 4 MW co-generation plants, FSC (Rarawai mill, Ba) – 5 MW and 4 MW co-generation plants, Tropik Woods Fiji Limited (9.3 MW plant) and Nabou Biomass Power Plant (10 MW) (EFL annual report 2016).

Moreover, EFL's thermal power plant uses fossil fuel operated diesel gensets for nearly all its electricity generation on islands other than Vitilevu. This is due to the fact that the demand for electricity cannot be met alone by hydropower scheme itself therefore it is catered by diesel generators (Chand 2013). Interestingly, there had been an extensive grid extension projects done by EFL at various sites from 2017 and this step forward ensures accessibility to modern energy and mitigating strategy for reducing fossil fuel use and sustainabe development. The areas with grid extension include: Naitasiri, Nasautoka grid extension in Wainibuka, Navuca and Sauva Grid Extension in Tailevu and Vunarewa and Naqali Settlements in Nadi (https://fijisun. com.fj/2018/01/11). However, this grid extension has not solved the problem of power demand but rather there is an increasing extensive demand for energy particularly during peak hours. The current generation fleet for 2016 is presented in Table 5.2 revealing that there is a pressing need for more hydropower stations in order to meet the targets of Nationally Determined Contributions, NDC's.

The Fiji government's Sustainable Energy for All (SE4All): Rapid Assessment and Gap Analysis have highlighted the importance of achieving Fiji's target of sustainable energy for all. The Sustainable Energy for All aims to provide a global initiative from all sectors in Fiji to achieve the following three objectives by 2030: (i) ensure universal access to modern energy services to all; (ii) double the global rate of improvements in energy efficiency; (iii) double the share of renewable energy in the global energy mix (Sustainable Energy for All (SE4All): Rapid Assessment and Gap Analysis Fiji).

Currently the access to electricity in Fijian households is approximately 85%. The last three consecutive census data for the years 1986, 1996 and 2007 are 48%, 67% and 89% respectively for the households that have access to electricity (Prasad et al. 2017). To meet electricity demand, EFL provides electricity either by grid connected or distributed generators such as diesel generators or biofuels generators or solar home systems or micro hydro systems.

By the year 2030, Fiji's energy sector targets to double its renewable energy capacity and in particular to increase its installed hydropower capacity by an additional 150 MW. According to Pacific Energy Update 2018, Fiji government has set the target of achieving 99% renewable energy generation by 2030 and 100% electrification rate by 2020. These investments may require approximately FJD $760 million over the coming decade. Moreover, the Fiji government has prioritized its actions in energy sector in various ways such as: (i) expanding the role of the private sector in power generation including the privatization of energy utility; (ii) increased the role of non- Energy Fiji Limited renewable energy via small-scale systems; (iii) restructuring regulatory arrangements to improve transparency and accountability, and to remove possible conflicts of interest (Pacific Energy Update 2018).

The Asian Development Bank (ADB) provides an extensive active technical assistance and helps develop an investment project. The key active technical assistance includes providing support for energy sector regulatory capacity and electrification investment planning. This will certainly support government of Fiji by developing: (i) the institutional capacity for regulation of the electricity sector and (ii) a sector investment planning framework. Moreover, the technical assistance provided has a positive impact on improved framework on Fiji's energy sector which include being resource efficient, cost-effective, and achieving environmentally sustainable energy sector (Pacific Energy Update 2018). More importantly, the technical assistance comprises of two major outputs: (i) capacity building for selected regulatory agencies and (ii) enhanced sector planning capacity at the relevant government department. The second output is very crucial and consists of detailed recommendations for establishing the policy framework for the identification, selection, and implementation of rural electrification investments, and capacity-building support for accelerated investments in Fiji's rural electrification program (Pacific Energy Update 2018).

5.1.4 Main Challenges and Gaps in Energy Sector

The power sector in Fiji faces enormous challenges with due respect to power generation through renewable sources. One of the major challenges in the power sector is being institutional in nature. EFL being the State-owned Enterprise (SOE) has not being offering attractive power purchase tariffs and had not been sufficient to attract private investors or Independent Power Producers (IPP's). As a result of this, Fiji Competition and Consumer Commission (FCCC) stepped in to regulate the tariff rate at an attractive rate of \$0.34/KWh. Secondly, the power generation project selection criteria offered by EFL had not been clear (Sustainable Energy for All (SE4All): Rapid Assessment and Gap Analysis Fiji).

The challenging business environment in Fiji provides another constraint in attracting private sector investment in energy projects. The major barrier is associated with private sector participation with regards to small-scale decentralised grid-connected renewable energy based generation. Currently, there is no feed-in tariffs, net metering provisions or incentive programmes to promote such generations by households and small and medium enterprises. To overcome and reduce these energy gap barriers, FDoE with support from Global Environmental Facility (GEF) and United Nations Development Programme (UNDP) has launched a new project called Fiji Renewable Energy Power Project (FREPP) (Sustainable Energy for All (SE4All): Rapid Assessment and Gap Analysis Fiji).

Furthermore, the Fijian government has reformed its policies on new IPP framework to increase the private sector participation in electricity supply at both small and large-scale through reform of regulatory aspects of the electricity sector (5-Year and 20-Year National Development Plan).

5.1.5 EFL's Future Power Development Projects

The FDoE had been monitoring the hydrology of several sites in Fiji as the major part of the hydro survey across the country. Based on these long-term hydro survey results and other factors such as terrain, road accessibility, water shed dealination, drought and rainfall patterns, feasibility studies are conducted. More to this, FDoE and EFL have already identified three feasible hydro potential sites on the main island of Vitilevu and one on Vanualevu at northern division. The Qaliwana Upper Wailoa Diversion Hydro Project is estimated to cost US$265 million (FJD $543 million) which includes hydropower and transmission connection. The potential output of this would be 44 MW and generation of 206 GWh. Secondly, the lower Ba Hydro Development Project will channel water from Nadarivatu and the potential output from this project would be 49 MW and generation would be 214 GWh. Currently, negotiations are ongoing with European Investment Bank in relation to this project (https://fijisun.com.fj/2015/04/11/fea-details-the-1.5bn-renewable-energyprojects/). A feasibility study is being carried out by the European Investment Bank to look into the possible establishment of two hydro power plants in Ba and upper Wailoa in Naitasiri (http://www.fijitimes.com/feasibility-study-for-hydro-power/) & (fhta.com.fj/feasibility-study-for-hydro-power/). Thirdly, the Japan International Cooperation Agency (JICA) study findings highlighted potential site at Waivaka river in Namosi and the estimated cost of the project is US$88.7 million (FJD $184 million) which includes hydropower and transmission (Japan International Cooperation 2015). The potential output of this would be 32 MW with generation of 67.6 GWh. In addition to this, EFL is awaiting the response from interested IPP's for the future power development projects (https://fijisun.com.fj/2015/04/11/fea-details-the-1.5bn-renewable-energy-projects/).

Other hydro power plans include exploring the options within Navua river basin. These projects are currently in the feasibility stage. The studies earlier indicate that the upper Navua river hydro power project could be of 48 MW capacity that could generate an average of 360,000 MWh of power a year. This would certainly result in savings from potential emissions of about 183,420 tons of CO_2 calculated using grid emission factor of 0.5095 tCO_2/MWh (Emission Reduction Profile Fiji 2013). Moreover, the Wailoa downstream hydro power project will be of planned capacity of 15 MW with annual emission reductions of 57,564 tons of CO_2 (Emission Reduction Profile Fiji 2013).

5.1.6   Energy Demand and Fossil Fuel Importation for Fiji

Like other Pacific Island Countries, Fiji heavily depends on imported petroleum-based fuels. Fluctuations in global oil supply not only affects energy security but also energy prices. Fiji had been thumped twice by the shortage of oil supply caused by two energy crisis in the 1970s – the Organisation for Economic Cooperation and Development (OECD) embargo and the Iranian revolution, which accounted for an average 8% increase in the price of energy – and the Iraq war in 2003, which caused an average 6.6% increase over 2001–2005 prices (IRENA 2015 cited in Kumar, 2011). International oil prices rose drastically between the period 2004–2008 thus increasing Fiji's energy expenditure. In 2008, Fiji has spent as much as 17% (FJD 744 million on fuel) importation of its gross domestic product (GDP) on energy up from 7% in 2003 (Fiji Islands Bureau of Statistics, various years; Reserve Bank of Fiji, various years).

Inflated oil prices have triggered small island states (SIDs) in the Pacific to invest in renewable energy technologies. The high oil prices had detrimental effect on electricity sector given the widespread reliance on generators that operate on diesel and heavy fuel oil (Dornan and Jotzo 2015). This had adverse effect on macroeconomic status and caused an increased electricity prices which threatened energy security of poor households. Exemption of tax duties and reduced prices of renewable energy technology has allowed route to more renewable -based electricity production over oil-based generation. Pacific island governments do support the investment in Renewable Energy Technology (RET) and are in search of potential donor agencies.

Fiji's first hydropower plant was built as early as the 1920s. However, there was no comprehensive survey of small scale hydropower. The majority of the hydropower was installed between the years 1980 and 1990 to supply the villages with mini-grids. In addition to this, the FDoE experienced difficulties in hydropower plant operations. The major challenges were associated with technical problems which include equipment breakdowns, poor management, inadequate community leadership and sedimentation exacerbated by difficult site access and the villager in charge of the plant operation lacked technical knowledge of operation (IRENA 2015).

The hydropower system is subjected to seasonal variation of annual hydrological cycles and extreme weather events, such as El Niño, El Niño Southern Oscillation or severe drought which eventually decreases reliability on hydropower production (IRENA 2015). Moreover, Fiji faces distinct wet and dry seasons of which 8 months of the year from May to December are deficient in precipitation particularly in the dry zone on the north western sides of the main islands (IRENA 2015). In this regard, Monasavu catchment resulted in critically low water levels at the Monasavu dam due to poor rainfall. However, during the months of November to April which coincides with tropical cyclone season, the level of precipitation is usually high particularly over the major islands of Fiji. During this period, the rainfall levels can reach as high as 6000 mm which accounts for up to 80% of the total annual rainfall (IRENA 2015 cited in Global Environmental Facility (GEF), 2009).

The increasing demands for energy and the continued concerns over the devastating effects of climate change have driven the shift in the energy from renewable sources. It is vital to have a closer look at all the possible renewable energy resources available to determine which if any, are suitable for development (Tarife et al. 2017). The various renewable energy options available are wind, solar, biomass and hydroelectric energy sources. Above all, hydropower is apparently the most common form of renewable energy option. Fiji as a Small Island Developing States (SIDS) needs to look at its available energy resources, electricity generation and consumption and consumption of imported fossil fuel to figure out the prominent threats and challenges to its energy economy. These energy issues need to be identified and concrete strategies enforced towards achieving its NDCs (Prasad et al. 2017). Fiji's NDC was developed to achieve the objective of the UNFCCC which is consistent with Green Growth Framework 2014 (GGF) and Sustainable Energy for All (SE4ALL) initiative of United Nations. The NDC targets are to be met by 2030 reducing the import of diesel and heavy fuel oil to 200 million litres, achieving 100% renewable electricity and reducing emissions by 30% (Prasad et al. 2017).

5.1.7 GIS Technologies for Hydropower Survey

Numerous studies reveal that a variety of methods and tools could be used to esti-mate the hydropower potential of a river. Most of the earlier studies concentrated on traditional methods using field investigations or classical approaches for hydro-power estimation whereas the recent study focuses on advanced tools of remote sensing and geographical information system. The use of traditional methods to survey sites with spatial context is more time consuming and costly hence this limits the scope of a study particularly in the remote areas with difficult terrains.

The use of classical approaches or simple rational calculations estimates power potential by converting river flows and elevation head at a single point, double point, or sub-drainage point (Zaidi and Khan 2018). More importantly, observation based hydropower assessment might miss some potential sites. The high potential sites are generally located in remote inaccessible mountainous areas with rough terrains and this makes the survey-based potential rather very challenging (Kurse et al. 2010). In such situations, to overcome these barriers to a large extent geographic information system (GIS) is adopted. These techniques are now becoming more apparent and accepted as it is cost and time effective. The first hydropower thematic GIS layers were developed and released to the public domain in 2004 in USA (Kurse et al. 2010). With the introduction of more innovative approach of GIS (modern compu-tational tool) and hydrological models it is quite uncomplicated to do resource assessment for hydropower. These spatial tools and modeling techniques with exist-ing terrain, hydrological data and varying climate had been employed in various countries as a tool to locate and select hydropower opportunities of different types (Tarife et al. 2017).

The Soil and Water Assessment Tool (SWAT) is used to model the input data such as soil map, map of forest protected areas, map of slope and map of fault lines used as Digital Elevation Model (DEM) and hydrological data including precipita-tion data, solar radiation, wind velocity, relative humidity and temperature to model output water runoff as output. The aim of this study is to identify the potential hydropower sites using Arc GIS version 10.5 and assess spatial and temporal avail-ability of water resources through the hydrological, SWAT 2012.10.21 modeling tool and estimate the hydropower potential using spatial analysis and simulation of SWAT model.

5.1.8 Study Location

This study assesses the potential hydropower sites on the two main islands of Fiji, namely Viti Levu and Vanua Levu (Fig. 5.2).

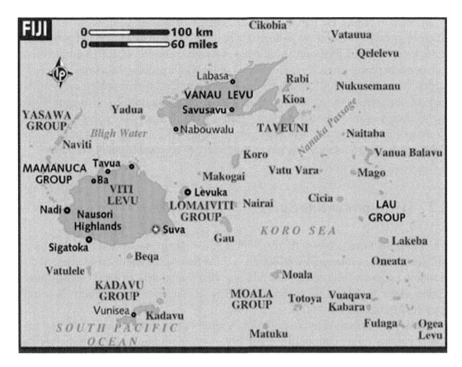

Fig. 5.2 Map of Fiji Islands. (Source: https://www.lonelyplanet.com/maps/pacific/fiji/)

5.1.9   Status and Investments in Hydropower Technology

There has been an increasing investment in various types of renewable energy technology to mitigate environmental scenarios such as rise in sea level and drastic change in climatic patterns due to emissions of green house gases. In order to curb alarming rise in green house gases, there had been proliferation in renewable energy. As a result, the Millennium Goals were set out in 2002 at the World Summit on Sustainable Development in providing access to modern and clean energy to all.

Currently, Fiji has achieved 85% in renewable energy technology and aims to become 100% renewable energy based by 2050. However, there is still a lot of gap in Fiji's renewable energy technology. There is still a greater capacity in renewable energy technology in Fiji such as solar photovoltaic, wind energy, hydro energy and biomass as climatic conditions are quite favourable or suitable to meet the requirements of these renewable energy technologies. Sustainable adoption of renewable energy technologies requires elimination of capacity gaps at local, national and global levels.

In order to reduce the gap in renewable energy technology, Fiji seeks assistance in green growth development particularly in the areas of hydro, solar and wind technologies. Moreover, Fiji's tariff rates have already being regulated to attract more independent power producers. This action will certainly allow the nation to meet its target. One of the major limitations in the development of renewable energy is the lack of infrastructure, expertise, funds and human capabilities. However, Fiji's hydro energy alone provides the largest share of total renewable energy that is fed to the national grid system which is approximately 60% and there is a great potential to use hydro to increase the percentage share of renewable energy in Fiji.

5.2   Hydropower Site Surveys in Fiji by FDoE

The FDoE had been constantly developing Fiji's energy sector through the use of indigenous renewable energy resources. Of the renewable energy resources, hydro resource has been fully researched and is a vital component of rural electrification option under the Rural Electrification Policy (REP) 2003 (Fiji's Hydro Potential Report Volume 1, 1995–2006).

FDoE has performed hydro preliminary surveys particularly during the dry season at various rural communities from 1995 to 2006 in order to study whether the site is feasible for developments. Parameters such as flow rate, catchment area, gross head, gross power and power available to each household were considered. When the site was found to be feasible during the initial preliminary survey, a long term monitoring for period of 3 years was carried out to fully assess the viability of the site in terms of the available flow rate (Fiji's Hydro Potential Report Volume 1, 1995–2006).

Normally, two methods employed in monitoring data are by installing a staff gauge and the villager will constantly read the water level or by installing a datalogger which automatically records the water level. And provided after 3 years the flow rate is viable then the site would undergo further developments before its final implementation. The following criteria needs were essential for selecting the appropriate site to measure the flow rate along the chosen creek. The chosen site must:

1. Have water flowing all throughout the year.
2. Be located at a high elevation and located not too far off from the village.
3. Not be utilized for any other purposes e.g. drinking water.

The FDoE has already carried out numerous surveys for micro hydro potential sites at various locations in Fiji at central, northern, western and eastern divisions. The survey conducted at central division includes surveys performed at Naitasiri villages, Namosi villages, Rewa village, Serua village and Tailevu villages. A total of 18 villages were surveyed for micro hydro potential sites at central division between time periods 1994 and 2007.

Moreover, surveys conducted in the northern division includes surveys performed at Bua villages, Cakadrove villages and Macuata villages. A total of 28 villages have been surveyed for micro hydro potential sites at northern division between time periods 1994 and 2006. Similarly, surveys conducted at western division include surveys performed at Ba villages, Nadroga-Navosa villages and Ra villages. A total of 30 villages have been surveyed for micro hydro potential sites at western division between time periods 1994 and 2006. Finally, survey conducted at eastern division includes surveys performed at Lomaiviti and Kadavu villages. A total of 25 sites have been surveyed for micro hydro potential sites at eastern division between time periods 1999 and 2006 (Fiji's Hydro Potential Report Volume I (1995–2006).

5.2.1 Methodology for the Identification of Suitable Hydropower Sites

5.2.1.1 GIS Modeling Technique

The modern calculation tools such as remote sensing, GIS and hydrological modeling are used to identify the potential or suitable hydropower sites. The practical representations of the parameters such as existing terrain, complex hydrological phenomenon, scientific assessment of drainage networks and climatic changes are possible with the development of spatial data and modeling techniques (Lakshmi and Sarvani 2018). Hence, the potential for hydropower development consequently lead to access to affordable energy. The hydropower potential is directly related to three very important parameters which include the geography of an area, amount of precipitation and soil saturation. Thus, GIS spatial tool can be specifically used to evaluate watersheds for runoff and determine the difference in elevation from where water is collected and the position of the turbine (Gerald et al. 2016).

GIS had been practically successively used over a number of years. Rojanamon et al. (2009) proposed new methods of GIS applications to select site(s) of small run-off-river hydropower. This had been quite challenging for site selection of small run-off-river hydropower particularly in rural and mountainous areas which required large amount of data and involved less participation by the local communities. Similarly, Das and Paul (2006) encountered difficulty in selection of potential hydropower site at the inaccessible tracts of Himalayan region. Moreover, Cuya et al. (2013) used the Arc-GIS based tool VAPIDRO-ASTE to assess hydropower potential in La Plata basin.

5.2.1.2 Digital Elevation Model (DEM)

GIS spatial tool is very supportive in hydrological modeling in terms of facilitating, managing and interpreting the hydrological data. GIS describes the topography of the area and this capability of GIS is used to develop Digital Elevation Model (DEM) (Hidayah and Indarto 2017). DEM is a digital representation of ground surface elevation. Thus DEM is specifically used for processing ground elevation values measured at the intersection of the horizontal grid lines. These elevation data grid is a type of raster data (an array of values measured at uneven locations spatially across the region). Thus, DEM is an essential tool required to determine the parameters such as flow direction, flow accumulation, flow length, slope and watershed (Hidayah and Indarto 2017).

5.2.1.3 Identification of Suitable Hydropower Site

The identification of hydro spots and flow determination at selected sites are the major requirements for the assessment of hydropower potential. The flowchart represented in Fig. 5.2 above shows the methodology used in assessing the hydropower of the watershed areas around Vitilevu. GIS spatial tool is used to identify and classify the theoretical hydropower potential sites around Vitilevu in Fiji Islands. Geospatial Arc GIS 10.5 based hydrological modeling is performed on raster cells using topographical and meteorological datasets and overlaid. Input datasets include Digital Elevation Model (DEM), fault lines, soil map, forests, and weather data such as precipitation. This spatial analysis involves firstly classifying the following:

- Fault lines with buffering region of 1 Km.
- Drainage (soils which are well and excessively drained with poor retention abilities).
- Slope (used slope below angle of 18° considering engineering aspects for construction-to minimise construction costs) derived from DEM.
- Protected forests cover.

These parameters were reclassified as rasters into suitable and unsuitable areas followed by overlaying of 10 years rainfall data to determine the hydro potential sites. Preliminary hot sport analysis is done on rainfall data sets to identify areas with frequent rainfall. Later watershed and historic rainfall data will be overlaid in DEM to determine the hydrospot areas to be utilised in analysis using SWAT tool (Fig. 5.3).

The methodology used to assess hydropower potential of the various sites in Vitilevu is presented through the flowchart below (Fig. 5.4):

5.2.2 Catchment Delineation

Catchment delineation is normally done to create boundary areas and drainage networks of watershed. These are extracted from DEM by creating boundary and the corrected DEM, flow direction and flow accumulation rasters will be used to develop a vector representation of catchments and drainage lines from selected points in SWAT (Kayastha et al. 2018).

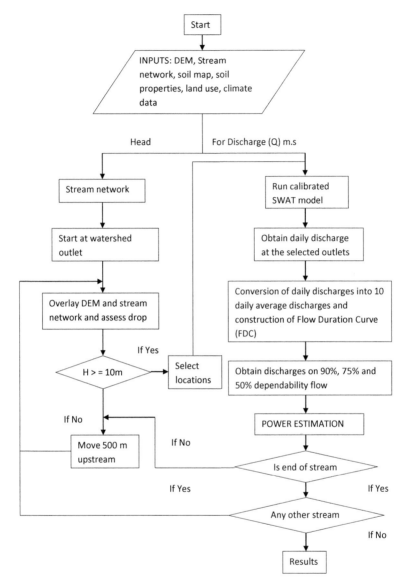

Fig. 5.3 Flow chart for assessment for hydropower potential. (Source: Kurse et al. 2010)

5.2.3   Selection of Hydrological Model

According to Bajracharya (2015), the hydrological models are classified as either lumped models, semi-distributed or distributed models. For lumped hydrologic models, the hydrological parameters are assigned at basin level and are assumed constant throughout the basin. Moreover, these models require less data but have

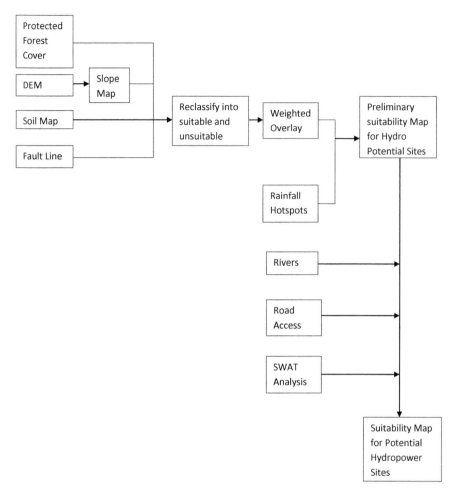

Fig. 5.4 Spatial analysis flow chart

limited representation of spatial variability of the natural system and this may restrict their applicability in integrated management purpose.

However, in semi-distributed models, the hydrologic parameters are partially allowed to vary at sub-basin level by dividing the basin into a number of smaller sub-basins, but the sub-basins with the same parameters are lumped together. Hence, semi-distributed model is a compromise between lumped and distributed models in terms of their representation in spatial variability and data requirement (Bajracharya 2015). The advantage of this model over lumped model is that it is more physically based and requires less input data than fully distributed models. Whereas in distributed models, the parameters are fully allowed to vary by splitting the given basin into smaller cells. This model provides the more realistic representation of the basin however; it poses the major constraint of utilising extensive data, computational time and resources (Bajracharya 2015).

While selecting the hydrologic model some of the factors such as: availability of data, study objectives, desired accuracy, cost of the software and cost related to the acquisition of the input data, processing speed of the computer at hand, user support need to be considered.

5.2.4   Assessment of Flow Rates

5.2.4.1   SWAT Hydrological Model

The Soil and Water Assessment Tool (SWAT) for Arc GIS 10.5 hydrological model is used to assess the flow rate of potential hydropower sites. Basically, SWAT for Arc GIS 10.5 is a continuous simulation watershed model which quantifies the impact of land management practices. The model has the capability to simulate surface runoff, percolation, return flow, erosion, nutrient loading, pesticide fate and transport, irrigation, ground water flow, channel transmission loses, pond and reservoir storage, channel routing, field drainage, plant water use and other supporting processes from small, medium and large watersheds (Kurse et al. 2010).

The SWAT model can be applied to a large un-gauged rural watershed with up to 100 sub-watersheds. These watersheds are subject to division into sub basins which accounts for land use and impact on soil properties. The model further sub-divides these sub basins into smaller homogenous units called Hydrologic Response Unit (HRU) (Kurse et al. 2010). The HRU report provides information about the distribution of watershed for land use, soil and slope. Spatial thematic map data and discrete data for specific locations are required for application of SWAT for Arc GIS 10.5 hydrological model. Spatial thematic map data includes: (i) soil map, (ii) slope map, (iii) map of fault lines, (iv) map of protected forests cover, (v) stream network data, and (vi) Digital Elevation Model. Thus, the input data are climatic and discharge data. The soil map is quite crucial in terms of determining the flow rate as it accounts for surface runoff and water absorptivity depending on soil type and soil property. More importantly, the soil properties include: (i) number of soil layers, (ii) soil texture- whether the soil is sandy, silt, clay and rock fragments, (iii) porosity fraction, (iv) moist bulk density, (v) saturated hydraulic conductivity and (vi) available water capacity (Kurse et al. 2010). Figure 5.5 represents the hydrologic cycle that works in SWAT model.

The SWAT model uses water balance principle. The following water balance principle is used:

$$SW_t = SW_0 + \sum_{i=1}^{t} \left(R_{day} - Q_{surf} - E_a - W_{seep} - Q_{gw} \right) \tag{5.1}$$

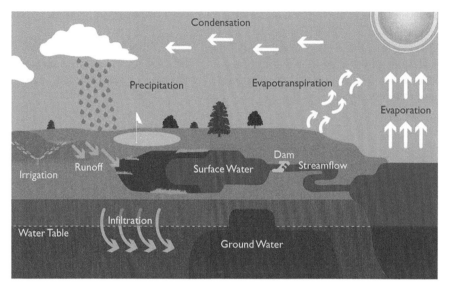

Fig. 5.5 Schematic representation of hydrologic cycle. (Source: http://nysgolfbmp.cals.cornell. edu/hydrologic-cycle/)

Where,

SW_t = Final soil water content (mm H_2O)

SW_0 = Initial water content on day i (mm H_2O)

t = Time (days)

R_{day} = Amount of precipitation on day i (mm H_2O)

Q_{surf} = Amount of surface runoff on day i (mm H_2O)

E_a = Amount of evapotranspiration on day i (mm H_2O)

W_{seep} = Amount of water entering the vadose zone from soil profile on day i (mm H_2O)

Q_{gw} = Amount of return flow on day i (mm H_2O)

Source: Bajracharya 2015.

Besides the spatial data, input data such as climatic data requires precipitation data, daily maximum temperature and daily minimum temperature. Finally, the model requires daily discharge data from various stream outlets for calibration and validation of the model.

5.2.4.2 Calibration, Validation and Evaluation of Model

The calibration process tunes the model parameters within recommended ranges to match the simulated output with the observed data. This action involves comparison of model that results with the recorded runoff data at selected outlets (Bajracharya 2015). This process requires model parameters to be adjusted in such a way that the simulated results are matched to the recorded flow pattern within some accepted criteria. Normally the calibration is done by trial and error manually or possibly done by automatic numerical optimization. Once calibration is completed then the model is validated. Validation process involves the comparison of model output with an independent observed dataset that has not been used in the calibration without further adjustment of model parameters (Bajracharya 2015).

A balanced split-sample approach could be used to perform calibration and validation of the SWAT model. According to Robert et al. 2008, model calibration is performed by using at least 4 years of data such as: measured daily stream flow, temperature and precipitation. This research project adopts the method used by Robert et al. 2008 utilised in calibrating, validating and evaluating the SWAT model. For instance, the SWAT model has to run for time period of 2019–2028. During the first 2 years (2019 and 2020) simulation will occur which allows the model to equilibrate to ambient conditions (Robert et al. 2008). Calibration of the model is based on the years 2023–2026 and the model will be validated using data from the 2 years prior (2021, 2022) and 2 years following (2027, 2028) the calibration period. This calibration period involves a wide range of climatic conditions such as: mean annual temperature, precipitation, peak stream discharge (flow rate) (Robert et al. 2008).

After calibration and validation of SWAT model it then gets configured and initially parameterized to run the model on a daily basis. The calibrated model is then assessed for the reliability of hydrologic predictions using two methods. The first is a traditional method where the relative error (RE) and Nash-Sutcliffe (NS) performance statistics is calculated from model output during the validation period (2021, 2022, 2027 and 2028) and compared to those generated during the calibration phase (2023–2026) for annual, seasonal and monthly time steps (Robert et al. 2008). Second method uses regression-based model to compare the estimated and measured stream flow values over seasonal and monthly time steps of the validation period (Robert et al. 2008).

5.2.4.3   Evaluation of Model

Finally, the models performance has to be evaluated. This is done to assess how close the models simulated values are with the observed values. Several statistical techniques such as coefficient of determination (R^2), Pearson's correlation coefficient (r), percent bias (PBIAS), root mean square error (RMSE), mean square error (MSE), mean absolute error (MAE), Nash-Sutcliffe efficiency (E_{NS}) and measure of relative error (RE) could be used to evaluate the performance of the model. The values of NS coefficient range from negative infinity to 1 and NS coefficients greater than 0.75 are considered "good", whereas values between 0.75 and 0.36 are "satisfactory" (Motovilov et al. 1999; Wang and Melesse, 2006 cited in Robert et al. 2008). Model predictions are no better than the mean of the observed data when NS coefficients are 0 or less (Robert et al. 2008).

Moreover, to gauge model performance, RE measures the percent difference between measured and simulated values over a specified time period. Smaller values of RE indicates better model performance (Robert et al. 2008). RE values less than 20% estimates that model is satisfactory (Robert et al. 2008).

5.2.4.4 Flow Duration Curve

The flow-duration curve (FDC) is a cumulative frequency curve that represents the percent of time that flow in a stream is likely to equal or exceed some specified value of interest (Bajracharya 2015). More specifically, the FDC shows the percentage of time the river flow is expected to exceed a design flow of some specified value or it shows the discharge of a stream that occurs or has exceeded some percent of time for example 70% of time. There are various applications of FDC which include: hydropower generation, river and reservoir sedimentation, water quality assessment and water-use assessment ((Bajracharya 2015). Stream flow data is required in hydropower design and hydro potential calculation which is obtained from the flow duration curve.

FDC also comparatively studies the flow characteristics of streams from one basin to another. The two important factors that determine the shape of FDC are the hydrologic and geologic characteristics of the watershed. Thus this FDC studies the hydrologic response of a watershed at various types of inputs. The shape of FDC significantly evaluates the stream and basin characteristics. The slope of curve at upper end represents the type of flood regime the basin will most probably have whereas the slope of curve at lower end indicates the ability of the basin to sustain low flows during dry seasons (Bajracharya 2015).

Daily, weekly or even monthly stream flow data can be used to prepare FDC and the exceedance probability is calculated as follows:

$$P_{rb} = 100 \times \left[M / \left(n_e + 1 \right) \right] \tag{5.2}$$

Where P_{rb} = the probability that a given flow will be equaled or exceeded (% of time)
M = the ranked position on the list (dimensionless)
n_e = the number of events for a period of record (dimensionless)
Source: Bajracharya 2015.

5.2.5 Potential Head Drop Estimation

Drop in the head along the river is required for the hydropower potential assessment. For the potential sites the drop in the head has to be estimated. There are various methods of estimating the head drop along the river course. The simplest method is by overlaying the DEM of the basin, sub-basin and river network shape file in order to obtain the raster value of upstream and downstream end point of each sub-basin river. Potential head drop of the river is the difference in raster value between the upstream and downstream end points of the river in a given sub-basin (Bajracharya 2015).

5.2.6   Hydropower Potential Calculation

Hydropower potential is the function of head drop discharge at certain flow exceedance. Theoretically, run of river hydropower potential is calculated by using Eq. 5.3.

$$P = \rho \times g \times Q \times H \tag{5.3}$$

Where,
P = Power generated in Watt (W)
ρ = Mass density of water (kg/m^3)
g = Acceleration due to gravity (m/s^2)
Q = Discharge (m^3/s)
H = Gross head drop (m)
Source: Bajracharya 2015

For example, if there are number of sub-basins in a given basin, the total power of the basin can be calculated by summing the potential of all sub-basins.

$$P = \sum_{i=1}^{n} \rho \times g \times Q \times H \tag{5.4}$$

Where,
i = *Sub*-basin number = i.............n
n = *Number* of sub-basins

The mass density of water ρ, is taken as 1000 Kg/m^3 and acceleration due to gravity as 9.81 m/s^2. The gross head elevation is taken as the difference between the headrace and the tailrace. Using the estimated head drop and the discharge, Q of any basin, theoretical hydro potential can be calculated.

5.2.7   Preliminary Results of Hydro Potential Sites Based on Suitability Map

5.2.7.1   GIS Based Preliminary Maps Representing Areas of Suitable Hydropower Sites

GIS technique was used to generate preliminary suitability map of hydro power potential sites on the main islands of Vitilevu and Vanualevu in Fiji through generated soil map, slope map, map of protected forests cover and map of fault lines.

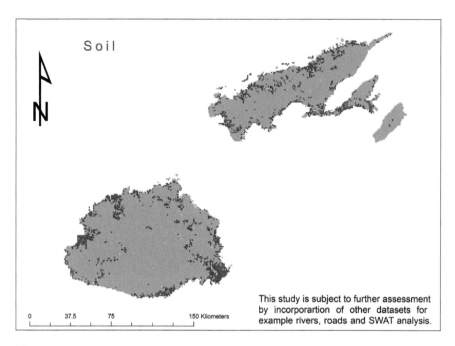

Fig. 5.6 Soil map of Vitilevu and Vanualevu Islands of Fiji

Figure 5.6 represents soil map of Vitilevu and Vanualevu Islands of Fiji with soils water retention properties. The areas marked red on the map represents soil having well drained and excessively drained properties with poor retention and greater run-off. This type of soil is very much suited for the development of hydropower station. On the other hand, the green regions on the map represents soils with greater water retention properties hence these regions are not suitable for development of hydro-power station.

Furthermore, Fig. 5.7 represents the map of fault lines and the areas marked in red are the regions or areas that are not suitable for construction of hydropower stations. These fault avoidance zones are areas created by establishing 1 Km buffer zones on either side of fault trace representing areas or sites not suitable for hydro-power development. The lower map in Fig. 5.7 represents the fault lines on the island of Vitilevu. The fault lines are contained within the land mass area but mostly concentrated on South Eastern, South West and South directions. In contrast to the upper half of Fig. 5.4 which shows the map of Vanualevu where the fault lines greatly extend into sea waters. As a result of this, it is extremely difficult to con-struct hydropower stations on the island of Vanualevu due to extensive fault lines reaching beyond land mass areas. Thus, there is a greater potential to set up new hydropower sites on the island of Vitilevu due to most of the land mass areas having less fault zones in comparison to Vanualevu. Furthermore, Fig. 5.8 repre-sents map of slope showing areas suitable for hydropower construction sites below an angle of 18°. Areas marked with red are the areas or regions not suitable for

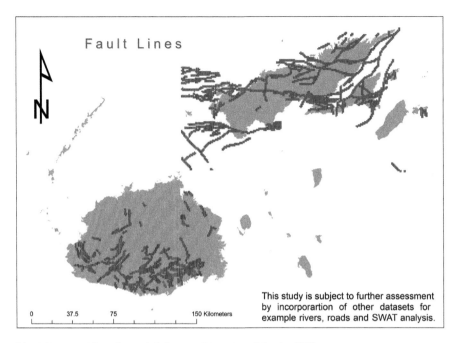

Fig. 5.7   Map of fault lines of Vitilevu and Vanualevu Islands of Fiji

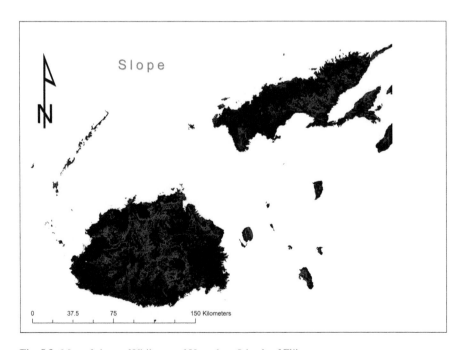

Fig. 5.8   Map of slope of Vitilevu and Vanualevu Islands of Fiji

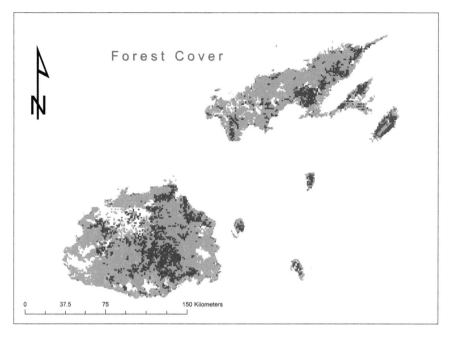

Fig. 5.9 Map of protected forest covers of Vitilevu and Vanualevu Islands of Fiji

construction of hydro power stations. Based on Fig. 5.8, approximately 70% of land mass area on the islands of Vitilevu and Vanualevu are suitable for hydropower construction. However, this is not possible in consideration to the fault lines and protected forest cover.

Moreover, Fig. 5.9 represents the map of protected forest cover. The regions marked as red are suitable areas for hydro power construction whereas non-suitable regions are marked in green. These green areas represent forest protected sites. Approximately 45% of the land mass area on the two islands has the potential for the construction of hydropower stations due to extensive protected forest cover areas. However, this study on determination of potential hydro power sites is subject to further assessment incorporating rainfall hotspot data, road accessibility, water shed delineation and the use of SWAT model to identify the various new hydro potential sites on the island of Vitilevu, Fiji.

5.3   Conclusion

The preliminary results of this study suggest that it is quite possible to construct hydro power stations at the potential sites on island of Vitilevu. Soil map in Fig. 5.6 reveals that very high proportion of the land mass areas of Vitilevu and Vanualevu are not suitable for the development of hydropower stations due to greater

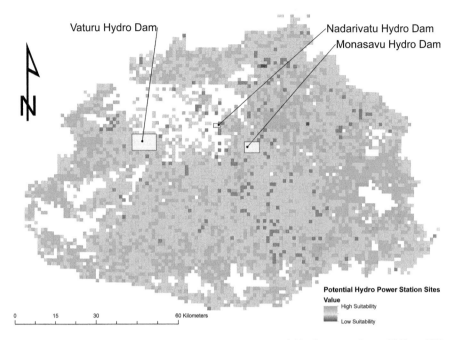

Fig. 5.10   Map of existing hydropower stations and new potential hydropower sites at Vitilevu, Fiji

water retention properties and poor water run-off. However, approximately 15% of land mass area on the island of Vitilevu has good water run-off and are the viable areas for construction of hydropower stations. The extensive fault lines and greater regions of protected forest cover prevent construction of hydropower stations on the island of Vanualevu. However, there is a greater potential for the construction of hydropower stations on the island of Vitilevu. The comparative analysis of overlayed maps by the use of GIS tool indicate that, approximately 40% of land mass area on the island of Vitilevu is quite suitable for the construction of hydropower stations.

Moreover, there is a strong correlation between areas of potential hydropower sites determined by GIS technique in this study and surveys done by the FDoE. Figure 5.10 shows existing hydro power stations and further supports that approximately 40% of land mass areas on Vitilevu is available as potential new hydro power sites. Work on construction of new hydro power stations may begin provided there is funding and technical assistance provided by agencies such as Asian Developmnt Bank (ADB) and Japan International Cooperation Agency (JICA). This will certainly boost Fiji's renewable energy potential contributing effectively to Fiji's climate change mitigation program and sustainable development.

As mentioned earlier, this study is subject to further assessment incorporating rainfall hotspot data, road accessibility; water shed delineation and the use of SWAT model to identify new hydro potential sites. The results of this research will be of great importance to decision makers in particular to government departments such as Department Infrastructure and Development, Department of Energy, Department of Environment and Ministry of Agriculture. Above all the use of hydrological modeling technique is quite relevant, effective, fast, and economical and generates maps of suitable areas of hydro potential sites in comparison to surveys which are performed physically.

Acknowledgements The climatic data such as precipitation and daily maximum and minimum temperatures were provided by Fiji meteorological centre, Namaka, Nadi. Spatial input data used to generate Figs. 5.6, 5.7, 5.8, and 5.9 were provided by Mr. Anish Maharaj, who also helped me in generating these maps using the GIS tool. And finally, thanks to the anonymous reviewer for providing the comments that improved the book chapter.

References

Ahl, R. S., Woods, S. W., & Zuuring, H. R. (2008). Hydrologic calibration and validation of SWAT in a snow-dominated Rocky Mountain watershed, Montana, U.S.A. *JAWRA Journal of the American Water Resources Association, 44*(6), 1411–1430. https://www.lonelyplanet.com/maps/pacific/fiji/. Accessed on 17 September 2019. http://nysgolfbmp.cals.cornell.edu/hydrologic-cycle/. Accessed on 17 September 2019.

Bajracharya, I. (2015). Assessment of run-of-river hydropower potential and power supply planning in Nepal using hydro resources. *Master of Science*, 1–131.

Carmela D. L. Pacific energy update 2018: 8.

Chand, D. (2013). Promoting sustainability of renewable energy technologies and renewable energy service companies in the Fiji Islands. International conference on sustainable energy engineering and application. *Energy Procedia, 32*, 55–63.

Cuya, P., Grace, D., Luigia, B., Popoescu, L., & Alterach, J. (2013). GIS-based assessment of maximum potential hydropower production in La Plata basin under global changes.

Das, S., & Paul, P. K. (2006). Selection of site for small hydel using GIS in the Himalayan region of India. *Journal of Spatial Hydrology, 6*(1), 18–28.

Department of Energy. *Sustainable Energy for All (SE4All): Rapid Assessment and Gap Analysis Fiji (n.d.)*, 1–58.

Dornan, M., & Jotzo, F. (2015). Renewable technologies and risk mitigation in small island developing states: Fiji's electricity sector. *Renewable and Sustainable Energy Reviews, 48*, 35–48.

Emission Reduction Profile Fiji. (2013).(pp. 5–22).

Energy Fiji Limited(2016). *Annual Report.*(pp.1–65).

Fiji Bureau of Statistics. Key statistics. Suva: Government Printery; 1995–2015.

Fiji Department of Energy (2009). Fiji's Hydro Potential Report Volume I (1995–2006). (pp. 1–182)

Gerald, C. K. C., David, N. S., & Elijah, K. B. (2016, September). Micro hydro potential modeling: Integrating GIS into energy alternatives for climate change mitigation. *Journal of Geoscience and Environmental Protection., 4*, 47–59.

Hidayah, E., & Indarto, W. S. (2017). Proposed method to determine the potential location of hydropowerplant: Application at Rawatamtu watershed, East Java. *Procedia Engineering, 171*, 1495–1504.

http://www.fijitimes.com/feasibility-study-for-hydro-power/. 25 June, 2018. Accessed on 6 Oct 2019.

International Renewable Energy Agency. (2015).*Fiji Renewables Readiness Assessment*. Adnan, Z.A. (pp. 1–60).

Kayastha, N., Singh, U., & Dulal, K. P. (2018). A GIS approach for rapid identification of Run-Of-River (RoR) hydropower potential site in watershed: A case study of Bhote Koshi watershed, Nepal. *Hydro Nepal, 23*, 48–55.

Kurse, B. C., Baruah, D. C., Bordoloi, P. K., & Patra, S. C. (2010). Assessment of hydropower potential using GIS and hydrological modelling technique in Kopili River basin in Assam (India). *Applied Energy, 87*, 298–309.

Lakshmi, S. D., & Sarvani, G. R. (2018). Selection of suitable sites for small hydropower plants using geo-spatial technology. *International Journal of Pure and Applied Mathematics, 119*(17), 217–240.

Ma, C.D.L. (2018). Pacific Energy Update 2018. Asian Development Bank. (pp. 1–22), from https://www.adb.org/sites/default/files/institutional-document/425871/pacific-energy-update-2018.pdf

Maraia V.S. (2018, January 11). *More Fijians to benefit from FEA power grid projects*. Stuff. Retrieved from https://fijisun.com.fj/2018/01/11/more-fijians-to-benefit-from-fea-power-grid-projects/

Ministry of Economy I. (2017). *28-Year and 20-Year National Development Plan Transforming Fiji*(pp. 1–24).

Prasad, R. D., & Raturi, A. (2017). Grid electricity for Fiji islands: Future supply options and assessment of demand. *Energy, 119*, 860–871.

Prasad, R. D., Bansal, R. C., & Raturi, A. (2017). Review of Fiji's energy situation: Challenges and strategies as a small island developing state. *Renewable and Sustainable Energy Reviews, 75*, 278–292.

Rachna L. (2015, April 15). *FEA details the $1.5bn renewable energy projects*. Stuff. Retrieved from https://fijisun.com.fj/2015/04/11/fea-details-the-1-5bn-renewable-energy-projects/.

Repeka N. (2018, June 25). *Feasibility of hydropower*. Stuff. Retrieved from fhta.com.fj/feasibility-study-for-hydro-power/.

Robert, A., Scott, W.W., & Hans, RZ. (2008). Hydrologic calibration and validation of SWAT in a snow-dominated rocky mountain watershed, Montana, USA. *Journal of the American Water Resources Association, 44*(6), 1411–1430.

Rojanamon, P., Chaisomphob, T., & Bureekul, T. (2009). Application of geographical information system to site selection of small run-of-river hydropower project by considering engineering/economic/environmental criteria and social impact. *Renewable and Sustainable Energy Reviews, 13*(9), 2336–2348.

Tarife, R.P., Tahud, A.P., Gulbe.n E.J.G., Macalisang, H.A.R.C.P., & Ignacio, T.T. (2017). Application of Geographic Information System (GIS) in hydropower resource assessment: A case study in Misamis Occidental, Philippines. *International Journal of Environmental Science and Development, 8*(7), 507–511.

The project for the effective and efficient use of renewable energy resources in power supply in Republic of Fiji final report vol. 1 executive summary February 2015.

Zaidi, A. Z., & Khan, M. (2018). Identifying high potential locations for run-of-the-river hydroelectric power plants using GIS and digital elevation models. *Renewable and Sustainable Energy Reviews, 89*, 106–116.

Chapter 6
Waste to Energy: Biogas from Municipal Solid Waste for Power Generation

Malvin Kushal Nadan

Abstract Anaerobic Digestion (AD) of the organic fraction of municipal solid waste (OFMSW) produces biogas which could be utlilized to produce energy, reducing waste which otherwise would have been landfilled. For the Pacific Island Countries (PICs), the use of incineration and anaerobic digestion is more viable due to the lower capital costs, complexity, operational costs and higher efficiencies compared to pyrolysis, gasification, plasma arc and other waste to energy technologies. However, the use of either technology for energy production would require further financial, social, technical and environmental analysis. The use of Waste to Energy (WtE) technologies for the production of energy helps greatly in achieving our Nationally Determined Contribution (NDC) goals in accordance with the Paris Agreement, which aims to reduce carbon emissions via sustainable biomass plantations and Waste to Energy by 212 $ktCO_2$/year. This chapter discusses the viability of electricity generation from biogas produced from the organic fraction of municipal solid waste disposed at the Vunato Disposal Site (VDS) in Fiji. The information and data gathered in this study will be useful in the implementation of waste to energy projects for electricity generation as a means of curbing our greenhouse gas emissions. The findings indicate that through the implementation of WtE technology at the Vunato Disposal Site, 258 kW (241 kWh/ton of MSW) and 1008 kW (637 kWh/ton of MSW) of power (electricity) could be produced respectively from AD and incineration of OFMSW and 825 kW (521 kWh/ton of MSW) from a combination of AD and incineration technologies. This would contribute towards carbon emission reduction of approximately 0.59 Gg, 2.29 Gg and 1.88 Gg annually from AD, incineration and combination of both the technologies respectively. However, for the implementation of WtE power plants certain barriers and challenges needs to be addressed and institutional frameworks needs to be put in place for the success of the WtE.

Keywords Waste to energy (WtE) · Anaerobic digestion · Incineration · Avoided carbon emissions

M. K. Nadan (✉)
School of Science and Technology, The University of Fiji, Lautoka, Fiji

© Springer Nature Switzerland AG 2020
A. Singh (ed.), *Translating the Paris Agreement into Action in the Pacific*,
Advances in Global Change Research 68,
https://doi.org/10.1007/978-3-030-30211-5_6

6.1 Introduction

There is a heavy reliance on the use of fossil fuels as a source of energy in Fiji, contributing 45.45% towards the electricity generation mix (Energy Fiji Limited (EFL) 2017); however, the use of fossil fuel has adverse implications with carbon emissions. Fiji in 2014 had emitted 1169.77 kt of CO_2; in comparison the Pacific Island neighbors Tonga, Samoa and Vanuatu had emitted 121.01 kt, 198.02 kt and 154.01 kt respectively (The World Bank 2019). It was also observed by EFL (2017) that there is an increase of 4% in the demand of electricity annually. Upgrading and retrofitting the existing generation source would not fully solve the issue of increasing demand, hence to cater for this other greener avenues needs to be explored.

Moreover, Fijian Government's commitment to the United Nations Framework Convention on Climate Change for the reduction of GreenHouse Gas emissions (GHG) into the atmosphere also needs to be upheld with investment on new and cleaner energy sources. The Fiji Government's NDC goal is to reduce CO_2 emissions by 30% from a BAU (business as usual) scenario in 2030, by 100% power generation from renewable energy (RE) and through energy efficiency. There are many sources of RE in Fiji and most of it has been investigated thoroughly for their feasibility and viability.

MSW management system aims to handle health, environment, aesthetic, land use resources, and economic concerns related to improper disposal of waste (Ouda et al. 2017). Waste to energy (WtE) technologies harness energy from waste and can be directed to generate electricity or can be used as a heat source. Five widely used and implemented WtE technologies are incineration, pyrolysis or gasification, plasma arc gasification, refused derived fuel (RDF) and bio-methanation or Anaerobic Digestion (AD) (World Energy Council [WEC] 2016).

According to WEC (2016), global waste to energy market was valued at US$25.32 billion in 2013, with thermal energy conversion technologies accounting for 88.2% of total market revenue. This trend is expected to maintain a steady growth with an estimated worth of US$40 billion in 2023 with a compound annual growth rate of (CAGR) of 5.5% from 2016 to 2023. The leading market in 2013 was Europe accounting for 47.6% of total market revenue and having China as the main contributor globally with its effort in doubling WtE technologies in the period 2013–2015 (WEC 2016).

WtE practices helps in reducing emissions of greenhouse gases to the atmosphere, therefore using gas generated from waste as a form of energy is actually a great way to combat global warming and climate change. Biogas plants lower the greenhouse effect by lowering methane emissions through apprehending this harmful gas and using it as fuel. Biogas generation helps cut reliance on the use of fossil fuels, such as oil and coal.

Contrary to other types of renewable energies, the biogas production process is natural, which does not require energy for the generation process. In addition, the raw materials used in the production of biogas are entirely renewable and readily available, as wastes and crops, will continue to grow. Manure, food scraps, and crop residue are raw materials that will always exist, which makes it highly sustainable option. Likewise, waste collection, and management, significantly leads to improvements in the environment, sanitation and hygiene.

This chapter aims to determine the electricity generation potential of biogas produced from municipal solid waste at Vunato Disposal Site in Lautoka and to carry out a comparative analysis with the generation potential of incineration of MSW in three different scenarios as below:

Scenario 1: Power generation from Biogas produced from OFMSW.
Scenario 2: Power generation from Incineration of OFMSW.
Scenario 3: Power generation from Biogas and Incineration of OFMSW.

In line with the NDC goals, avoided carbon emissions for each scenario were also determined based on the avoided fossil fuel usage through the implementation of a WtE as described in the three scenarios.

6.2   Biogas Production and Technology

Waste to energy technologies utilizes either solid or liquid waste to produce energy. Many technologies have been researched and some have been developed commercially not only as a solution to waste management but also imposing as an option for energy security. With the increasing urban population, the need for WtE technologies will be vital for local municipalities. WEC (2016) had further stated that biological WtE; such as Anaerobic Digestion (AD), treatment is becoming commercially viable.

Anaerobic digestion is a biological process in which a group of micro organisms' breakdown organic matter in the absence of oxygen and produce biogas which is made up of methane (CH_4), carbon – dioxide (CO_2) and traces of other gases. It has been in use for decades in India and China for gas production but recently has started to be commercialized for the generation of biogas from wastes. The digestion process occurs in four stages: hydrolysis, acidogenesis, acetogenesis, and methanogenesis (Fig. 6.1).

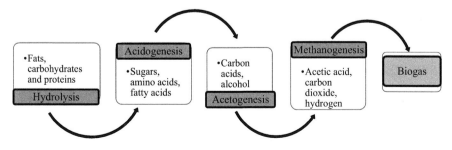

Fig. 6.1 Anaerobic digestion pathway

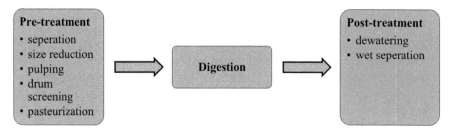

Fig. 6.2 Unit processes of biogas production from MSW

The ideal conditions for anaerobic digestion to occur are:

- Near neutral pH level
- Constant temperature, either at thermophilic (50–60 °C) and mesophilic (30–40 °C)
- Constant feeding rate and an appropriate
- Carbon to nitrogen (C/N) ratio

The production of biogas is highly dependent on these parameters and significant changes in them can inhibit the amount of biogas produced and the methane concentration. In addition, a higher total solid concentration could also lead to an inhibition in biogas production. Biogas production from MSW has to undergo unit processes to ensure the production of quality biogas (Fig. 6.2).

The Global Methane Initiative (GMI), an international public-private initiative that advances cost-effective, near-term methane abatement and recovery and use of methane as a clean energy source in three sectors: biogas, coal mines, and oil and gas systems, during their subcommittee meeting in Florianopolis, Brazil on 14th March 2014, confirmed that 100 m³ of biogas could be produced from 1 ton of MSW. This was based on model assumptions and published digester data from various sources as described below:

- Institute for Global Environmental Strategies (IGES) – *118 m³* biogas per tonne of MSW organics
- Regional Information Service Centre for South East Asia on Appropriate Technology (RISE-AT) – *100–200 m³* biogas per tonne MSW organics
- California Integrated Waste Management Board – *112 m³* per tonne MSW organics (biogas yields from 14 full-scale digesters in Europe treating a variety of wet MSW types)

The use of biogas to generate electricity requires efficient biogas production systems. Vandevivere et al. (2003) categorized the common MSW AD technologies as follows:

- One-stage Continuous Systems
- Two-stage Continuous Systems
- Batch Systems

In a single stage all the reactions take place in a single reactor while in a two-stage system the reactions take place in at least two reactors. Because of the simple design and less costs involved in maintenance of the one stage system; it is preferred industrially over the two-stage system. Around 90% of the digesters in Europe operate using the one stage principle (Vandeviere et al. 2003). In batch systems, digesters are filled with fresh wastes and allowed to go through all degradation steps, similar to a landfill.

According to a report presented by the University of California (2008), one of the largest single stage systems is located in Groningen, Netherlands, which produces around 0.10–0.15 m³/kg of biogas from wet source-separated waste, with a weight reduction of 50–60%. Other single stage systems, Dranco (Dry Anaerobic Composting) system, is reported to have biogas yields in the range of 0.103–0.147 m³/kg and the Valorga system can produce in the range of 0.080–0.16 m³/wet kg.

Biogas can be used as a fuel for combustions engines, which convert it to mechanical energy (used to power an electric generator to produce electricity). The electrical efficiency of biogas can be in the range of 2 kWh/m³ (Electrigaz 2017; Yingjian et al. 2011). Surroop and Mohee (2012) has reported an efficiency of 222 kWh/ton of biowaste. Organic Waste Systems (having built 31 plant in Europe and Asia) which markets the Dranco systems has reported that this system can produce electricity in the range of 0.17–0.35 MWh/ton of MSW (Organic Waste Systems 2013).

6.3 Electricity Generation Capacity of Biogas Produced from OFMSW at Vunato Disposal Site

The Vunato Disposal Site caters for Municipal Solid Waste collected by the Lautoka City and Nadi Town Councils, hotels in the Lautoka and Nadi area and nearby islands. The site is located approximately 2 km from the Lautoka Farmers Market and has a land area of 16 ha, divided into 6 sections for waste disposal and a recycling yard, surrounded by mangroves and a grave yard seperated by a 4.5m high bank (Lautoka City Council [LCC] 2012).

Waste disposed at the site is either collected by the municipal councils or is disposed in person; in either case the weight of the waste is recorded prior to disposal. Waste management at VDS is carried out by the Health Department of the Lautoka City Council. The respective municipal councils' collects wastes from households twice per week, which is placed on curb side. Curbside collection service is free, which is covered in the city and town rates.

VDS has an open aerobic evaporation method for waste management, except for market waste which is collected and is turned into compost and sold as fertilizer. No other intermediate treatment occur except for incinertion of medical waste at the Lautoka Hospital; infectious medical waste is treated at the hospital except for the sharp objects, which is treated in Suva, separate from other infectious medical wastes (LCC 2012).

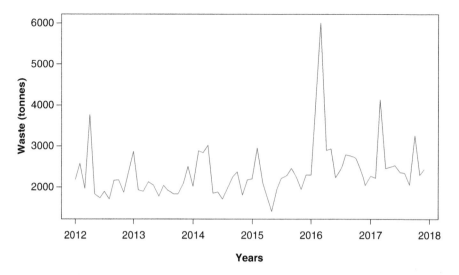

Fig. 6.3 Waste disposed at Vunato disposal site 2012–2017

Municipal Solid Waste disposed at Vunato includes of food, business, market and green, drain, street and park waste, also including factory, construction and hotel waste. According to LCC on average 60 tons of waste is disposed at Vunato daily of which approximately 42% is household waste as shown in Fig. 6.3.

Waste disposal pattern at VDS (Fig. 6.3) shows that the waste disposed is highly affected by cyclones and hurricanes during which period the amount of waste disposed increases. The spikes of wastes disposed in Fig. 6.3, in the years 2012, 2014 and 2016 has been due to hurricanes/cyclones Evan, Lusi and Winston respectively.

The waste disposed at VDS is also highly dependent on the population of the area, based on the 2017 Census data; the urban population for Lautoka and Nadi was 142,621. The MSW generation rate in the two municipalities is approximately 90 kg/person/year or 240 g/person/day.

6.3.1 OFMSW of Waste Disposed at Vunato

The process of generating electricity from biogas is highly dependent on the composition of organic waste. The Organic Fraction of the Municipal Solid Waste (OFMSW) at Vunato was determined using the methods described in "ASTM D 5231–92 Standard Test Method for the Determination of the Composition of Unprocessed Municipal Solid Waste."

Waste collected was divided into 12 components; paper/corrugated, newsprint, plastic, yard waste, food waste, wood, diaper, other organics, ferrous, aluminum, glass and other inorganics.

The components of the waste were determined from the household wastes which were placed on the roadside for collection by the Lautoka City Council. Household wastes from different areas around the Lautoka City were collected and analyzed, the process was carried over a week.

OFMSW of household waste at Vunato was determined to be 69.63%, which included of 46% food waste, 8% plastic and 15% paper (Fig. 6.4). It was determined that approximately 46.32% of waste (food waste) could be used for AD due to its composition. Food and kitchen waste contributed 66.53% towards the OFMSW with 21% and 12% of paper and plastics respectively and 0.3% of other organics (Fig. 6.5). The Lautoka City Council has a waste collection activity organised for the collection

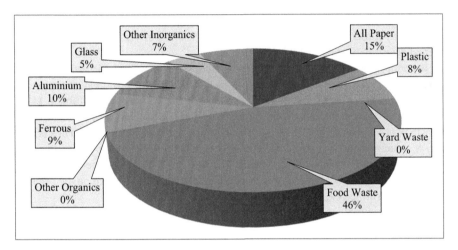

Fig. 6.4 MSW composition fraction of Household Waste disposed at Vunato disposal site

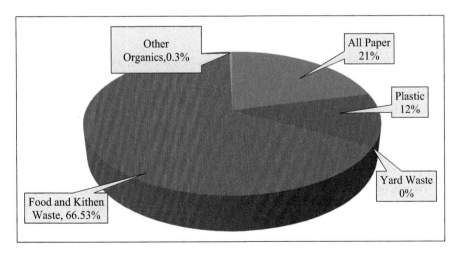

Fig. 6.5 Organic fraction of MSW disposed at Vunato disposal site

Table 6.1 Organic weights of the different MSW disposed at Vunato disposal site as employed by 2017 waste disposal data

Waste component	Annual disposed – 2017 (tons)	Weight of different OFMSW components (tons)				
		Food and kitchen waste	Green waste	Plastics	Paper	Other organics
Household waste	12,957	6001	0	1078	1911	310
Market waste	611	611	0	0	0	0
Hotel waste	1421	658	0	118	210	34
Green waste	13	0	13	0	0	0
Drain, park and street waste	4696	2175	0	390	692	112
Total	**19,698**	**9445**	**13**	**1586**	**2813**	**456**

of yard and garden wastes due to which the composition of yard waste in the experimental determination of OFMSW is zero.

In addition to kitchen and food waste from households; waste collected from markets, hotels, drain/street and green waste was also taken into consideration for use in the AD process for the production of biogas for electricity generation (Table 6.1).

The analysis on the electricity generation potential of biogas produced from OFMSW at VDS employs 2017 data provided by Lautoka City Council. The current MSW management system at VDS involves collection, weighing at the site and classification and disposal at the landfill. The weight off each classified waste component by LCC was broken down into its different components based on the organic fraction determined experimentally.

The organic fraction for hotel waste was assumed to be the same as the organic fraction of household waste.

6.3.2 Estimation of Electricity Generation Potential of Biogas Produced from OFMSW at Vunato: Scenario 1

There are many technologies available for the extraction of energy from waste. However, only AD and incineration of OFMSW was accounted for the generation of electrcity at VDS, The two WtE technologies was studied based on three scenarios as below:

Scenario 1: Power generation from Biogas produced from OFMSW
Scenario 2: Power generation from Incineration of OFMSW
Scenario 3: Power generation from AD and Incineration of OFMSW

The use of incineration and anaerobic digestion was chosen based on lower capital costs, complexity, operational costs and higher efficiencies compared to pyrolysis, gasification and plasma arc technologies.

Biogas can be produced through the digestion of biodegradable materials of MSW such as food waste, yard waste, vegetable waste, kitchen waste etc. Biodegradable waste is the waste that can be decomposed and will be broken down

Table 6.2 Determination of weight of organic mass that is suitable for anaerobic digestion at Vunato disposal site

Waste component	Annual disposed (tons)	Weight used for anaerobic digestion for the production of biogas (tons)
Household waste	12,957	6001
Market waste	611	611
Hotel waste	1421	658
Green waste	13	0
Drain, park and street waste	4696	2175
Total	**19,698**	**9445**

into carbon dioxide, water, methane or simple organic molecules by the action of micro-organisms in reasonably less time (Reddy et al. 2017).

Table 6.2 below shows the mass of the different wastes (household, hotel, market and green and drain/street waste) that could be used for anaerobic digestion.

Based on the amount of waste that was disposed at VDS in 2017; approximately 9445 tons of organic waste would be available for AD annually.

In order to determine the amount of biogas that could possibly be produced from MSW at VDS, the "rule of thumb" for biogas production from MSW as suggested by the Global Methane Initiative (GMI) was adopted. According to GMI, on average 100 m^3 of biogas could be produced for every ton of MSW digested. With a total of 9445 tons of organic waste disposed at Vunato annually, which can be used for AD and with a biogas yield of 100 m^3/ton of OFMSW; a total of 944,500 m^3 of biogas could be produced annually, equivalent to 2587 m^3/day. There are several other operational and technology-specific variables that may affect the biogas production rate for AD for which a sensitivity analysis was also conducted to ensure the sustainability of electricity generation potential of biogas produced from OFMSW at VDS.

Many technologies exist in the market for the conversion of biogas into electricity; gas engines being the most widely used technology (Surroop and Mohee 2012). Due to the use of the gas engines, lower heating value of methane is taken into consideration to determine the amount of power produced from biogas. The following equation was used to determine the power generation capacity of biogas produced from OFMSW.

Where,

$$Power\,Capacity\,(MW) = \dot{m}_{OFMSW} \times \gamma \times \theta \times \rho_{CH_4} \times LHV_{CH_4} \times \eta_{el} \times 0.2778$$

$\dot{m}_{OFMSW}$ = mass flow rate of OFMSW (kg/h)
γ = biogas yield (m^3/kg)
θ = methane fraction $(\%)$
ρ_{CH_4} = density of methane $(tons/m^3)$

Table 6.3 Parameters used to determine the amount of power produced from biogas produced from OFMSW at Vunato disposal site

Parameters	Approximate values	Source
Mass flow rate of OFMSW	9445 tons/year (1078 kg/h)	Lautoka City Council
Biogas yield	100 m³/ton	Global methane initiative
Methane fraction	65%	Biogas (2019)
Density of methane	0.717 kg/m³	Swedish Gas Centre (2012)
Lower heating value of methane	50 MJ/kg	Engineering toolbox (2003)
Efficiency	37%	
Power output (MW)	**0.258**	

LHV_{CH_4} = lower heating value of methane (*MJ/kg*)
η_{el} = efficiency of the biogas engine (%)

The amount of power produced from biogas depends on the methane content of the gas and not solely on the size of the digester and the feedstock. Typically, biogas contain around 55–75% of methane, 25–50% of carbon-dioxide and traces of other gases (Biogas 2019). The biogas generated at VDS is assumed to have 60% methane content and with a yield of 100 m³/ton as stipulated by the Global Methane Initiative as a thumb rule for the production of biogas from MSW. The energy content of biogas was taken as 50 MJ/kg (Engineering Toolbox 2003) with the efficiency of the biogas engine as 37% (Table 6.3).

It can be seen from Table 6.3 that approximately 260 kW of power could be generated from biogas produced from OFMSW at the VDS. The amount of power produced per unit mass of OFMSW was found to be 241 kWh/ton of biowaste and the amount of power produced per unit volume of biogas is 2.41 kWh/m³. A total of 2,277,600 kWh of electrcity or 8200 GJ of electrical energy can be produced annually from the production of biogas through the anaerobic digestion of OFMSW at the Vunato Disposal Site.

As stated by Surroop and Mohee (2012), that 222 kWh of electricity can be produced from AD of a ton of biowaste, Yingjain et al. (2011) and Electrigaz (2017) states that 2 kWh of electricity can be produced per cubic meter of biogas and as reported by University of California, Department of Biological and Agricultural Engineering (2008) that on average 0.15–0.32 MWh of electricity could be produced from a ton of biowaste; in comparison the estimated electricity from biogas at Vunato Disposal Site is very much feasible.

6.3.2.1 Sensitivity Analysis

A sensitivity analysis was conducted to determine the uncertainty in the power output from biogas at VDS. The uncertainty in the power output is attributed to the different input parameters which changes over time; such as the mass flow rate, organic fraction, biogas yield and the fraction of methane in biogas.

The variations in the input would determine the output power for the biogas generator; the changes in the output power due to these would significantly affect the viability of a biogas power generation plant at VDS.

The effect of these changes on the power production were studied as follows:

Case 1 – Change in Organic fraction (30–50%)
Case 2 – Change in Biogas yield (0.080–0.16 m³/kg)
Case 3 – Change in methane fraction (50–70%)

The range for the three variables was taken from various literatures and the change in the power output was calculated to determine the impact of each variable on the output power.

Case 1: Change in the Organic Fraction of MSW
Anaerobic digestion involves the breakdown of organic matter by bacteria in the absence of oxygen; for this reason, percentage of digestible organic component of MSW is very important in the production of biogas. The organic fraction of MSW is greatly influenced by the amount and composition of waste that is being disposed (Table 6.5), seasonal variation, lifestyle of the people in the area and the demography.

The organic fraction of MSW at Vunato can be between 30% and 55% at any given time (Kumar 2013) with the experimentally determined digestible OFMSW at VDS being 46.32% (9445 tons out of 19,698 tons yearly).

Table 6.4 shows that 20% change in the organic fraction of municipal solid waste has power outputs in the range of 160 kW to 300 kW. The analysis shows that on average with every percent of change in the OFMSW, the power output would increase on average by 5 kW.

Together with the change in the organic fraction, the composition of the organic fraction significantly contributes towards the biogas yield as shown in Table 6.5.

Case 2: Changes in the Biogas Yield
Fiji being a tropic island nation, the constituents of MSW would be greatly affected by the seasons. With the different seasons, the food intake and the supply of fresh

Table 6.4 Power output from biogas at Vunato disposal site with change in the organic fraction of MSW

Parameters	Approximate values		Source
	Minimum	Maximum	
Mass flow rate of MSW	19,698 tons/year		Lautoka City Council
Organic fraction	30%	55%	Kumar (2013)
Biogas yield	100 m³/ton		Global methane initiative
Methane fraction	65%		Biogas (2019)
Density of methane	0.717 kg/m³		Swedish Gas Centre (2012)
Lower heating value of methane	50 MJ/kg		Engineering toolbox (2003)
Efficiency	37%		
Power output (MW)	**0.160**	**0.296**	

Table 6.5 Biogas and methane yields from different substrates found in MSW. (Biteco 2013)

Substrate	Dry basis %	Organic dry substance %	Biogas yield m³/ton	Methane CH₄%
Vegetable waste products	15.0	76.0	57.0	56.0
Onion	9.6	94.0	80.3	65.0
Onion peel	82.4	67.0	267.8	65.0
Carrot	11.9	88.3	73.3	52.0
Cauliflower	9.6	92.7	59.2	56.0
Fresh pumpkin	8.4	91.5	50.9	55.8
Food wastes with the low level of fat content, wet	14.4	81.5	75.4	59.8
Food wastes with the high level of fat content	18.0	92.3	126.5	62.0
Fresh potato	26.0	93.4	177.1	51
Miscellaneous food wastes	40.0	50.0	120.0	60.0

Table 6.6 Power output from biogas at Vunato disposal site with change in the biogas yield

| Parameters | Approximate values | | Source |
	Minimum	Maximum	
Mass flow rate of OFMSW	9445 tons/year		Lautoka City Council
Biogas yield	0.080 m³/kg	0.16 m³/kg	University of California, Department of Biological and Agricultural Engineering (2008)
Methane fraction	65%		Biogas (2019)
Density of methane	0.717 kg/m³		Swedish Gas Centre (2012)
Lower heating value of methane	50 MJ/kg		Engineering toolbox (2003)
Efficiency	37%		
Power output (MW)	**0.206**	**0.413**	

fruits and vegetables would change the OFMSW and thus the biogas yield. For the determination of power generation due to changes in the biogas yield, yields in the range of 0.080 m³/kg to 0.16 m³/kg was assumed as stated by University of California, Department of Biological and Agricultural Engineering (2008).

The analysis on the change in the biogas yield through the AD of OFMSW showed that changes in the biogas yield of 0.08 m³/kg increases the power production by 207 kW as seen in Table 6.6. This states that for every 0.01 m³/kg increase in biogas production, approximately 26 kW of extra power is produced.

Case 3: Changes in Methane Fraction of the Biogas
The generation of electricity from biogas is largely dependent on the methane fraction of the biogas, which can be in the range of 50 to 70% according to Biogas (2019). The power capacity of AD of OFMSW with changes in the methane fraction of the biogas as specified was determined as seen in Table 6.7.

Table 6.7 Power Output from biogas at Vunato disposal site with changing methane content

Parameters	Approximate values		Source
	Minimum	Maximum	
Mass flow rate of OFMSW	9445 tons/year		Lautoka City Council
Biogas yield	100 m³/ton		Global methane initiative
Methane fraction	50%	70%	Biogas (2019)
Density of methane	0.717 kg/m³		Swedish Gas Centre (2012)
Lower heating value of methane	50 MJ/kg		Engineering toolbox (2003)
Efficiency	37%		
Power output (MW)	**0.198**	**0.278**	

For a biogas power plant at VDS, changes in the methane fraction of 50% to 70% would produce power of 198 kW and 278 kW respectively. This states that for every 1% increase in the methane fraction of biogas produced, power output would increase by 4 kW.

The sensitivity analysis revealed that with fluctuations in the key parameters contributing towards the power generation capacity of biogas, the power capacities did not change significantly and on average would produce 258 kW. The average power is almost the same as the power capacity of the plant.

6.4 Comparison of Electricity Generation Between Anaerobic Digestion and Incineration Technologies for OFMSW at Vunato

6.4.1 Power Generation Through MSW Incineration: Scenario 2

Incineration refers to the combustion of OFMSW in excess oxygen with temperatures in excess of 800 °C (Ouda et al. 2017) to obtain heat energy to drive a generator through a steam turbine producing power. Harmful flue gases are treated and then released in the atmosphere. According to Ouda et al. (2017) the by – product of the incineration process is bottom ash which consists of silicon, iron, calcium, aluminum, sodium and potassium in their oxide state which are present within a range of 80–87% by mass.

The overall efficiency of this technology is between 20% and 25%, after losses in the technology as studied by Ouda et al. (2017) and Swapna (2012) and can lower the volume of the waste by 90% (Gupta and Mishra 2015). Incineration remains to be the most integral part of MSW management in many countries, with unit process of the technology as shown in (Fig. 6.6). According to WEC (2016) Japan is dominating the Asia-Pacific market utilizing approximately 60% of its waste, however

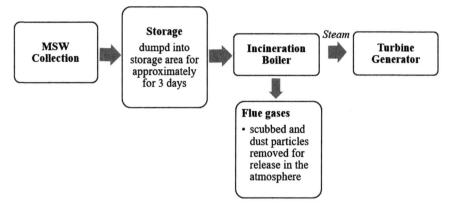

Fig. 6.6 Process flow of MSW incineration in China. (Xu et al. 2016)

the increase in waste generation will see a growth in the waste to energy technologies being implemeted.

As stated by Xu et al. (2016) the characteristics of MSW in China are; unsorted with low calorific values, high moisture content (45–55%) with the organic fraction being around 40–60% and is dumped and stored to reduce humidity and increase calorific value. Municipal Solid Waste at the Vunato Disposal Site was found to have similar characteristics with the organic fraction determined experimentally to be around 66% and having a moisture content of 40% as studied by Swapna (2012).

One ton of organic MSW can produce around 2.5 ton of steam equivalent of 21 kW of power (Anshar et al. 2014). They had further stated that one ton of MSW would be equivalent to 0.3 tons of coal or 0.2 tons of oil.

The following equation was used to determine the energy produced from the combustion of OFMSW at VDS.

$$Energy\,Content\,(MJ) = m_{OFMSW} \times CV \times MC \times \eta_{el}$$

Where:

m_{OFMSW} = dry mass of OFMSW (*kg*)
CV = calorific Value of OFMSW component (*MJ/kg*)
MC = moisture content
η_{el} = efficiency of the generation unit

For the determination of the energy available from the combustion of MSW at VDS, the calorific values of the different waste components was adopted from Swapna (2012). The calculations in Table 6.8 shows that from the combustion of MSW disposed at Vunato, a total of 127236 GJ of heat energy could be extracted. The electrical energy available however is dependent on the system efficiency of

Table 6.8 Determination of energy content from OFMSW through incineration at Vunato disposal site

OFMSW component	Mass of OFMSW component (tons)	% Moisture content	Dry mass of OFMSW component (tons)	Calorific value (MJ/kg)	Energy content (GJ)
Food and kitchen waste	9445	66	3211	17.42	55,935
Green waste	13	66	4.42	17.42	77
Plastic waste	1586	29	1126	41.50	46,729
Paper waste	2813	47	1490	16.44	24,495
Total energy content (GJ)					**127,236**

conversion of heat energy into electricity. For determing the annual electricity generation capacity from the combustion of MSW at VDS, the efficiency of conversion of heat energy to electricity was assumed to be 25% as stated by Ouda et al. (2017) and Swapna (2012). The calculations showed that through the incineration of MSW disposed at VDS approximately 8835833 kWh of electricity (31809 GJ of energy) could be generated annually with a power plant capacity of approximately 1 MW.

6.4.2   Power Generation Through Anaerobic Digestion and Incineration of OFMSW: Scenario 3

While determining the organic fraction of MSW it was evident that not all organic wastes could be used for AD as seen in Tables 6.1 and 6.2, as some of these are not digestible like plastics and paper. Due to the availability of less input in the AD process and lower energy generation capacity the power from the AD of MSW is less compared to incineration. However, according to Chaya and Gheewala (2007) incineration of MSW has a higher greenhouse gas (GHG) emission incomparison with the AD process.

In order to minimize the high GHG emission from the incineration of MSW disposed at VDS, an evaluation on the power production capacity from the possible implementation of both the technologies at VDS was conducted.

Based on the experimental evaluation of the OFMSW at VDS, a total of 13,857 tons of organic waste is approximately available for WtE conversion; however due to the nature of the waste, it needs to be divided into two streams of inputs for WtE conversion i.e. food and kitchen waste (9445 tons) for anaerobic digestion and green, plastic, paper and other organic wastes (4412 tons) for incineration as seen in Table 6.9.

The calculations show that a total of 2,277,600 kWh of electrcity could be generated annually through anaerobic digestion and 4,951,388 kWh of electricity could be produced annually through the incineration process (Table 6.9). The implementation of dual waste to energy plant (AD and incineration) at Vunato would approxi-

Table 6.9 Determination of electricity production from anaerobic digestion and incineration

OFMSW component	Mass of OFMSW component (tons)	WtE conversion technology used for processing	Electricity produced (kWh)
Food and kitchen waste	9445	Anaerobic digestion	2,277,600
Green, plastic paper waste and other organics	4412	Incineration @ 25% efficiency	4,951,388
Total	**13,857**		**7,228,988**

Table 6.10 Power output for the different scenario in using WtE

Scenario	Waste processed (tons)	Electricity production (kWh/year)	Power plant capacity (kW)	Electricity production per ton of waste (kWh/ton)
1. *Anaerobic digestion only*	9445	2,277,600	260	241
2. *Incineration only*	13,857	8,835,833	1008	637
3. *Anaerobic digestion and incineration*	13,857	7,228,988	825	521

mately generate a total of 7,228,988 kWh of electricity annually with a power plant of capacity of 825 kW. In order to achieve this, LCC will have to campaign and create an awareness to notify the public in ensuring that the waste that is disposed by the public for collection by LCC for disposal is sorted. In addition, a separating unit might have to be installation, which will separate the waste and feed into the respective processing plant.

The amount of energy that could be produced from waste that is disposed at Vunato was analysed for three scenarios; anaerobic digestion, incineration and anerobic digestion with incineration. The findings as shown in Table 6.10, shows that the incineration has the highest power generating potential over anaerobic digestion and a combination of anaerobic digestion with incineration. The three scenarios presented shows a viable option for not only power generation but also for MSW management, which otherwise would have been landfill creating health and environmental impacts.

The use of both anaerobic digestion and incineration technologies in a single system for WtE, though will reduce its negative impacts but will not be feasible. The initial costs involved in setting up the plant would be higher as two different types of technologies would be installed, moreover the operational costs in running the plants would be higher. In order to ensure that the setting of the plant is viable, electricity would need to be sold at a higher price to obtain breakeven.

As discussed by Ouda et al. (2017) the use of either technology for WtE would require a further financial, social, technical and environmental analysis. This needs to be taken with considerable emphasis on the government's commitment towards the NDC goals.

6.5   Avoided Carbon Emissions

Energy generation through the utilization of MSW helps in the reduction of GHG emissions through preventing uncontrolled GHG emissions from landfills and by generating energy that could displace the use of fossil fuels. The study in this section looks at CO_2 emission reductions from the latter, where CO_2 emissions was determined based on the Tier 1 Approach of the stationary combustion of fuel sources according to the 2006 Intergovernmental Panel on Climate Change (IPCC) Guidelines for National Greenhouse Gas Inventories, Volume 2, Chapter 2.

The guidelines states that for every tera joule (TJ) of fossil fuel burnt approximately 74100 kg of CO_2 is released (0.26 kg of CO_2 for every kWh of electricity produced through the combustion of fossil fuels) (IPCC 2006). In comparison according to US Energy Information Administration (2018), 161.3 Pounds of CO_2 is emitted per million British thermal units (Btu) of energy or 22.6 pounds per US gallon of diesel fuel. This is the same as 0.24 kg of CO_2 emissions for every kWh of electricity produced through the diesel fuel. CO_2 emissions from the combustion of fossil fuel was determined by multiplying the fuel consumption with the fuel emission factor as stated in the IPCC (2006).

According to the study it was found that incineration of MSW reduces emissions of 2,297,316 kg of CO_2 (2.29 kt CO_2), anaerobic digestion 592,176 kg of CO_2(0.59 kt CO_2) and 1,879,536 kg of CO_2 (1.88 kt CO_2) reduction if a dual plant consisting of AD and incineration technology is set up at VDS as seen in Table 6.11.

Although electricity production from MSW, using either AD or incineration is viable, the environmental impacts associated with each process can be the deciding factor in implementing the technology. Incineration has a high global warming potential than anaerobic digestion (Bolin et al. 2009). Due to the wet nature of the waste, combustion would be difficult (Chaya and Gheewala 2007) and would use more energy internally than compared to anaerobic digestion.

The carbon emissions in this study do not take into consideration CO_2 emissions from transportation of MSW to the plant, use of the fossil fuel in the WtE process, landfill of rejected MSW, fugitive methane loss, management of digestate and AD facility construction and operation. In order to achieve the net carbon emissions a full life cycle analysis of the WtE process would need to need undertaken. Life cycle analysis conducted on AD and incineration of MSW by Chaya and Gheewala

Table 6.11 Avoided carbon emissions through the implementation of waste to energy technologies at Vunato disposal site

Scenario	Electricity production (kWh/year)	Avoided carbon emission (Gg/year)
1. *Anaerobic digestion only*	2,277,600	0.59
2. *Incineration only*	8,835,833	2.29
3. *Anaerobic digestion and incineration*	7,228,988	1.88

(2007) and Bolin et al. (2009) suggests that production of biogas is a better option than incineration of MSW.

6.6 Barriers and Challenges

The availability of renewable energy resources in Fiji is vast and to increase its share in the electricity generation mix and reduce consumption of fossil fuel, these resources needs to be explored and developments in the sector needs to be enhanced. Regardless the growth in the RE sector has been slow due to barriers and challenges faced by the industry.

(i) *Financial and market barriers*

High capital cost and unavailability of financing are the most prominent barriers for RE dissemination in Fiji, especially for a new RE technology like AD. In addition, the readily available cheaper RE technologies suppresses the advantages of AD technology. The competitiveness of the AD technology in the market would be difficult without considering the economical, social and environmental needs. In this regard it is important to note that lack of experience and understanding of AD technologies among financial institutions and investors leads to low participation of national financiers and may increase the cost of capital.

(ii) *Regulatory and institutional barriers*

The government plays an important role in the implementation of any RE technology. Although the Draft Fiji National Energy Policy and the Draft Strategic Action Plan was reported in 2013 but has not been endorsed as of end of 31st December 2018. The two documents provide an in-depth review and strategies on how RE share could be increased. Due to the financial and technical capabilities, it is challenging for municipal councils to implement WtE technologies.

(i) *Technical and Infrastructural barriers*

The lack of technical services and infrastructure decreases the potential of RE implementation. The need to retain experts in the country is very crucial for the industry, as most of them tend to migrate.

(ii) *Socio-cultural barriers*

The need for creating public awareness on AD and RE is very crucial as well. There are limited public awareness on RE technologies, advantages of RE in daily life, having adequate fuel sources for ever, accessing low cost energy and awareness of social and environmental impact of non-renewable energy sources. For AD, awareness needs to be created for the public in terms of separating waste at the source.

6.7 Discussion

This chapter looks at the possibility of using MSW disposed at VDS to generate electricity. The prime objective is to determine the generating potential using anaerobic digestion of OFMSW at VDS, however the use of incineration technology for electricity generation and a combination of AD and incineration was also looked at and a comparative analysis of the power generation capacity of each technology was deduced.

A study on the composition of waste at VDS showed that approximately 60 tons of MSW is disposed at VDS daily with 69.63% organic fraction. Based on the 2017 disposal data, 13,857 tons of organic waste was available for WtE conversion, of which 9445 tons of waste could be used for AD resulting in an approximate biogas production of 944500 m^3/annum. With the biogas produced 2277600 kWh of electricity could be generated annually aggregating 241 kWh/ton of waste and having a power plant capacity of 260kW. A sensitivity analysis of the AD was also conducted to determine the viability of this technology, if implemented at VDS. The analysis looked at fluctuations in key parameters for the production of biogas and electricity generation; OFMSW flow rate, biogas yield and the methane fraction of the biogas. With a change of 30–55% in the OFMSW flow rate the power capacity was determined to be in a range 160 kW and 296 kW. The biogas yield fluctuations in the range of 0.08 m^3/ton and 0.16 m^3/ton of MSW produced power in the range of 206 kW and 413 kW and finally changes in the methane fraction of 50% and 70% has a power capacity of 198 kW and 278 kW. The range for the analysis was taken from literature and this showed on average 258 kW of power could be generated with the changes in key parameters.

In comparison, implementation of an incineration power plant would result in a power plant with a capacity of 1008 kW aggregating 637 kWh/ton from 13,857 tons of OFMSW. A combination of AD and incineration power plant would produce 825 KW of power with 521 kWh/ton of MSW. The construction of a power plant at VDS would require a socio-techno assessment.

In addition, the amount of CO_2 emissions, through thermal electricity generation, that will be prevented through the implementation of the WtE technology was also determined. According to the NDC of Fiji, 212 kt of CO_2 emissions is projected to be reduced annually through the implementation of RE for power generation and energy efficiency. Through the implementaion of WtE power plants at VDS CO_2 emissions could be reduced by 0.59, 2.29 and 1.88 Gg from AD of OFMSW, incineration of OFMSW and a dual power plant with a combination of AD and incineration technologies respectively.

The use of WtE technologies to generate electricity does not only help in the reduction of carbon emissions, but also in MSW management. Landfill as practiced in Fiji, poses a threat to the environment and also is a health issue.

Barriers and challenges in the implementation of a WtE power plant at VDS would also need to be addressed, which would require reforms in polices and institutional framework, creating public awareness, ensuring expertise are available for the operations of the facility and securing financial support either from government or through a Private Public Partnership (PPP).

6.8   Conclusion

Fiji being an island nation with reliance on fossil fuel for its energy needs, the development of renewable sources would be an answer to energy security. The use of MSW to generate electricity has an added advantage of MSW management and also helping in GHG mitigation. The study shows that with the amount of MSW disposed at VDS, 260 kW of power could be generated through AD and 1008 kW through incineration. A sensitivity analysis of the AD process shows that with change in the key parameters the power capacity remains in the range of 160–413 kW, with an average of around 250 kW. The implementation of WtE technologies at VDS contributes towards achieving the NDC goals, with reductions in carbon emissions of 0.59, 2.29 and 1.88 Gg for AD, incineration and a combination of AD and incineration respectively.

References

Anshar, M., Ani, F. N., & Kader, A. S. (2014). The energy potential of municipal solid waste for power generation in Indonesia. *Jurnal Mekanikal, 37*(2), 42–54. Retrieved from https://jurnalmekanikal.utm.my/index.php/jurnalmekanikal/article/view/41/40.

Biogas. (2019). Retrieved January 4, 2019, from Wikipedia: https://en.wikipedia.org/wiki/Biogas

Biteco. (2013). *Out of gas from different types of substrates.* Retrieved from http://www.biteco-energy.com/biogas-yield/

Bolin, L., Lee, H.M., Lindahl, M. (2009). *LCA of biogas through anaerobic digestion from the organic fraction of municipal solid waste (OFMSW) compared to incineration of the waste.* Proceedings of EcoDesign 2009: 6th International Symposium on Environmentally Conscious Design and Inverse Manufacturing, 6–9th December, Sapporo, Japan.

Chaya, Q., & Gheewala, S. H. (2007). Life cycle assessment of MSW to energy schemes in Thailand. *Journal of Cleaner Production, 15*(15), 1463–1468. Retrieved from https://www.researchgate.net/publication/222565900_Life_Cycle_Assessment_of_MSW-to-energy_Schemes_in_Thailand.

Electrigaz. (2017). *Biogas FAQ.* Retrieved from http://www.electrigaz.com/faq_en.htm

Energy Fiji Limited. (2017). *FEA annual report 2016.* Retrieved from http://efl.com.fj/wp-content/uploads/2017/08/fea-annual-report-2016.pdf

Engineering ToolBox. (2003). *Fuels – Higher and lower calorific values.* Retrieved from https://www.engineeringtoolbox.com/fuels-higher-calorific-values-d_169.html

Gupta, S., & Mishra, R. S. (2015). Estimation of electricity generation from waste to energy using incineration technology. *International Journal of Advance Research and Innovation, 3*(4), 631–634. Retrieved from http://www.ijari.org.

Intergovernmental Panel on Climate Change (IPCC) 2006. (2006). In H. S. Eggleston, L. Buendia, K. Miwa, T. Ngara, & K. Tanabe (Eds.), *IPCC guidelines for National Greenhouse gas Inventories, prepared by the National Greenhouse gas Inventories Programme.* Japan:

IGES. Retrieved from https://www.ipccnggip.iges.or.jp/public/2006gl/pdf/2_Volume2/V2_2_Ch2_Stationary_Combustion.pdf.

Kumar, P. (2013). *Brief country analysis paper for Fiji*. Paper presented at the 4th Regional 3R forum in Asia, Ha Noi, Viet Nam.

Lautoka City Council. (2012). Solid Waste Management (SWM) Master Plan for Lautoka City Council.

Organic Waste System. (2013). *Biogas plants: References*. Retrieved from http://www.ows.be/biogas-plants/references/

Ouda, O. K. M., Razam, A. S., Al-Waked, R., Al-Asad, J. F., & Nizami, A. S. (2017). Waste to energy potential in the western province of Saudi Arabia. *Journal of King Saud University – Engineering Scinces, 29*, 212–220.

Reddy, N. S., Satyanarayana, S. V., & Sudha, G. (2017). Biogas generation from biodegradable kitchen waste. *International Journal of Environment, agriculture and Biotechnology, 2*(2), 689–694. Retrieved from 10.22161/ijeab/2.2.15

Surroop, D., & Mohee, R. (2012). Technical and economic assessment of power generation from biogas. *International conference on environmental science on technology*. Singapore. Retrieved from http://www.ipcbee.com/vol30/022%2D%2DICEST2012_N30008.pdf

Swapna, S. (2012). *An energy balance analysis of MSW for energy production in Fiji*. MSc thesis, University of the South Pacific, Fiji.

Swedish Gas Technology Centre. (2012). *Basic Data on Biogas*. Retrieved from http://www.sgc.se/ckfinder/userfiles/files/BasicDataonBiogas2012.pdf

The World Bank. (2019). *CO2 emissions (kt)*. Retrieved from https://data.worldbank.org/indicator/EN.ATM.CO2E.KT?end=2014&locations=FJ&start=1990&view=chart&year_high_desc=false

U.S Energy Information Administration. (2018). *How much carbon dioxide is produced when different fuels are burned?* Retrieved from https://www.eia.gov/tools/faqs/faq.php?id=73&t=11

University of California, Department of Biological and Agricultural Engineering. (2008, March). *Current anaerobic digestion technologies used for treatment of municipal organic solid waste. Report presented to the California Integrated Waste Management Board*. Retrieved from https://www2.calrecycle.ca.gov/Publications/Download/1099

Vandevivere, P., De Baere, L., & Verstraete, W. (2003). Types of anaerobic digester for solid wastes. In J. Mata-Alvarez (Ed.), *Biomethanization of the organic fraction of municipal solid wastes* (pp. 111–140). IWA Publishing, London, United Kindom. Retrieved from http://hdl.handle.net/1854/LU-210258.

World Energy Council. (2016). *World energy resources 2016*. Retrieved from https://www.worldenergy.org/wp-content/uploads/2016/10/World-Energy-Resources-Full-report-2016.10.03.pdf

Xu, S., He, H., & Luo, L. (2016). Status and prospects of municipal solid waste to energy technologies in China. In O. P. Karthikeyan (Ed.), *Recycling of solid waste for biofuels and biochemicals*. Singapore: Environmental Footprint and Eco-design of Products and Processes. https://doi.org/10.1007/978-981-10-0150-5_2.

Yingjain, L., Qi, Q., Xianzhu, H. E., & Jiezhi, L. (2011). Energy use project and conversion efficiency analysis on biogas produced in breweries. *Industrial Energy efficiency (IEE)*. Conference proceedings of the World Renewable Energy Congress, Linkoping, Sweden. Retrieved from http://citeseerx.ist.psu.edu/viewdoc/download?doi=10.1.1.452.56&rep=rep1&type=pdf

Chapter 7
Viability of Commercial On-Shore Wind Farm Sites in Viti Levu, Fiji

Lionel Joseph and Ramendra Prasad

Abstract The Energy Fiji Limited's (EFL's) first major wind project, commissioned in late 2007 at Butoni, Sigatoka has been unable to perform to expectations. Among the reasons given is inadequate information on the wind resource potential at the selected site. This stresses the need for a thorough resource evaluation before a wind project is actually implemented. This chapter carries out a wind resource assessment of two sites in Viti Levu using 10-min interval data over a 2 year period obtained from the Fiji Meteorological Services.

Considering the mean wind speed, Rakiraki seems to be a more viable option for commissioning a wind farm over Yaqara with a higher wind speed of 6.19 m/s. However, this information is not enough for determining the best site. Thus, the capacity factor (CF) values for two different turbine models were computed by deriving the best probability distribution function (PDF) equation. Out of the three PDFs (Weibull, Gamma and Lognormal) that were tested, the best fit to the observed data was provided by the Weibull distribution. Other criteria, including power output, annual energy production (AEP), cost saved (FJD), cost of electricity (COE) produced and payback period were all evaluated as part of a techno-economic assessment to select the better site out of the two.

The results attained revealed that the Vestas V27 model for Rakiraki had the higher CF of 0.304. As for the output power and energy, the Vergnet 275 model for Rakiraki produced the higher values of 73.41 kW and 610,896.38 kWh, respectively. In general, the COE (FJD/kWh) and the payback period for Rakiraki were lower than those for Yaqara. However, for Rakiraki the Vergnet 275 model witnessed a slightly higher COE and payback period over the Vestas V27 model. Moreover, it is clear that Rakiraki is the more viable site for commissioning a wind farm over Yaqara (with AEP estimate of 22,603.17 MWh for the 37 Vergnet turbines). Also, while Vergnet would be more suited for Rakiraki in terms of AEP, cost saved and emission savings of 15,374.5 tCO_2 (when replaced with IDO) and 15,272.4 tCO_2 (when replaced with HFO), it also needs to be considered that the COE and payback period of the same model would be slightly higher over Vestas.

L. Joseph (✉) · R. Prasad
School of Science and Technology, Department of Science, The University of Fiji,
Lautoka, Fiji

© Springer Nature Switzerland AG 2020   151
A. Singh (ed.), *Translating the Paris Agreement into Action in the Pacific*,
Advances in Global Change Research 68,
https://doi.org/10.1007/978-3-030-30211-5_7

This research would be of fundamental importance to policy makers as an aid to formulating a well-planned policy relating to wind energy, thus enabling the usage of a broader range of RE resources.

Keywords Wind energy · Feasibility assessment · Economic evaluation · EFL · PDFs · Nationally determined contributions (NDC)

7.1 Introduction

Conventional fossil fuel resources such as coal, oil, and natural gas have constantly been a major energy source since the industrial revolution and the rapid economic growth throughout the world is leading to an increased demand in this form of energy (Leung and Yang 2012). The two main global energy crises include: depletion of fossil fuel reserves prior to the increase in energy demand with the increasing population over the years and secondly, excess dependence on fossil fuel derived energy leads to increased accumulation of GHGs in the atmosphere, which is the leading contributor towards global warming and climate changing impacts. With this, there is an increasing interest in global emission reduction and energy conservation (Wang et al. 2015; Yesilbudak et al. 2013).

Use of clean renewable energy and energy efficiency offers the most effective solution to world energy problems. Renewable energy resources including hydro, wind, solar, traditional and modern biomass, biofuels, and geothermal is responsible for supplying 19.2% of the world's final energy consumption, while 2.5% is nuclear power and greater portion of 78.3% comes from fossil fuels (REN21 2016). Recently, wind energy and wind technology development has become a very interesting topic for both researchers and developers in developing nations (Khahro et al. 2014). At the end of 2015, the global wind power generation capacity reached 435 GWs, which is just 7% of the total global power generation capacity (GWEO 2016).

The global wind power production capacity has been growing at a rapid pace, with a mean growth of an approximate 30% per annum (Leung and Yang 2012). China is the leading wind power producer and it was in 2010 when China surpassed the US by adding 16.5 GW over that year to get a total of 42.3 GW, which was an increase in 64% from 2009 (GWEC 2011a). Although China started to use wind energy in the 1970s, wind power generation in China began in earnest from 2006 when "The Renewable Energy Law of China" was established, leading to a doubling of the installed wind power capacity since that year (Leung and Yang 2012). Policy played a crucial role in China surpassing the US, and it is predicted that with proper policies in place, 20% of the total electricity will be generated by wind power by 2030 (GWEC 2011b). Therefore, to implement successful wind power projects in the Pacific Island countries (PICs), it is very important to put into practice what the leading countries are doing.

According to Sharma and Ahmed (2016), the developing PICs such as Fiji largely rely on fossil fuels for transportation and power generation, which makes them the main victims when petroleum price rises. For instance, in Fiji, 45.45% of the total electricity generation comes from industrial oil and heavy fuel oil (EFL 2016). In Fiji, the main renewable power source is hydropower, contributing 53.05% of the total grid electrification. A little support is given to the Energy Fiji Limited (EFL) by the independent power producers (IPPs), namely Tropik Wood Industries Limited (TWIL) and Fiji Sugar Corporation (FSC), which contributes to about 1.11% of the power generation mix. This statistic is for the year 2016 (EFL 2016), where the remaining 0.39% of the energy mix was provided by wind energy. Fiji has only one wind farm (Butoni wind farm), which has a total rated capacity of 10 MW and was installed in the year 2007. It is evident that it has not been able to reach close to its rated capacity but this should not discourage the developers and investors from further investing in wind technology to meet future power demand and reduce fossil fuel derived fuel imports.

For predicting wind energy resources at any particular site, it is very important to know the wind speed and direction of that site (Sharma and Ahmed 2016). Literature shows that over the years since the installation of Butoni wind farm, about 19 other sites in Fiji have been investigated for their feasibility as potential sites for new wind farms. These sites include: Benau (Kumar and Nair 2013), Qamu (Kumar and Nair 2014), Vadravadra (Gau Island) (Singh 2015), Bligh waters (Dayal 2015), Kadavu and Suva (Peninsula) (Sharma and Ahmed 2016), and Wainiyaku (Taveuni) (Kumar and Nair 2012). In addition, a study by Kumar and Prasad (Kumar and Prasad 2010) selected the following sites for wind resource assessment: Suva (Laucala bay), Ba, Tavua, Savusavu, Lautoka, Labasa, Nadi, Ucu Point, Nausori, Rakiraki, Monasavu, and Nabouwalu. While not all sites would be expected to be suitable for wind farm installation, it is quite possible that they included a few feasible sites which could be further enumerated after mutual comparisons. It is worthwhile to note that most of these studies also incorporated economic analyses together with wind resource analysis, which is a very important aspect of wind energy feasibility assessments.

However, it has to be noted that the wind resource assessments of the 12 sites done by Kumar and Prasad (Kumar and Prasad 2010) relied on NASA data, which is not very accurate for feasibility studies. In comparison the current study obtained the wind climate data (i.e., wind speed and wind direction) from the Fiji Meteorological (MET) Services which is much more reliable because of the method of collection. In addition, the present work includes detailed economic assessment as well as assessing the amount of greenhouse gases (GHGs) saved by replacing conventional fuel sources like Heavy Fuel Oil (HFO) and Industrial Diesel Oil (IDO) with wind energy. These additional aspects of methodology make this study novel in its nature and scope.

According to Kantar et al. (2018), statistical distributions are generally used to characterize the wind speed data. The choice of wind speed distribution is fundamental for the accurate approximation of the wind energy potential of a particular site. Weibull Distribution, a two-parameter distribution is widely used in the

literature concerning wind energy and this is due to its flexible and easily comput-
able nature (Kantar et al. 2018; Mohammadi et al. 2016; Akgül et al. 2016; Alavi
et al. 2016a, b; Morgan et al. 2011; Adaramola et al. 2014). Apart from Weibull
distribution, numerous models have been used, including the single parameter exponential (Alavi et al. 2016a), Rayleigh (Kantar et al. 2018; Alavi et al. 2016b; Morgan
et al. 2011) and Lindley models (Kantar et al. 2018), the two-parameter inverse
Weibull (Akgül et al. 2016), Gamma, inverse Gaussian, Lognormal, Log-logistic
and Nakagami distribution (Kantar et al. 2018; Alavi et al. 2016a, b; Morgan et al.
2011; Mohammadi et al. 2017) and the three-parameter Generalized Extreme Value
and Birmum Saunders models (Alavi et al. 2016a; Mohammadi et al. 2017). Clearly
there are various distribution models to choose from, each with their pros and cons.

Note that these distribution models have parameters that need to be estimated for
determination of wind speed distribution. For higher levels of accuracy, a correct
estimation of distribution parameters is crucial. Various parameter estimation methods have been used in the past and compared in terms of different criteria. For
instance, Mohammadi et al. (2016) and Akgül et al. (2016) used maximum likelihood estimation method (MLE) and modified maximum likelihood estimation
method (MMLE) to estimate the parameters of Weibull model. Alavi et al. (2016b)
have utilized the maximum likelihood method and method of moments (MOM) to
estimate parameters of Weibull, Gamma, Lognormal and Rayleigh distributions and
it was found that the MLE method performed perfectly in estimating the Weibull
parameters to fit the wind speed distribution. Also, Kantar et al. (2018) claim that
MLE method is preferable when the sample size (n) is more than 100 and thus in
this study we have used MLE to estimate the parameters of Weibull, Lognormal and
Gamma distribution models.

Moreover, economic evaluation and the type of turbine model to choose for the
best performance are critical. Economic analysis is fundamental for any renewable
energy project to ensure that the investment gives a net profit. Most literature on
wind energy is based on Weibull analysis, while some talk about capacity factor as
being the main variable and others consider economic assessment as the key factor
(Khraiwish Dalabeeh 2017). This work has linked all these devices in an effort to
evaluate the most feasible solution for the selected sites. The objective of this chapter is therefore to evaluate the wind energy potential for two selected sites (Rakiraki
and Yaqara). These are both located in the Western region of Viti Levu, which is one
of the two major islands of Fiji. The study will use the wind speed data for 2 years
(2016–2017) attained from Fiji Meteorological Services to perform an economic
evaluation on the selected wind turbines. This information will help the Fijian government and the other energy planners and policy makers to make more informed
decisions while investing in wind energy technologies.

7.2   Background

7.2.1   Energy Situation in Fiji

The Republic of the Fiji is a self-governed island nation having a population of approximately 880,000 people with high population density distributed in the two main islands, namely Viti and Vanua Levu (ADB 2015). Fiji is a developing country, and its dependence on petroleum derived fuel is high. Like other SIDS, it is highly affected by sudden changes in oil price (OECD 2016). This in turn forces the Fijian government to heavily subsidize electricity distribution (Michalena et al. 2018). However, to minimize the environmental burden of HFO and diesel dependence, the ratified Nationally Determined Contributions (NDC) has set an ambitious target of 81% of renewable energy share by 2020 and meeting their entire electricity demand through renewable energy sources by 2030 (Timilsina and Shah 2016; "Government of Fiji" 2015). Initially, the renewable energy target was 90%, which was to be achieved by 2015 (Betzold 2016). This figure was unfortunately not met in 2015, as only 48% of the total demand could be met by renewable sources (EFL 2015). Meeting the current target seems quite possible because Fiji has the renewable resources required, but proper policies and regulatory frameworks need to be in place to utilize these resources effectively.

Also, as stated by Timilsina and Shah (2016), setting an ambitious target is a good indication of Fiji to promote its renewable resources. Other than oil derived fuel, hydropower has been contributing a significant share in the total primary energy supply as revealed in Fig. 7.1. Other renewable energy sources, especially solar and wind are unreliable power sources at times due to their inherently stochastic nature (Yang et al. 2009). This factor makes these sources uneconomical when

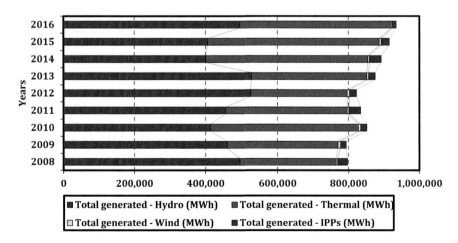

Fig. 7.1 Overview of Fiji's electricity generation mix for grid-connected electrification over the years. (EFL 2016)

compared to the conventional energy sources (i.e. coal, gas, large hydro) (Timilsina and Shah 2016). However, promoting sources like wind or even solar can only be effective through successful implementation of the policies and measures in place. For instance, policies will lower the costs of renewable energy technologies through research and development, economy of scale, market maturity, and increased competition amongst the suppliers (Timilsina and Shah 2016). REN21 (2016) strongly asserts that energy efficiency improvements, smart management, and the right mix of RE technologies can provide a reliable and an affordable electricity supply.

7.2.2 Wind Power Generation in Fiji

This section will discuss the operational status of Fiji's only wind farm (i.e. Butoni wind farm) over the years since installation.

7.2.2.1 Butoni Wind Farm Installation

Fiji's only significant wind farm, which is operated by EFL, feeds directly into the grid for Viti Levu that has a total installed generation capacity of around 236.5 MW (IRENA 2015). The farm was commissioned in August 2007 and is situated in Sigatoka, in the south-western part of the island of Viti Levu. According to the report by IRENA (2015), initially Pacific Hydro (a private sector) formed partnership to develop this project with EFL. However, due to concerns about the commercial viability of the project, Pacific Hydro pulled out in 2003, leaving EFL to continue with this project alone.

The farm comprises of 37 Vergnet turbines (model GEV-MP 275) with a total rated capacity of 10 MW (Kumar and Weir 2008). Each turbine is designed for maximum output at a hub height of 55 m. These turbines are horizontal axis wind turbines (HAWT) with 2-bladed downwind rotor and a two-speed generator. In addition, these turbines have a teetering hub with a hydraulic pitch control. Also, the turbines have a two-stage planetary gear box and two-stage asynchronous squirrel cage generators with a hydraulic active yawing mechanism, which automatically turns it in the direction of the wind.

According to Kumar and Weir (2008), for operation in Fiji, the tower hinges at the base is one of the most crucial features of these turbines. This is so that the turbine can be laid nearly horizontal to prevent the destructive force of the wind during cyclonic period (IRENA 2015). This tilting mechanism is considered advantageous in rural remote regions. Furthermore, this cyclone proof design was installed by Vergnet in New Caledonia (a cyclone prone Island), which was quite successful. In accordance to Zieroth (2006), New Caledonia is very successful with wind energy projects.

7.2.2.2 Status of Power Generation

The average power generated per year via Butoni wind farm fluctuates over the years, with the least power generated in 2007 and the most generated in 2009 as shown in Fig. 7.2. The reason of low electricity production in 2007 was due to its late operation and hindrance in the performance caused by Category 3 Severe Tropical Cyclone Gene. During the cyclone period, the wind turbines were lowered in order to prevent damages to wind turbine components. The maintenance work required some time, which affected power generation. However, the annual electricity production increased later on till 2009.

The year 2009 marked the highest electricity production. During that year there were two Category 2 Tropical Cyclones, named: Mick and Lin. Despite that, Butoni wind farm managed to produce the highest electricity since the cyclones did not directly affect the site. Moreover, during the cyclone period the wind turbines were lowered to the ground to prevent damage. During that time maintenance work were carried out, so when the wind turbines were in a functional position later on, it managed to work efficiently. After 2009, there was a decline in power generation till 2011. This could be most probably due to lower wind speed during those years.

Moreover, there was a sudden increase in power production in 2012. This could be due to high average wind speed during that period of the year. On the contrary, there was a significant decline in the power output till 2014. During 2013 and 2014, there were a series of cyclone [Category 5 severe tropical cyclone Ian (2013), Category 2 tropical cyclone Kofi, Category 3 severe tropical cyclone Lusi (2014)] in Fiji and due to this the wind turbines were lowered to prevent any damage; however, one wind turbine was damaged permanently while two other turbines were also under maintenance. This resulted in lowered electricity output. The two turbines lowered for maintenance purpose is still not operational to date as was discovered after a field visit to the site in 2017. In 2015, there was a slight increase in the electricity production prior to favorable average wind speed during that year. Lastly, the

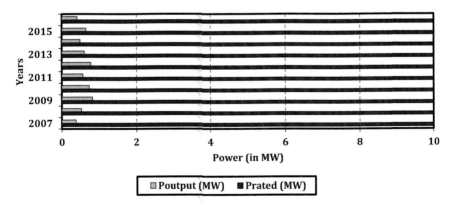

Fig. 7.2 Variations in the average annual power output (MW) of Butoni wind farm over the years. (EFL 2015)

graph (Fig. 7.2) shows that the average annual power production in 2016 was also very low. Some reasons for this could be: only 35 out of 37 turbines were functional, annual wind speed was low, and maintenance work might have been under process.

7.3 Methodology

7.3.1 Study Site and Wind Speed Statistics

Wind speed data for two different sites, namely Rakiraki and Yaqara in Western Fiji as depicted in Fig. 7.3 are used for this study. The data were obtained from Fiji Meteorological Services. The data used are 10-min interval wind speed sampled for a period of 2 years from January 2016 till December 2017, collected via the use of anemometers installed at a height of 10 m at geographical locations specified in Table 7.1. The percentage of missing data are furnished in Table 7.2 below; however,

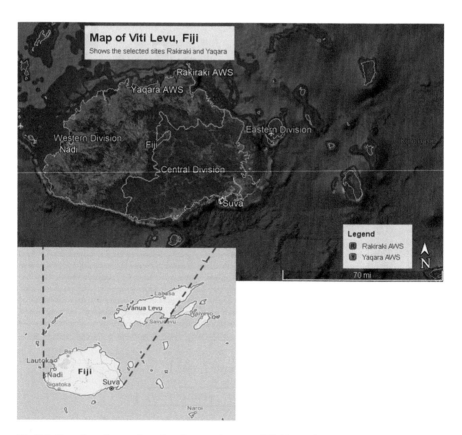

Fig. 7.3 Location of two selected stations on the map of Viti Levu, Fiji Islands

Table 7.1 Geographical information for Rakiraki and Yaqara

Selected site	Latitude (°S)	Longitude (°E)	Height above MSL (m)
Yaqara (site 1)	17.43290	177.97745	33
Rakiraki (site 2)	17.34037	178.22138	15

Table 7.2 Information on data availability recorded at 10 min intervals

Selected site	Data period	Total dataset(n)	Expected dataset	Absent dataset	Percent of absent dataset
Yaqara	1/01/2016– 31/12/2017	105,219	105,264	45	0.043%
Rakiraki	1/01/2016– 31/12/2017	96,842	105,264	8422	8.000%

Table 7.3 Descriptive statistics of the wind speed data used for Yaqara and Rakiraki

Statistical indices	Minimum	Maximum	Mean	Median	S.D	Skewness	Kurtosis
Yaqara	0.010	16.910	4.107	4.010	1.955	0.370	2.651
Rakiraki	0.020	21.620	6.190	6.320	3.003	0.124	2.421

there was no urgent need in imputation of these absent values as this study mainly focused on learning about the wind speed during the years of data availability rather than forecasting future wind speed values.

Table 7.3 presents some descriptive statistics including minimum, maximum, mean, median, standard deviation, Skewness and Kurtosis of the used wind speed data for two selected stations. As noticed, Rakiraki has the higher mean wind speed with the value of 6.190 m/s, which gives a good indicator for a probable future wind farm set-up and Yaqara has the lower mean wind speed with the value of 4.107 m/s. Moreover, both the stations have the coefficient of Kurtosis being significantly lower than 3; hence, a distribution with kurtosis <3 is called platykurtic. Thus, compared to a normal distribution, its central peak is lower and broader, and its tails are shorter and thinner. Additionally, the coefficient of Skewness is a positive value, which denotes that the distribution would be distorted towards the y-axis of the graph.

7.3.2 Probability Distribution Curves of Wind Speed Using MLE

Generally, according to Alavi et al. (2016a) plots depicting probability distribution of wind speed is of indispensable significance to characterize the wind speed behaviour, evaluate the wind energy potential and performance of wind energy systems. In this study, three PDFs are utilized to describe the wind speed frequency distributions. These distribution functions include Weibull, Gamma and Lognormal. The

Table 7.4 Parameters of PDFs estimated using MLE method

PDFs	Parameters (MLE)	Yaqara	Rakiraki
Weibull	Shape (k)	2.229761	2.163397
	Scale (c)	4.654635	6.993114
Gamma	Shape (k or α)	3.669335	3.183620
	Rate (β)	0.889682	0.512754
Lognormal	Meanlog (μ)	1.274472	1.660798
	Sdlog/shape (σ)	0.598277	0.655208

maximum-likelihood estimation (MLE) method was utilized via the "fitdistrplus" package in RStudio software to compute the parameters defining each PDF as seen in Table 7.4. The MLE is typically considered as a very robust technique in determining the values of the parameters that maximize the probability of the used data (Conradsen et al. 1984). For instance, for a larger sample size, MLE is more efficient than other known estimation methods such as method of moments (MOM), and it generates a lower mean squared error (MSE).

7.3.3 Error Evaluation for Best Curve Fit

7.3.3.1 AIC and BIC

Akaike information criterion (AIC) and Bayesian information criterion (BIC) are utilized in this study to determine which PDF out of the three tested gave the best fit to the observed wind speed data for the two sites. AIC (Akaike 1973) and BIC (Schwarz 1978) are the two most commonly used model selection criteria based on model parameters using the MLE method employed in this study. Lower values of AIC and BIC indicate a better PDF model and the values are summarized in Table 7.5 revealing Weibull as the best PDF out of the three.

7.3.4 Wind Direction Analysis

In accordance to Sharma and Ahmed (2016), the wind direction assessment is an integral part of resource assessment as it reveals the prevailing wind direction from which the wind energy can be harnessed effectively. Wind speed alone falls under the univariate category and the bivariate category of wind speed and direction forms a better way of conducting feasibility studies to make more accurate and reasonable assumptions. Wind speed and direction data for both sites are used to construct a wind direction radar chart, also known as wind rose chart, which indicates the most dominant and least dominant wind direction over the 2 years used for this research.

Table 7.5 Error evaluation values of AIC and BIC

Test	Yaqara (n = 105,219)		Rakiraki (n = 96,842)	
	AIC	BIC	AIC	BIC
Weibull	433470.7	433489.8	484945.3	484964.2
Gamma	439518.8	439537.9	494436.0	494455.0
Lognormal	458697.5	458716.6	514609.7	514628.6

The "openair" package in Rstudio was used to make the plotting of wind rose charts possible.

7.3.5 Selected Best Fit PDF for Capacity Factor (CF) Equation

The error evaluation criteria (AIC and BIC) used in Sect. 7.3.3 made it clear that Weibull PDF gave the best fit to the observed wind speed data. Hence, this section will show how the Weibull PDF equation is used to derive the capacity factor (CF) formula.

The wind speed near the ground surface changes with height. The wind speed near the ground surface is often lower and is disturbed due to obstacles present; while the wind speeds at high heights (i.e. height of wind turbines) is higher with a laminar flow motion. Furthermore, the measured wind speed used for this study is measured at a height of 10 m, being the height of the Fiji MET tower. The most common expression for the variation of wind speed with height is the power law (Akpinar and Akpinar 2005; Celik 2004; Ahmed Shata and Hanitsch 2006), represented as:

$$\frac{v_2}{v_1} = \left(\frac{h_2}{h_1}\right)^{\alpha} \tag{7.1}$$

where, v_1 = mean wind speed at h_1 = 10 m and v_2 is the wind speed to be calculated at h_2. The power law exponent α, depends on factors such as surface roughness and atmospheric stability (Khraiwish Dalabeeh 2017). Moreover, similar to the wind speed v_2 at required height h_2, the Weibull parameters (k and c) can be calculated using the expressions (Al-Abbadi 2005; Ahmed Shata and Hanitsch 2008; Schallenberg-Rodriguez 2013):

$$k_h = k_{10}\left[1 - 0.0881 Ln\left(\frac{h}{10}\right)\right]^{-1} \tag{7.2}$$

$$c_h = c_{10}\left(\frac{h}{10}\right)^\alpha \tag{7.3}$$

$$\alpha = \left[0.37 - 0.0881\ln\left(c_{10}\right)\right] \tag{7.4}$$

where k_{10} and c_{10} are the shape and scale factor at height = 10 m and k_h and c_h are parameters at height = h.

Therefore, integrating the best fitted Weibull PDF with a typical wind turbine power curve one can calculate the theoretical capacity factor using the equation (Abed and El-Mallah 1997; Kaldellis and Zafirakis 2013) given below:

$$CF = \frac{\exp\left[-\left(\frac{v_c}{c}\right)^k\right] - \exp\left[-\left(\frac{v_r}{c}\right)^k\right]}{\left(\frac{v_r}{c}\right)^k - \left(\frac{v_c}{c}\right)^k} - \exp\left[-\left(\frac{v_f}{c}\right)^k\right] \tag{7.5}$$

In the above equation v_c = cut-in speed (m/s), v_r = rated speed (m/s) and v_f = cut-out speed (m/s).

Using the CF equation and the rated power of the selected turbines (P_{rated}) in kW, the average output power (P_{av}) can be established using the formula (Lydia et al. 2014) given as:

$$P_{e.av} = CF \times P_{rated} \tag{7.6}$$

The average number of wind turbines (ANWT) needed for a specific wind farm site can be attained via:

$$ANWT = \frac{P_{L.av}}{P_{e.av}} \tag{7.7}$$

In the equation, $P_{L\,av}$ is the average load requirement for a given site. Hence, assuming the availability of the wind turbine as being 95%, the annual energy production (AEP or E_{out}) can be equated using:

$$AEP = CF \times 0.95 \times P_{rated} \times t \tag{7.8}$$

where t is the operating time taken for this study in hour/year, taken to be 8760 h.

7.3.6 Technical Aspects

The wind turbines and their models differ in many forms, mainly in form of their rated power capacity (P_r), the height and the architecture in which it effectively harnesses the wind speed. It is very crucial to select the best turbines to ensure it is economical in the long-term period and does not get damaged during cyclonic seasons. Butoni wind farm has employed the use of Vergnet 275 turbines due to its cyclone proof nature. This model can be tilted down during times of extreme events, including hurricane and tropical cyclones and also during times of maintenance and repair. Pacific Island territories, including New Caledonia, Vanuatu and Samoa has also installed similar models. For the study sites, the turbine models need to have rated power capacity in the range of 225–275 kW because turbines with higher rated power capacity require higher rated wind speed than what is achievable from the sites under study. The other technical specifications are presented in Table 7.6.

7.3.7 Economic Evaluation

Moreover, to compare the economic values achieved for these sites, the assumptions used in (Sharma and Ahmed 2016) was used for economic analysis since it is a recent paper. The present value of costs (PVC) and the key assumptions are listed as follows:

$$PVC = I + C_{omr}\left[\frac{1+i}{r-i}\right] \times \left[1-\left(\frac{1+i}{1+r}\right)^t\right] - S\left(\frac{1+i}{1+r}\right)^t \tag{7.9}$$

- The operational maintenance and repair cost (C_{omr}) are taken to be 25% of the cost of the turbine/annum (i.e. [machine price/lifetime] × 0.25).
- Cost of civil work (C_{cw}) is estimated to be 20% the price of turbine.
- Investment (I) incorporates the turbine price and its civil work and grid integration costs.
- The inflation rate (i) and the interest rate (r) are considered to be 5.2% ("Inflation Rate – Fiji", 2018) and 15%, respectively.
- The scrap value (S) is assumed to be 10% of the turbine price and civil work.
- The lifespan (t) of the turbine is considered to be 20 years.

Table 7.6 Technical specifications of two selected wind turbine models

Turbine model	Type/no. of blades	P_r(kW)	Rotor diameter (m)	V_c (m/s)	V_r (m/s)	V_f (m/s)	Hub height (m)
Vergnet GEV MP (275)	HAWT/2	275	32	3.5	16.0	20.0	55.0
Vestas V27 (225)	HAWT/2	225	27	3.5	13.5	25.0	32.5

Table 7.7 Costs associated with the selected turbines

Turbine model	Vergnet 275		Vestas V27	
	Per turbine	Per 37 turbines	Per turbine	Per 37 turbines
Cost of turbine (FJD)	550,800	20,379,600	489,600	18,115,200
C_{omr} (FJD)	6885	254,745	6120	226,440
C_{cw} (FJD)	110,160	4,075,920	97,920	3,623,040
I (FJD)	660,960	24,455,520	587,520	21,738,240
i (%)	5.2	5.2	5.2	5.2
r (%)	15	15	15	15
S (FJD)	66,096	2,445,552	58,752	2,173,824
t (years)	20	20	20	20
PVC (FJD)	711,290.49	26,317,748.13	632,258.21	23,393,553.77

- Vergnet GEV MP 275 and Vestas V27 225 wind turbine models are chosen for comparison prior to their suitability for Fiji and also the rural remote regions. The model specifications are shown in Table 7.6.
- The cost of turbine (C_t) is estimated from (Kumar and Nair 2013) (Table 7.7).

Other criteria used for economic evaluation included; cost/kWh of the technologies, cost saved by the government by reduction of fossil-derived fuel imports and the payback period of the considered projects, given by the equations enlisted below:

$$\frac{Cost}{kWh} = \frac{PVC}{20 \times AEP(kWh)} \quad (7.10)$$

$$Costsaved = AEP(kWh) \times 0.34(FJD / kWh) \quad (7.11)$$

$$PBP = \frac{PVC}{AEP(kWh) \times 0.34(FJD / kWh)} \quad (7.12)$$

7.3.8   Avoided Emissions

The projected greenhouse gas emissions (GHG) saved from replacing Heavy Fuel Oil (HFO) or Industrial Diesel Oil (IDO) with wind energy has also been incorporated in this study. Firstly, the average electricity (kWh) generated by using 1 kg of HFO and IDO (used by EFL) was determined. This was done by analyzing the EFL annual reports from 2009 to 2016. It was found that 1 kg of HFO generated 4.63 kWh electricity and 1 kg of IDO generated 4.68 kWh electricity, which reveals minimum difference between the two fuel sources. The IPCC guideline on GHG emissions from stationary combustion was used for determining the emissions of CO_2, CH_4 and N_2O (IPCC 2006). The net calorific values and GHG emission factors are shown in Table 7.8. The tier 1 equation used for determining the GHG emissions is:

Table 7.8 Shows the net calorific value and GHG emission factor of HFO and IDO fuel sources (IPCC 2006)

Fuel type	Net calorific value (TJ/Gg)	Emission factor (kg of GHG/TJ)		
		CO_2	CH_4	N_2O
HFO	40.4	77,400	3	0.6
IDO	43	74,100	3	0.6

$$Emissions_{GHG, fuel} = Fuel\ Combustion_{fuel} \times Emission\ factor_{GHG, fuel} \qquad (7.13)$$

Where:

$Emissions_{GHG, fuel}$ = emissions of a given GHG by type of fuel (in kg GHG)
$Fuel\ Combustion_{fuel}$ = amount of fuel combusted (in TJ)
$Emission\ factor_{GHG, fuel}$ = default emission factor of a given GHG by type of fuel (in kg GHG/TJ)

The total GHG emission is given by:

$$Emissions_{GHG} = \sum_{fuels} Emissions_{GHG, fuel} \qquad (7.14)$$

7.4   Results and Discussions

The study has employed mathematical models and calculations for the two selected sites in the Western Fiji, namely; Yaqara and Rakiraki and for two types of wind turbines which are rendered suitable for Fiji, prior to its cyclonic nature. The 10-min interval wind speed data for the two locations over a period of 2 years have been attained and analyzed to conduct a feasibility study. The important statistics for the two sites are summarized in Table 7.3 and can be seen that Rakiraki has a greater mean wind speed of 6.19 m/s over Yaqara, with an average wind speed of 4.107 m/s that is quite close to the wind speed of the Butoni region (i.e.~4 m/s). The maximum speed reaches 16.91 and 21.62 m/s for Yaqara and Rakiraki, respectively. This reveals that for Rakiraki, the maximum wind speed can be slightly higher than the cut-out speed of the Vergnet GEV MP (275) turbine. However, this should not be a major concern if the turbines are lowered to safety because extreme wind speeds are only present during cyclonic periods.

Figure 7.4 is a time-series plot for the two sites and illustrates a clearer picture of the wind speed at the chosen sites. The plots however do not represent any clear increasing or decreasing trend, but it does show seasonality, cyclic patterns and few uneven anomalies. The seasonal increase and decrease in wind speed indicate that the wind speed is often not constant but varies from time to time. This is understandable as wind speed is a climatic variable, which is dependent on various other factors. Furthermore, it is this same varying nature of wind speed that gives rise to intermittency constraints that deters most of the investors away.

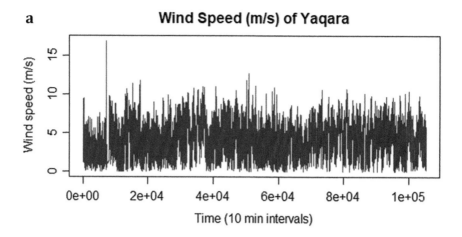

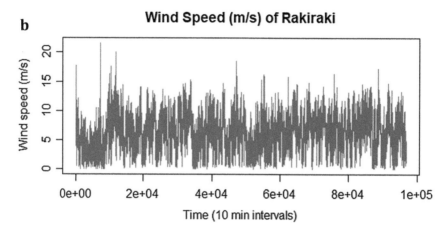

Fig. 7.4 Time series plots of (**a**) Yaqara and (**b**) Rakiraki sites. Ten minutes interval data are used for both the sites from 01/01/16 to 31/12/17, where Yaqara has a total of 105,219 data points and Rakiraki has a total of 96,842 data points as represented by the x-axis of the graphs. For example, 2e+04 indicate 20,000th data point

Moreover, wind speed alone cannot accurately determine the most feasible site for commissioning a wind farm. Wind direction is something else that needs to be considered as well. The direction of wind changes often and it becomes a key aspect to know which direction gets more effective wind speed than the others. For instance, Sharma and Ahmed (2016) claims that wind direction analysis indicates the prevailing wind direction from which the energy can be harnessed and thus a dominant sector is preferred over a dispersed energy distribution. Figure 7.5 displays both the sites wind distribution in four parts with discrete 90° intervals.

The wind rose chart for Yaqara (Fig. 7.5a) indicates that the prevailing wind direction analyzed over the 2 years fall dominantly somewhat under the South-East

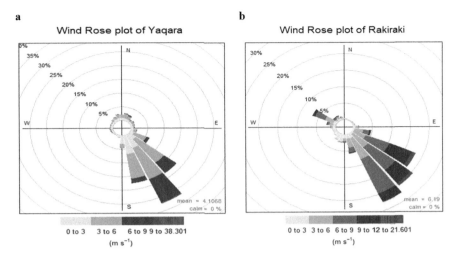

Fig. 7.5 Wind rose plots of (**a**) Yaqara and (**b**) Rakiraki

(SE) direction. The common wind speed witnessed under this direction falls in the range of 3–9 m/s, with 4.107 m/s being the mean wind speed. The same dominant direction of South-East is common to Rakiraki (Fig. 7.5b) as this is the direction attracting nearly all the preferred wind speeds being in the range of 6–12 m/s. The minor diversion from this dominant direction may be due to the presence of local boundaries and mountainous regions since the prevailing wind direction for South Pacific region is South-East.

An integral part of this study was to use the probability distribution functions (PDFs), namely; Weibull, lognormal and gamma in this case to understand the wind distribution probabilities of each site. Cumulative distribution functions (CDFs) were also plotted. The CDFs give the same information as the PDFs but are depicted in a slightly different manner. Figure 7.6a, b shows the PDFs and CDFs of Yaqara and Rakiraki, respectively. For Yaqara, the maximum probability of the wind speed being in the range of 3–5 m/s is close to 0.19, while for Rakiraki, it is nearly 0.12 for wind speed in the range of 6–8 m/s. The purpose for choosing these three PDFs is because of their common use, with Weibull distribution being the most used for wind statistical studies. The need for PDFs in wind energy feasibility study simply lies in the fact that it has several parameters. The PDFs used in this study, each have two parameters. These parameters then can be used in useful mathematical equations and models to find important information in regards to wind energy as seen in the Eqs. (7.1, 7.2, 7.3, 7.4, and 7.5). Although the parameters found using MLE method are for 10 m height, the Eqs. (7.1, 7.2, 7.3, and 7.4) can determine the parameters at any desirable height. The main aim of this research was to choose the PDF that is having the best fit with the observed wind speed of the sites and using the AIC and BIC criteria as summarized in Table 7.5 shows that Weibull is the best PDF for this study.

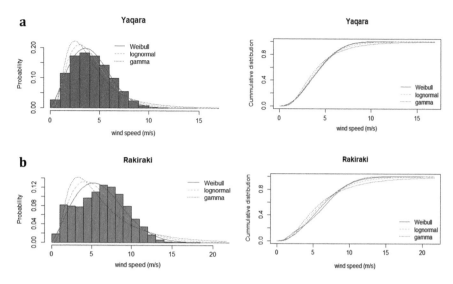

Fig. 7.6 Probability distribution functions (PDFs) and cumulative distribution functions (CDFs) of (**a**) Yaqara and (**b**) Rakiraki

The shape (k) and scale parameters (c) at desired heights of 55 m for Vergnet 275 and 32.5 m for Vestas V27 models were determined using the Eqs. (7.2 and 7.3) and then the capacity factor (CF) for the selected turbines for the two chosen sites were modeled using Eq. (7.5). Capacity factor is crucial when it comes to renewable energy technologies, especially the ones affected by intermittency issues, such as wind and solar. It is revealed in Fig. 7.7a that the CF of Vergnet 275 and Vestas V27 in Yaqara and Rakiraki are (0.101; 0.105) and (0.267, 0.304), respectively. The CF of both the turbine models for Yaqara is less than the CF of the same turbines selected for Rakiraki. This is simply because Rakiraki has more wind resource to be utilized per year. Another unique observation made here is that the CF for Vergnet model is lower than Vestas, despite Vergnet having a higher installed capacity of 275 kW/turbine. This can be explained by the given rated velocity (V_r) of the two turbines. Vergnet has a V_r of 16 m/s, while Vestas has a V_r of 13.5 m/s as shown in Table 7.6. This means that the mean wind speed of both the sites is closer to 13.5 m/s than a much higher speed of 16 m/s; thus, making it easier for Vestas model to produce close to its rated power easily than Vergnet.

Figure 7.7b, c shows the power (kW) and electrical energy (kWh) produced by the turbines, respectively. Both the graphs are considering output power and energy per turbine and are used for basic comparison purpose. The figures show that the power output by Vergnet and Vestas turbines in Yaqara and Rakiraki are 27.88; 23.53 kW and 73.41; 68.44 kW, respectively. Given the rated power of Vergnet being 275 kW and Vestas being 225 kW, the output power achievable by the turbines for the sites are quite low, which is often the case for wind technology due to intermittency constraints. For energy output, Vergnet and Vestas turbines in Yaqara and Rakiraki have potential of producing (232,057.6; 195,841.42 kWh and 610,896.38;

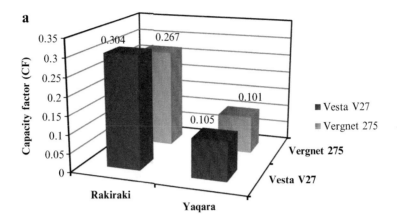

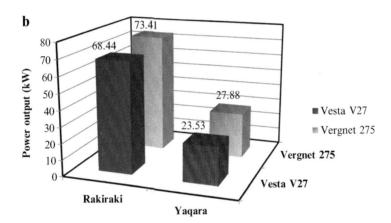

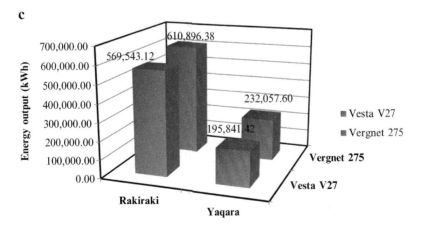

Fig. 7.7 Comparison of the sites; Yaqara and Rakiraki using two selected turbine models; Vestas V27 and Vergnet 275 in terms of (**a**) capacity factor, (**b**) power output (kW) and (**c**) energy output (kWh). Comparisons are for a single turbine at each location

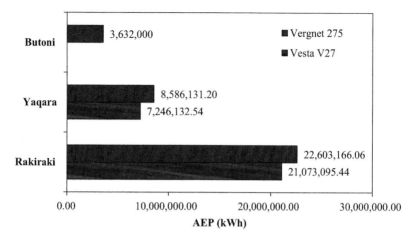

Fig. 7.8 Comparison of Annual Energy Production (AEP) between Butoni, Yaqara and Rakiraki, considering 37 turbines

569,543.12 kWh), respectively. Thus, from the analysis the preferred turbine in terms of power and energy output is Vergnet and the feasible site out of the two would most likely be Rakiraki.

The above figures (Fig. 7.7a–c) are compared in terms of a single turbine, but a more realistic scenario for considering wind farms would be comparison that entails more than one wind turbine. This section has considered a total of 37 turbines for comparison purpose. The reason for using 37 turbines is to make a valid comparison of the selected sites with Butoni wind farm. Butoni wind farm that initially started with 37 turbines is considered a virtual failure and investors are often de-motivated to make any similar investments. Figure 7.8 shows an annual energy production (kWh) comparison of the sites Yaqara and Rakiraki with Butoni. Since Butoni has all Vergnet turbines, comparison relating Vestas model was unfortunately limited. Therefore, it is rather clear that Rakiraki is the more viable site for commissioning a future wind farm that is more likely to succeed.

Also, Vergnet turbine would be more suited for Rakiraki as it yields more energy than the Vestas model and more cost is saved by not having to import diesel fuel as seen in Fig. 7.9.

To evaluate the feasibility of a wind energy application at a specific site, there can be two additional criteria that can be applied in order to choose the more viable site and proper turbine model. These include, the minimum cost of unit electricity (COE) produced (FJD/kWh) and the payback period, which can be computed using the PVC values summarized in Table 7.7. Over time the COE of wind technology has been going down and can be considered a cost-effective technology in the future. On the other hand, payback period needs to be considered as it is very important for an investor to recover and reap the profit before the turbine's lifetime, which often spans around 20 years. Figure 7.10a, b presents an interesting result that shows a slightly lower COE for Vergnet for the site Yaqara (0.153 over 0.161 FJD/kWh),

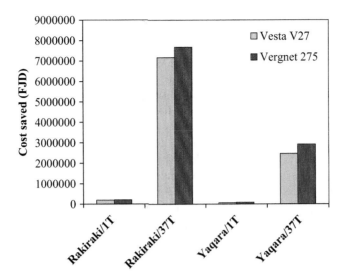

Fig. 7.9 Cost saved (FJD) by employing single wind turbine and 37 wind turbines of two models in the selected sites

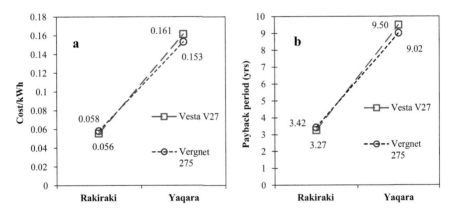

Fig. 7.10 Presents the (**a**) Cost/kWh (FJD/kWh) and (**b**) Payback period (years) for Rakiraki and Yaqara using Vestas V27 and Vergnet 275 wind turbine models

while the COE for Vergnet for the site Rakiraki is slightly higher (0.058 over 0.056 FJD/kWh). The difference in these costs; however, are very small, almost negligible. Furthermore, the same result is obtained in terms of the payback period, where Vergnet model for Yaqara has a low payback period and the same model for Rakiraki has a slightly higher payback period. The graphs clearly depict that Rakiraki is the best site over Yaqara. For turbine model selection, both the models have their pros and cons and it can be slightly difficult to debate on which model to choose from. For instance, when it comes to energy generation to meet higher demand, then

Vergnet model is preferred over Vestas due to its higher rated capacity. On the contrary, when it comes to COE and payback period (Fig. 7.10a, b) show that the more viable site (i.e. Rakiraki) entails a lower COE and payback period for the Vestas model, making it a preferable option over Vergnet.

Renewable energy sources, such as wind offers a cleaner alternative for electricity generation. This section discusses how much GHG (CO_2, CH_4 and N_2O) emissions can be saved by switching from HFO and IDO to wind energy. Here the fossil energy used in installing and maintenance work is not accounted as it is negligible. Butoni is used as a baseline (from actual energy generation data) showing how much GHG would have been released if HFO or IDO were used. It can be said that although Butoni plays a minor role in energy generation, but it needs to be credited for certain level of emission reductions. The comparison is shown in Fig. 7.11. In general, it can

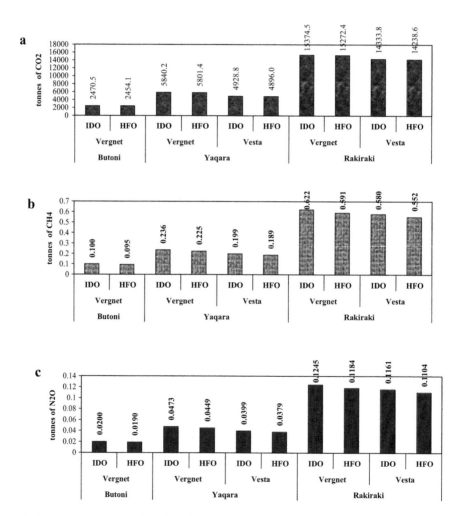

Fig. 7.11 Presents tonnes of (**a**) CO_2, (**b**) CH_4 and (**c**) N_2O emissions that would incur if IDO or HFO fuel sources are used in the given sites over wind energy

be seen that the use of IDO slightly results in higher emissions over HFO, while no emissions would be released via the use of wind technology. As for the wind technology used, Vergnet model has tendency to reduce more emissions due to its higher rated capacity over Vestas model. Moreover, wind farm set-up in Rakiraki will reduce more GHG emissions per year due to high energy generation potential of the site. It can be noted that the theoretical wind farm in Rakiraki using Vergnet model results in the most emissions saved per year, with values of 15,374.5 tCO_2, 0.62 tCH_4 and 0.125 tN_2O (when replaced with IDO) and 15,272.4 tCO_2, 0.59 tCH_4, 0.118 tN_2O (when replaced with HFO). When using the Vergnet wind turbine, the total emissions saved from all three sites are: 23,685.2 tCO_2, 0.958 tCH_4, 0.19 tN_2O when replaced with IDO and the total emissions saved over HFO are: 23,527.9 tCO_2, 0.911 tCH_4, 0.18 tN_2O.

Fiji's NDC Roadmap has its own ambitious targets to reduce 30% of the nation's GHG emissions by 2030. The milestones are divided into short, medium and long-term actions. For short-term target, the average annual GHG planned to be mitigated is 42,000 tCO_2 in total, with 21,000 tCO_2 from solar and 21,000 tCO_2 from biomass. For medium-term target, the average yearly GHG planned to be mitigated is 158,000 tCO_2 in total, with 17,000 tCO_2 from solar PV, 86,000 tCO_2 from hydro and 55,000 tCO_2 from biomass. Lastly, the long-term target plans to mitigate 227,000 tCO_2 in total, with 34,000 tCO_2 from solar PV, 137,000 tCO_2 from biomass/biogas and 55,000 tCO_2 via grid extension. Thus, NDC plan of Fiji has disregarded wind energy not knowing that it has a potential of mitigating over 23,000 tCO_2 per year that is 56.4% under the short-term plan, 14.9% under the medium-term plan and 10.4% under the long-term target.

7.5   Conclusions

Wind resource assessments of two selected sites in Fiji, namely Yaqara and Rakiraki are presented. The overall mean wind speed for Yaqara and Rakiraki were found to be 4.107 m/s and 6.19 m/s, respectively. The wind speed attained for the two sites pre-dominantly falls in the low-medium wind speed regime. Also, the dominant wind direction for both the sites was South-East, corresponding to the trade winds.

Out of the three probability density functions used, Weibull had the best fit with observed data and therefore, its parameters were used to model the capacity factors of the two different turbine models used in this study. It was learnt that the CF values of both Vergnet and Vestas models for Yaqara was less than the CF of the same turbine models selected for Rakiraki. Hence, the output power by Vergnet and Vestas turbines in Yaqara and Rakiraki were 27.88; 23.53 kW and 73.41; 68.44 kW, respectively. Additionally, Vergnet and Vestas turbines in Yaqara and Rakiraki had the potential of producing 232,057.6; 195,841.42 kWh and 610,896.38; 569,543.12 kWh output energy, respectively.

For further feasibility evaluation of the sites, two additional criteria, the minimum cost of unit electricity (COE) produced (FJD/kWh) and the payback period were also applied. Overall, both the COE and payback period values for Rakiraki were lower than that of Yaqara. However, in terms of the turbines, the COE and payback period of the Vergnet model for Rakiraki resulted in slightly higher values over the Vestas V27 model. Thus, the techno-economic analysis showed that distinct wind turbine models would extract different amounts of energy from the wind at different sites, which can significantly influence the final COE produced and the payback time to reap the profits. This study provides an important example of a pre-feasibility study that can be applied to other suitable sites of Fiji as well. Despite, the lack of performance of the first wind farm in Fiji, wind energy must still be given a high priority for development and utilization for power generation in this country.

In summary, it is clear that Rakiraki is the more viable site for commissioning a future wind farm over Yaqara. Also, Vergnet turbine would be more suited for Rakiraki in terms of energy generation to meet higher demand of the Fijians and cost saved. Rakiraki is also theoretically the better site in terms of emissions saved per year. The use of Vergnet turbine can result in emission savings of: 15,374.5 tCO_2, 0.62 tCH_4 and 0.125 tN_2O (when replaced with IDO) and 15,272.4 tCO_2, 0.59 tCH_4, 0.118 tN_2O (when replaced with HFO).

Acknowledgements The authors LJ and RP are thankful to the Fiji Meteorological Office first and foremost for providing the data that made the calculations possible.

References

Abed, K., & El-Mallah, A. (1997). Capacity factor of wind turbines. *Energy, 22*(5), 487–491. https://doi.org/10.1016/s0360-5442(96)00146-6.

Adaramola, M., Agelin-Chaab, M., & Paul, S. (2014). Assessment of wind power generation along the coast of Ghana. *Energy Conversion and Management, 77*, 61–69. https://doi.org/10.1016/j.enconman.2013.09.005.

ADB. (2015). *ADB basic statistics 2014*. Retrieved from http://www.adb.org/publications/basic-statistics2014

Ahmed Shata, A., & Hanitsch, R. (2006). Evaluation of wind energy potential and electricity generation on the coast of Mediterranean Sea in Egypt. *Renewable Energy, 31*(8), 1183–1202. https://doi.org/10.1016/j.renene.2005.06.015.

Ahmed Shata, A., & Hanitsch, R. (2008). Electricity generation and wind potential assessment at Hurghada, Egypt. *Renewable Energy, 33*(1), 141–148. https://doi.org/10.1016/j.renene.2007.06.001.

Akaike, H. (1973). Information theory and an extension of the maximum likelihood principle. In H. N. Petrov & F. Csaki (Eds.), *Second international symposium on information theory*. Budapest: Akademiai Kaido.

Akgül, F., Şenoğlu, B., & Arslan, T. (2016). An alternative distribution to Weibull for modeling the wind speed data: Inverse Weibull distribution. *Energy Conversion and Management, 114*, 234–240. https://doi.org/10.1016/j.enconman.2016.02.026.

Akpinar, E., & Akpinar, S. (2005). An assessment on seasonal analysis of wind energy characteristics and wind turbine characteristics. *Energy Conversion and Management, 46*(11–12), 1848–1867. https://doi.org/10.1016/j.enconman.2004.08.012.

Al-Abbadi, N. (2005). Wind energy resource assessment for five locations in Saudi Arabia. *Renewable Energy, 30*(10), 1489–1499. https://doi.org/10.1016/j.renene.2004.11.013.

Alavi, O., Mohammadi, K., & Mostafaeipour, A. (2016a). Evaluating the suitability of wind speed probability distribution models: A case of study of east and southeast parts of Iran. *Energy Conversion and Management, 119*, 101–108. https://doi.org/10.1016/j.enconman.2016.04.039.

Alavi, O., Sedaghat, A., & Mostafaeipour, A. (2016b). Sensitivity analysis of different wind speed distribution models with actual and truncated wind data: A case study for Kerman, Iran. *Energy Conversion and Management, 120*, 51–61. https://doi.org/10.1016/j.enconman.2016.04.078.

Betzold, C. (2016). Fuelling the Pacific: Aid for renewable energy across Pacific Island countries. *Renewable and Sustainable Energy Reviews, 58*, 311–318. https://doi.org/10.1016/j.rser.2015.12.156.

Celik, A. (2004). A statistical analysis of wind power density based on the Weibull and Rayleigh models at the southern region of Turkey. *Renewable Energy, 29*(4), 593–604. https://doi.org/10.1016/j.renene.2003.07.002.

Conradsen, K., Nielsen, L., & Prahm, L. (1984). Review of Weibull statistics for estimation of wind speed distributions. *Journal of Climate and Applied Meteorology, 23*(8), 1173–1183. https://doi.org/10.1175/1520-0450(1984)023<1173:rowsfe>2.0.co;2.

Dayal, K. (2015). *Offshore wind resource assessment, site suitability and technology selection for Bligh Waters Fiji using WindPro*. Master of Science, Uppsala University.

EFL. (2015). *EFL annual report 2014*. Suva, Fiji. Retrieved from http://www.efl.com.fj

EFL. (2016). *EFL annual report 2015*. Suva, Fiji. Retrieved from http://www.efl.com.fj

Global Wind Energy Outlook (GWEO). (2016). Retrieved from http://www.gweo.net

Government of Fiji. (2015). *Fiji's intended nationally determined contribution*. Retrieved from: http://www4.unfccc.int/ndcregistry/PublishedDocuments/Fiji%20First/FIJI_iNDC_Final_051115.pdf

Inflation Rate – Fiji. (2018). Retrieved from https://tradingeconomics.com/fiji/inflation-cpi

IPCC. (2006). *Guidelines for national greenhouse gas inventories: Volume 2. Energy: Stationary combustion (chapter 2)*. [online] Available at: https://www.ipcc-nggip.iges.or.jp/public/2006gl/vol2.html. Accessed 07 Feb 2019.

IRENA. (2015). *Fiji renewables readiness assessment*. Retrieved from http://www.irena.org/publications/2015/Jul/Renewables-Readiness-Assessment-Fiji

Kaldellis, J., & Zafirakis, D. (2013). The influence of technical availability on the energy performance of wind farms: Overview of critical factors and development of a proxy prediction model. *Journal of Wind Engineering and Industrial Aerodynamics, 115*, 65–81. https://doi.org/10.1016/j.jweia.2012.12.016.

Kantar, Y., Usta, I., Arik, I., & Yenilmez, I. (2018). Wind speed analysis using the extended generalized Lindley distribution. *Renewable Energy, 118*, 1024–1030. https://doi.org/10.1016/j.renene.2017.09.053.

Khahro, S., Tabbassum, K., Soomro, A., Dong, L., & Liao, X. (2014). Evaluation of wind power production prospective and Weibull parameter estimation methods for Babaurband, Sindh Pakistan. *Energy Conversion and Management, 78*, 956–967. https://doi.org/10.1016/j.enconman.2013.06.062.

Khraiwish Dalabeeh, A. (2017). Techno-economic analysis of wind power generation for selected locations in Jordan. *Renewable Energy, 101*, 1369–1378. https://doi.org/10.1016/j.renene.2016.10.003.

Kumar, A., & Nair, K. (2012). Wind characteristics and energy potentials at Wainiyaku Taveuni, Fiji. *Management of Environmental Quality: An International Journal, 23*(3), 300–308. https://doi.org/10.1108/14777831211217503.

Kumar, A., & Nair, K. (2013). Wind Power Potential at Benau, Savusavu, Vanua Levu, Fiji. *International Journal of Energy, Information And Communication, 4*(1), 51–62.

Kumar, A., & Nair, K. (2014). Wind energy potential, resource assessment and economics of wind power around Qamu, Navua, Fiji. *International Journal of Wind And Renewable Energy, 3*(1), 2277–3975.

Kumar, A., & Prasad, S. (2010). Examining wind quality and wind power prospects on Fiji Islands. *Renewable Energy, 35*(2), 536–540. https://doi.org/10.1016/j.renene.2009.07.021.

Kumar, A., & Weir, T. (2008). Wind power in Fiji: A preliminary analysis of the Butoni wind farm. *International solar energy society conference*, Sydney, Australia.

Leung, D., & Yang, Y. (2012). Wind energy development and its environmental impact: A review. *Renewable and Sustainable Energy Reviews, 16*(1), 1031–1039. https://doi.org/10.1016/j.rser.2011.09.024.

Lydia, M., Kumar, S., Selvakumar, A., & Prem Kumar, G. (2014). A comprehensive review on wind turbine power curve modeling techniques. *Renewable and Sustainable Energy Reviews, 30*, 452–460. https://doi.org/10.1016/j.rser.2013.10.030.

Michalena, E., Kouloumpis, V., & Hills, J. (2018). Challenges for Pacific Small Island developing states in achieving their nationally determined contributions (NDC). *Energy Policy, 114*, 508–518. https://doi.org/10.1016/j.enpol.2017.12.022.

Mohammadi, K., Alavi, O., Mostafaeipour, A., Goudarzi, N., & Jalilvand, M. (2016). Assessing different parameters estimation methods of Weibull distribution to compute wind power density. *Energy Conversion and Management, 108*, 322–335. https://doi.org/10.1016/j.enconman.2015.11.015.

Mohammadi, K., Alavi, O., & McGowan, J. (2017). Use of Birnbaum-Saunders distribution for estimating wind speed and wind power probability distributions: A review. *Energy Conversion and Management, 143*, 109–122. https://doi.org/10.1016/j.enconman.2017.03.083.

Morgan, E., Lackner, M., Vogel, R., & Baise, L. (2011). Probability distributions for offshore wind speeds. *Energy Conversion and Management, 52*(1), 15–26. https://doi.org/10.1016/j.enconman.2010.06.015.

OECD. (2016). *Geographical distribution of financial flows to developing countries 2016: Disbursements, commitments, country Indicators.* Retrieved from http://www.oecd.org/dac/geographical-distribution-of-financial-flows-to-developing-countries-20743149.html

REN 21. (2016). *Renewables 2015 –global status report.* Retrieved from http://www.ren21.net/wp-content/uploads/2015/07/GSR2015_KeyFindings_lowres.pdf

Schallenberg-Rodriguez, J. (2013). A methodological review to estimate techno-economical wind energy production. *Renewable and Sustainable Energy Reviews, 21*, 272–287. https://doi.org/10.1016/j.rser.2012.12.032.

Schwarz, G. (1978). Estimating the dimension of a model. *The Annals of Statistics, 6*(2), 461–464. https://doi.org/10.1214/aos/1176344136.

Sharma, K., & Ahmed, M. (2016). Wind energy resource assessment for the Fiji Islands: Kadavu Island and Suva peninsula. *Renewable Energy, 89*, 168–180. https://doi.org/10.1016/j.renene.2015.12.014.

Singh, R. (2015). A pre-feasibility study of wind resources in Vadravadra, Gau Island, Fiji. *Energy And Environmental Engineering, 3*(1), 5–14.

The Global Wind Energy Council (GWEC). (2011a). *PR China.* Retrieved from http://www.gwec.net/index.php?id=125

The Global Wind Energy Council (GWEC). (2011b). *United States.* Retrieved from http://www.gwec.net/index.php?id=121

Timilsina, G., & Shah, K. (2016). Filling the gaps: Policy supports and interventions for scaling up renewable energy development in Small Island developing states. *Energy Policy, 98*, 653–662. https://doi.org/10.1016/j.enpol.2016.02.028.

Wang, J., Qin, S., Jin, S., & Wu, J. (2015). Estimation methods review and analysis of offshore extreme wind speeds and wind energy resources. *Renewable and Sustainable Energy Reviews, 42*, 26–42. https://doi.org/10.1016/j.rser.2014.09.042.

Yang, H., Wei, Z., & Chengzhi, L. (2009). Optimal design and techno-economic analysis of a hybrid solar–wind power generation system. *Applied Energy, 86*(2), 163–169. https://doi.org/10.1016/j.apenergy.2008.03.008.

Yesilbudak, M., Sagiroglu, S., & Colak, I. (2013). A new approach to very short term wind speed prediction using k-nearest neighbor classification. *Energy Conversion and Management, 69*, 77–86. https://doi.org/10.1016/j.enconman.2013.01.033.

Zieroth, G. (2006). *Feasibility of grid connected wind power for Rarotonga, Cook Islands.* SOPAC Bulletin.

Chapter 8
Solar Energy for Power Generation in Fiji: History, Barriers and Potentials

Ravita D. Prasad and Atul Raturi

Abstract In the last 5 years, there has been rapid growth in "behind the meter" solar photovoltaics (solar PV) installations for several commercial companies around the main island of Fiji, Viti Levu. In total, around 4 MW of solar PV is installed with some grid-connected solar systems planned and many off-grid solar system planned by Fiji Department of Energy with funding from Fijian government and overseas donor agencies. This chapter reviews solar PV developments in Fiji and discusses the future development plans that are documented in publically available domains. Some barriers and challenges are also discussed for slow deployment of solar PV. The theoretical potential of solar PV power generation was found to be around 170 GWh/year which would result in around 150,000 metric tonnes of carbon dioxide avoided emissions. Using Long Range Energy Alternative Planning System (LEAP), grid electricity model was constructed and a range of new renewable energy technologies were used for future electricity generation with addition of solar PV. It was found that 90 MW of new solar PV on Viti Levu's grid, 5 MW in Vanua Levu's grid and 4 MW in Ovalau's grid, would have the potential of generating 167 GWh by 2030 with around 148,000 metric tonnes of avoided carbon dioxide emissions.

Keywords LEAP · Solar energy · Emission reduction · Long-term planning

R. D. Prasad (✉)
Faculty of Science, Technology and Environment, The University of the South Pacific, Suva, Fiji

College of Engineering, Science and Technology, Fiji National University, Nasinu, Fiji
e-mail: ravita.prasad@fnu.ac.fj

A. Raturi
Faculty of Science, Technology and Environment, The University of the South Pacific, Suva, Fiji
e-mail: atul.raturi@usp.ac.fj

© Springer Nature Switzerland AG 2020   177
A. Singh (ed.), *Translating the Paris Agreement into Action in the Pacific*,
Advances in Global Change Research 68,
https://doi.org/10.1007/978-3-030-30211-5_8

8.1 Introduction

Fiji has good solar insolation. Using 1983–2005 NASA data (NASA 2017), average annual insolation on a horizontal surface in Fiji is 5.4 kWh/m^2/day with a standard deviation of 0.6 kWh/m^2/day (see Fig. 8.1). During the mid-year, solar insolation reaches the lowest point of 4.0 kWh/m^2/day while high solar insolation (around 6 kWh/m^2/day) occurs from October to February.

Solar photovoltaic (solar PV) systems are gaining popularity globally and likewise for Fiji. Globally, the price of solar PV has dramatically decreased over the last decade, resulting in an increase in new solar PV installation for electricity generation. Fiji's solar PV generation on grid was nil before 2010. However, from 2012, there has been an increase of solar PV installations on the roof-tops of several commercial companies leading to around 3.6 MW of installed solar PV. Some reasons for this increase are given below.

Firstly, the unit capital cost of solar PV decreased to around FJD3,100–3500/kW. Secondly, one major company (Sunergise) that began their operations in 2012 in Fiji has an attractive business model that encourages businesses to have roof-top solar PV grid connected systems. How does this model work? Sunergise provides and funds everything required in a solar project; from the design of the system, the cost of solar PV panels to installation, monitoring, maintenance and insurance of the system (Berno 2017). It does not charge its customers any upfront capital cost. The customer of Sunergise then signs a Power Purchase Agreement (PPA) for 10–20 years and pays monthly for the electricity generated by solar PV system at a rate which is less than what the user will pay if they buy electricity from the grid utility company (Berno 2017). This company has also done off-grid

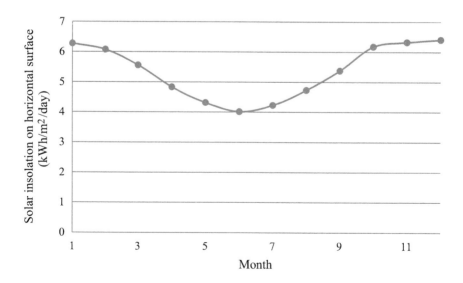

Fig. 8.1 Fiji's average solar insolation for horizontal surface. (Data Source: NASA 2017)

installations for island resorts and some villages. Thirdly, two other major installers (Clay Energy and Complete Battery Solutions (CBS)), have gained popularity amongst government departments and energy stakeholders in installing solar PV thus providing confidence to new investors to uptake solar PV projects. Lastly, the availability of donor agency funds such as Leonardo Di Caprio funds for solar and Masdar funds for solar villages have resulted in some villages having around 100–200 kW of solar PV mini-grid systems installed.

The main objective of this chapter is to estimate the avoided carbon dioxide emissions when solar PV is fully utilised or utilised on a large scale. The next section reviews the developments made in solar PV in Fiji. Section 8.3 presents development plans documented for solar PV in publically available reports while Sect. 8.4 discusses the relation of solar PV with sustainable development. Section 8.5 discusses barriers and challenges to the rapid deployment of solar PV. Section 8.6 assesses the electricity generation potential using solar PV in Fiji followed by an estimation of avoided greenhouse gas (GHG) emissions. Section 8.7 discusses strategies to increase renewable energy share in electricity generation. Section 8.8 concludes the chapter.

8.2 Review of Solar PV Development in Fiji

8.2.1 Solar Home System (SHS)

Solar PV has been in use in Fiji for almost three decades. One of the first use of solar PV was in solar home system (SHS) that provided electricity to power basic appliances in rural households where grid electricity was not reachable. Currently, there are two types of SHS installed in Fijian homes. Type I SHS has two 50 W solar panels, a 100 Ah battery, DC lights and charge controller (one by 10 A) while Type II SHS has two 135 W solar panels, a 200 Ah battery, DC LED lights, charge controller (one 20 A) and a 300 W inverter (Nand and Raturi 2015). According to Fiji Department of Energy (FDoE) statistics, there were 5808 SHS by end of 2014 (Valemei 2014) (see Fig. 8.2). Out of this, 60% is Type I SHS, which were installed in 2013 and before. From 2014, Type II SHS is being installed. According to (Nand and Raturi 2015) by 2015–2016, there will be a total of 9000 SHS in Fiji. FDoE is currently working to replace Type 1 SHS with Type II SHS. This change is being done since Type I SHS had problems with those households where more electrical appliances were connected to the system than recommended. Type II SHS is a bigger system that can cater for relatively more electrical appliances including AC appliances.

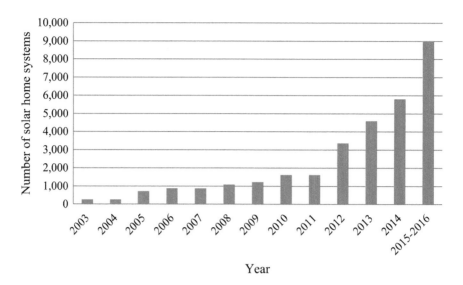

Fig. 8.2 Number of solar home systems installed in Fiji annually since 2003. (Data Source: Nand and Raturi 2015; Valemei 2014)

8.2.2   Solar PV in Off-Grid Island Resorts

There are a number of island resorts in Fiji, which have over the past decade installed solar PV systems with battery storage for supplying electricity with diesel genera- tors as back-up. The largest system to date is Six Senses Fiji Resort on Malolo Islands in the Mamanuca Group that has a 1 MW solar PV system with 4 MWh of Lithium ion battery storage system (SEANZ 2017). Other resorts include Turtle Island resort with 240 kW solar PV system with 120 kVA of diesel generator as back-up (Syngellakis et al. 2016), Tokiriki Island Resort with 198 kW of solar PV system with a 162 kW of standalone battery inverter and 1000 MWh of battery bank and Coral View Island Resort with 50 kW solar PV.

8.2.3   Solar PV Mini-grids for Remote Islands

A mini-grid comprises of solar PV modules with inverter plus battery storage and diesel generators as back-up (Fig. 8.3). In addition to SHS for households, the use of such solar PV mini-grids with diesel generator back-ups is gaining prominence. One of the recent projects is the Masdar project that helped establish solar PV/diesel generator mini-grids on three different islands; Rotuma, Lakeba and Kadavu. These mini-grids do not have any battery storage and have an array of ground mounted solar PV(Table 8.1 and Fig. 8.4). These systems are a bit different from the typical village mini-grid shown in Fig. 8.3, with the main difference being a central ground based solar PV. In addition, there are no wind turbines installed or battery storage.

Fig. 8.3 An example of village solar/wind/diesel hybrid system. 1. PV generator, 2. PV inverter, 3. Battery inverter, 4. Storage battery, 5. Diesel generator, 6. Wind turbine. (Source: SMA 2010)

Table 8.1 Mini-grid specifications for Lakeba, Rotuma and Kadavu islands in Fiji. (Source: Caruso 2016)

	Lakeba	Kadavu	Rotuma
Project PV capacity (kW)	150	225	150
Diesel capacity (kW)	138	184	138
Module type	494 × JA-310 Poly	726 × JA-310 Poly	494 × JA-310 Poly
Inverter	7 × Sunny Tripower 20000TL	10 × Sunny Tripower 20000TL	7 × Sunny Tripower 20000TL
Racking	Schletter – ground mount	Schletter – ground mount	Schletter – ground mount
Fuel save controller	SMA fuel save controller 2.0	SMA fuel save controller 2.0	SMA fuel save controller 2.0
Diesel type	3 × Hatz low load diesel twin pack	4 × Hatz low load diesel twin pack	3 × Hatz low load diesel twin pack
Diesel controller	6 × Woodward easYgen – 3200	8 × Woodward easYgen - – 3200	6 × Woodward easYgen – 3200

The overall electrical energy produced from these mini-grid is projected to reduce Fiji's annual diesel consumption by 0.259 million litres resulting in a reduction of 722 tonnes of GHG emissions (Masdar 2015). Forty percent of the daily demand is met by the solar mini-grid on each island (Engerati 2015). The annual energy output from Kadavu solar-diesel mini-grid system is 319 MWh, while it is 217 MWh/year each for Lakeba and Rotuma. These systems were installed by Sunergise with financial support from the Abu Dhabi Development Fund (ADFD).

Another mini-grid system commissioned in July 2016 is at Nasoki village in Moala Island. It is a solar hybrid system that supplies electricity to 53 households, 12 streetlights, a church, a school and a community hall (Sunergise and ClayEnergy 2017).

Fig. 8.4 Ariel view of solar PV array installation on Kadavu island. (Source: Caruso 2016)

8.2.4 Solar PV for Nursing Stations and Telecommunications

Solar PV also supplies electricity to nursing stations that are in remote areas not connected to national grid. There are a total of approximately 13 kW of solar PV installed at multiple remote nursing stations (CBS 2018).

For telecommunications to cover nearly all locations in Fiji, telecommunication repeater towers have to be built in remote locations where grid power access is not possible. Clay Energy have installed dedicated solar PV systems at 31 different locations including outer islands (Sunergise and ClayEnergy 2017).

8.2.5 Solar PV Grid Connected System

A total of 3.6 MW of grid connected solar PV is installed on Viti Levu (in 2018) (see Table 8.2). All these systems have been installed by Clay Energy and Sunergise in the last 6 years and are mainly roof-top installations. According to the annual reports of Energy Fiji Limited (EFL), there has been some solar electricity generated from 1998 to 2007 by solar PV system that was commissioned in November 1997 (FEA 2016). In 1998, this system generated around 12 MWh of electricity and was doing well for almost 6 years. From 2005 to 2007, solar generation significantly decreased and after 2007 there were no on-grid solar PV generation recorded by EFL annual reports.

Table 8.2 Grid connected solar PV installations around Fiji

Company	Type	Installed capacity (kW)	Annual energy production (MWh)
Western Viti Levu			
Terraces Apartment Resort Denarau	Roof top	72.5	120
Radisson Blu Resort Denarau	Roof top	420	600
Rooster Poultry	Roof top	515	745
RB Patel Supermarket Jet Point	Roof top	125	139
RB Patel Supermarket West Point	Roof top	46	73.5
Port Denarau Marina	Roof top	68	
Port Denarau Marina Retail complex	Roof top	250	376
Drasa Sai School	Roof Top	5	7
Central Viti Levu			
USP Lower campus	Ground based	45	56
USP Renewable energy training Center		6	
RB Patel Center Point Supermarket	Roof top	135	167
MarkOne Apparel	Roof top	275	350
Coca-Cola factory, Kinoya	Roof top	1100	1383
Shreedhar motors Ford Dealership	Roof top	69	88
Shreedhar motors Subaru Showroom	Roof top	55	
International secondary school Suva	Roof top	88	112
Performance Floatation Development	Roof top	147	188
ANZ Pacific Operations Service Center	Roof top	140	180

Data Source: PRDR (2016), Sunergise 2017), Sunergise and ClayEnergy (2017), and SunnyPortal (2018)

8.2.6   Solar PV for Electric Vehicle Charge

The University of the South Pacific, Laucala Campus in Suva has an electric vehicle charging station that is powered by solar PV. This is the first electric vehicle charging station in the country and is currently working as a demonstration and research station (Datt et al. 2015). Performance of this charging station is being studied to build more of similar and improved versions as Fiji plans to move towards electric mobility.

8.2.7   Solar PV for Streetlights and Jetty Lights

Fiji Roads Authority (FRA) has installed 25 solar streetlights in Nailaga Village in Ba (FRA 2018). This is the first solar streetlight project undertaken by FRA. In addition, Fiji Department of Energy funded the installation of solar lights on jetties at several remote locations (Valemei 2014).

8.3 Development Plans

EFL has solar PV development plans for the future. According to its 2016 annual report, plans are underway for the development of a 5 MW solar PV farm in Western Viti Levu (FEA 2016). This solar PV system will have 1–2 h of battery back-up as well (FEA 2016).

For rural electrification, FDoE with aid from GIZ, plans to install solar PV hybrid systems in locations where national grid electricity cannot reach. This hybrid system will be composed of 20 kW solar PV, 50 kVA diesel generator and a battery storage (FDoE 2018). This mini-grid system will be able to provide electricity to 60–70 households and will be replacing the diesel generator only systems that are currently supplying electricity to villages in remote areas.

There are other donor funded projects for installation of solar PV systems in rural areas. One such example is the Leonardo DiCaprio Foundation that is assisting in the building of a solar PV/battery hybrid system on Vio island, off the coast of Lautoka (GoF 2018). This is a mini-grid system where 46 households will be connected. This will have a business model similar to SUNACCESS that Sunergise uses for providing rural electrification in remote locations (Sunergise and ClayEnergy 2017).

8.4 Solar PV and Sustainable Development

Solar PV technology is one of the most promising technologies in achieving sustainable development for the developing countries. The use of solar home systems in rural areas has enabled Fiji to achieve 96% of electricity access to the total number of households as one studies the preliminary data from 2017 household census survey (FBoS 2018). Electricity in urban areas is considered part of everyday life and not much importance is given to where it comes from or whether it would be available tomorrow or not. However, the picture in remote islands or remote rural areas is different. Some remote villages are not provided with electricity "24/7" but have electricity from diesel generators for a few hours only in the evening. When diesel supply runs out, villages have to procure diesel from the main islands which is transported via shipping vessels.

At times, when transport is unavailable for transporting fuel from main islands to remote sites, homes are left without power for almost 2–3 weeks. Solar home systems do not have this issue. In addition, electricity is provided for extended hours in the evening which helps students in their studies and also assists women in their household chores. SHS also improves communication and entertainment through the use of TV, computers, mobile charging and radios. Hence, remote villagers prefer solar home systems to be installed at their premises.

Solar lights at jetties aid mariners for their productive activity during night-time. Solar PV electricity generation at nursing stations in remote islands aids in refrigeration of medical supplies. Apart from these, there are some projects done by

University of the South Pacific students under the guidance of Dr. Raturi where solar PV electricity with diesel as back-up is providing electricity to power refrigerators in villagers (IRENA 2017). These refrigerators are used to store the fish which is then sold to the market which increases the income of people in remote areas. Hence, solar PV is a means for increasing the economic activity.

8.5   Barriers and Challenges

Solar PV has many advantages such as it has no moving parts and therefore does not require extensive operation and maintenance; solar resource is free and abundant at most locations in Fiji.

For Fiji, the current installation cost of rooftop solar PV grid connected system is around 3100–3500 FJD/kW. Six years ago the cost was around 13,000 FJD/kW (Stolz 2012). Hence, cost of solar PV can no longer be considered as a barrier for solar PV development for electricity generation. Even though the cost has decreased, there is no solar PV in EFL's generator portfolio at present. Some possible challenges and barriers are discussed below for uptake of grid-connected solar PV in Fiji.

8.5.1   Institutional

In the past, the Fiji Electricity Authority (FEA) was acting as the regulator in addition to being the sole power utility in Fiji. However, with the new Electricity Act 2017 this has now changed. FEA is now corporatized to EFL where its regulatory role will be given to an independent body. However, currently (2018) the details are still being finalized.

8.5.2   Natural Disasters

Nasau in Koro island had a 6 kW system where 60 households, a school and teacher's quarters were connected with electricity. This system was operational only for 1 month after which Category 5 Tropical Cyclone Winston completely destroyed most of the components of this system (Sunergise and ClayEnergy, 2017). However, there are other systems on mainland that have withstood tropical cyclones wrath over the past years and no major damage have been reported. Hence, the civil works of a solar system needs to be well within standards to withstand high winds during cyclone season in Fiji; November to April.

8.5.3 Technical

Several studies done on the performance of grid with solar PV penetration reports that 20–30% of the grid demand can be catered by solar PV without any grid stability issues. However, for greater penetration of variable power source such as solar and wind, grid storage technologies need to be considered as well as placement of solar PV generators around the country. According to Katz et al. (2015) distributing solar PV generators will reduce the variability of solar PV output significantly. With the addition of large solar PV on grid in future years can lead to "duck curve" as observed by California grid system (Lazar 2016). Hence, grid operators need to consider the over generation during mid-day and then in the evenings when the solar generation goes down there is a ramp up (additional generation) needed from conventional power stations to meet the demand in the evenings. If planned properly the excess electricity generation from solar PV can be stored in some form of grid storage system for example, battery storage and pumped hydro storage. There is a need for detailed techno-economic study on storage options for Fijian central grid.

However, for mini grid system in smaller remote islands, solar PV with diesel and battery storage would make the system reliable.

8.5.4 Absence of Attractive Tariff for Solar PV Electricity Suppliers to Grid

Currently, around 15 cents Fijian/kWh is paid to producers who sell solar PV generated electricity to grid. However, this rate is low in comparison to rates that other IPPs are paid. This is a limiting factor to potential IPPs to use solar PV in selling electricity to the grid. Net metering can also be considered for domestic systems.

8.5.5 Lack of Streamline Processes for Starting Up a New Business

There are different government departments that are involved in getting license to start a new business and they have their own requirement and application processing times. For example, one renewable energy company shared its experience when starting its business in Fiji in (Berno 2017) and stated that it took 8 months to register the company. But Berno (2017) also reports that now it takes much less time to register a new company.

Overall, Fiji is learning from its experiences in sustainable energy projects; technical and institutional, in addition to bringing in consultants to develop framework to increase IPP participation in power sector and make recommendations on improving their current way of doing business (ITP 2014, 2015).

8.5.6 Financing Opportunities for New Sustainable Energy Companies

With Fiji's NDCs of reducing emissions by 30% and making the electricity sector 100% renewable, huge investments of around USD1.7 billion will be needed (GoF and GGGI 2017). In order to start any new business, initial capital is required. However, if a company starts from scratch, then it would need loans or subsidies to start the business.

Currently, concessional loans are offered by Import Substitution and Export Finance Facility for Reserve Bank of Fiji facilitated by Fiji Development Bank for new investments in renewable energy, sustainable public transport, export and import substitution. In addition, government of Fiji is giving 5-year tax holiday to any taxpayer who invests in sustainable energy projects. Further to this, zero duty applies to importation of sustainable energy goods. Recently, Fiji's 2019 budget announced 250% tax deduction is allowable on any expenses incurred by companies investing in research and development in renewable energy and ICT under the Incentive Package for Research and Development.

However, these may not be enough for new businesses who are starting from scratch. They need capital to start up a sustainable energy project. Hence, aid from global and regional development banks would benefit new businesses.

8.6 An Assessment of Solar Electricity Generation Potential for Fiji and Its Potential for Avoided Emission

8.6.1 Method 1: Theoretical Solar Electricity Generation Potential and Its Avoided Emissions

The electric power generation from a solar panel (E_S) can be calculated from panel area and solar irradiance on the panel over time, Eq. (8.1) given by Noguchi et al. (2013).

$$E_S = \int \left[H_S(t) \cdot K_{1S} \cdot K_{2S} \cdot \eta_S \cdot A_S \right] dt \tag{8.1}$$

Where Table 8.3 gives the parameters for estimating solar electrical energy potential.

According to FBoS (2010), there are 721 primary schools, 172 secondary schools, 4 teacher training schools, 69 vocational training schools, 17 special schools and 3 Universities in Fiji. For solar energy potential calculation from schools, 40% of the total number of schools (that is 400 schools) with Fiji's average solar insolation was used for calculation. Data for number of small offices were obtained from EFL's list of commercial customers at each location. From this, 30% of the commercial customers were taken to estimate solar generation potential by commercial sector. A total 31 factories, which are high demand consumers of EFL (counted as industrial customers by EFL listing), were considered for estimating solar generation potential from Suva area.

Of all the hotels that are grid connected, 20 are listed as industrial customers for EFL. These 20 hotels as well as 30 island resorts that are off-grid (have their own generation system mainly diesel) were taken for calculation of solar potential using Nadi's average monthly solar insolation. Table 8.4 gives the annual solar electrical energy output from different locations calculated using Eq. 8.1. Total annual energy output potential from solar PV is estimated to be 170 GWh. These estimations are based on solar installation on existing structures roof-tops that are assumed to be in good condition and are able to cater for the area of panels as given in Table 8.3.

The total solar energy output can be increased if ground installation are done in areas that receive high solar insolation and there are no land tenure issues. Fiji's land area is 18,333 km^2 out of which Viti Levu and Vanua Levu (two main islands of Fiji) covers 87%. As seen from roof-top solar PV applications, around 0.6 km^2 of total roof-area is required with total installed capacity of 100 MW, Table 8.4. In addition, WBG (2016) shows that Fiji's solar power potential ranges from 1022 to 1667 kWh/kW$_p$/year depending on the location,(see Fig. 8.5). Hence, considering the large land area in Viti Levu and Vanua Levu, land based solar installations can be done near locations with demand depending on the solar resource and land availability for installations.

Table 8.3 Parameters considered for solar PV generation potential

Parameter	Symbol	Unit	Other information
Electric power generation	E_S	kWh	
Surface area of panel	A_S	m^2	House: 20 m^2 for 3 kW using 190 W BP4190T solar panel size (1.587 m by 0.79 m)
			Office: 33 m^2 for 5 kW
			School: 66 m^2 for 10 kW
			Factory: 263 m^2 for 40 kW
			Hotel: 460 m^2 for 70 kW
Conversion efficiency	η_S		0.18
Temp. correction coefficient	K_{1S}		0.9
Total correction coefficient	K_{2S}		0.8
Geographical condition			These solar installations are assumed to be on roof-tops of existing structures

Table 8.4 Solar PV generation potential for different regions in Fiji for households and commercial customer

Town	30% of urban households	Annual energy output (GWh)	30% of EFL commercial customer	Annual energy output (GWh)	Annual energy output from 31 factories (GWh)	Annual energy output from 50 hotels (GWh)	Annual energy output from 400 schools (GWh)
Suva	5141	25	1352	11			
Lautoka	3133	17	506	4			
Ba	1112	6	202	2			
Labasa	1677	8	271	2			
Levuka	264	1	27	0			
Nadi	2537	14	571	5			
Savusavu	422	2	92	1			
Sigatoka	577	3	182	2			
Nausori	2856	14	231	2			
Lami	1232	6	69	1			
Tavua	143	1	69	1			
Nasinu	5247	26	265	2			
TOTAL	24,341	123	3834	32	2	6.1	6.7
Area of roof-top (km²)	0.5		0.1		0.008	0.02	0.03
Total installed capacity (MW)	73		19		1	4	4

8.6.1.1   Potential Avoided Emissions

One hundred and seventy GWh of annual electricity is estimated to be generated from solar PV roof-top applications with 100 MW of installations. In addition, there are more theoretical potential of ground based solar PV electricity generation but this has not been estimated in this work as land area for each location has to be considered with other factors as well. Avoided emissions was calculated using IPCC default emission factors for diesel generators used in stationary combustion, Table 8.5.

The main logic behind avoided emissions calculation was that, if solar PV is not generating, then a diesel generator would generate this electricity. The efficiency of a diesel generator, η_D, was taken as 30%.

Hence, for the 170 GWh of solar generation annually, avoided emission is calculated using Eqs. (8.2) and (8.3).

Fig. 8.5 Photovoltaic power potential in Fiji. (Source: WBG 2016

Table 8.5 Default emission factors for diesel generator. (Source: IPCC 2006)

Emission source	Emission factor kg/ TJ of energy consumed
Carbon dioxide (CO_2)	74,100
Methane (CH_4)	3
Nitrous oxide (N_2O)	0.6

$$\text{Energy}_{\text{consumed by diesel}} = \text{Annual solar energy output } (MWh) \times \frac{1GJ}{0.278MWh} \times \frac{1}{\eta_D} \quad (8.2)$$

$$\text{Avoided emissions} = \text{Emission factor} \times \text{Energy}_{\text{consumed by diesel generator}} \quad (8.3)$$

Using the energy consumed by diesel generator and emission factors, the avoided CO_2 emission is approximately 151,000 metric tonnes/year while the CH_4 and N_2O emissions are 6 and 1 metric tonne respectively per year. Hence, methane and nitrous oxide emissions are negligible in comparison with carbon dioxide emissions.

8.6.2 Method 2: Solar Electricity Generation Potential with Other Renewable Energy Generation Using LEAP Tool

This is a dynamic method where considering the current demand and forecasting this demand in future based on GDP and population average annual growth rate (AAGR), existing and future generation technologies are dispatched to meet the growing demand considering the hourly system load curve and maintaining a reserve margin of 40%.

Assessment of electricity generation potential uses Long Range Energy Alternative Planning System (LEAP) to develop a model for grid connected electricity system. The planning period is from 2016 to 2030. LEAP is an energy accounting tool that is widely used by around 150 countries for carrying out energy-environment analyses that is useful for their emission inventory or making policy decisions. However, one limitation of LEAP is that it does not have a module to simulate grid storage systems. Hence, for this work grid storage is not considered.

8.6.2.1 Fiji's Grid Electricity Model

At present, Energy Fiji Limited (EFL) is responsible for providing grid electricity generation to four different islands (Viti Levu, Vanua Levu, Ovalau and Taveuni) where each one of them have their own grid network and power generation stations. The demand, transmission and distribution and generation modules for each of these islands were made in LEAP. Data used in LEAP modelling was primarily sourced from EFL annual reports, Fiji Bureau of Statistics and other reports. The data requirement for each module and output are shown in Fig. 8.6. Future electricity demand increase is forecasted using linear regression model for domestic and non-domestic demand where domestic demand just depends on population growth rate which is assumed to be 0.6%/annum and non-domestic demand depends on population growth rate as well real GDP growth rate that is taken as 3% per annum.

8.6.2.2 New Solar PV with New Hydro, Biomass, Wind and Geothermal Electricity Generation Technologies

In order to obtain zero emissions from grid electricity sector and to have a diverse energy supply options, solar PV, hydropower, biomass power plant, wind farms and geothermal power plants are considered in this work. New capacities are added at each location as shown in the timeline in Fig. 8.7. There are lots of generation technologies added on to Viti Levu grids as this is the largest grid network in Fiji with 93% of the total grid electricity demand from this island. Altogether, 108 MW of new biomass power plant, 80 MW of new hydro, 90 MW of new solar PV, 40 MW of new wind and 2 MW of new geothermal was added to the Viti Levu grid. For

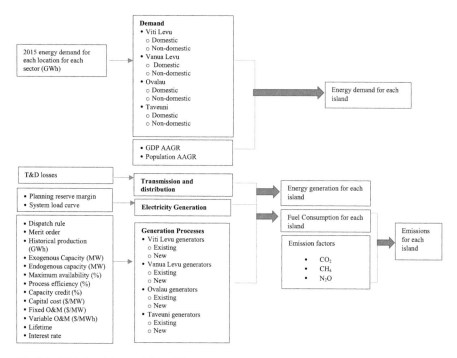

Fig. 8.6 Grid electricity model for Fiji

Vanua Levu, 5 MW of new hydro, 5 MW of new biomass and 5 MW of new solar PV was added. Ovalau Island does not have any published reports on hydro potential or geothermal potential. Hence, 4 MW of new solar PV, 2 MW of new biomass power plants and 1 MW of new wind farm was added. Taveuni island that has recently come under EFL operations has 0.7 MW of new hydro commissioned in 2017 with 2 MW of diesel generators. This capacity is currently enough to cater for future demand increase as the current peak demand for the islands is 0.25 MW (ParlimentFiji 2018).

With around 8 m² of area needed for 1 kW of solar PV installation, an area of 0.72 km² would be required to install in Viti Levu. These installations can be rooftop or ground mounted with grid customers acting as prosumers or installations can be a utility scale where EFL or IPPs are responsible for generation and transmitting to the grid.

To cater for future demand increase and with 10% transmission and distribution losses, Fig. 8.8 shows the annual generation till 2030. Annual generation in 2015 was 934 GWh, which increases to 1500 GWh to meet the growing demand. There is an average 3.3% annual growth rate in generation. In comparison with historical grid electricity generation, which is dominated by hydro and diesel/residual fuel oil generation (Fig. 8.9), future grid electricity generation is quite diverse. By 2030, 11% of the total generation is from solar PV, 8% of the total generation is from wind, 46% from hydro, 33% from biomass and 0.4% from geothermal.

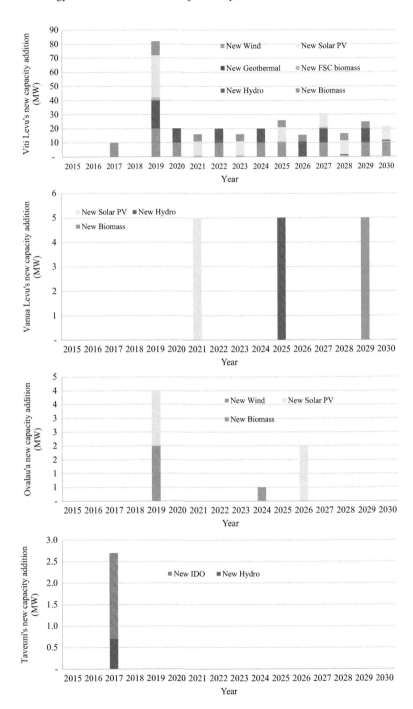

Fig. 8.7 Timeline of new capacity addition to the grid in the four islands

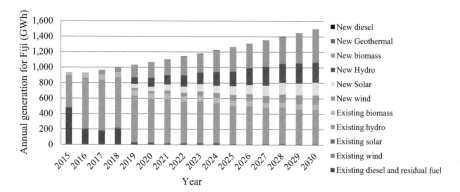

Fig. 8.8 Future annual generation from modelling for the period from 2015 to 2030

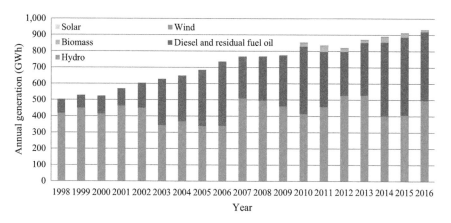

Fig. 8.9 Historical grid annual electricity generation from 1998 to 2015. (Data Source: FEA 2016)

8.6.2.3 Potential Avoided Emissions

When adding new renewable energy generation technologies, 100% renewable energy generation is achieved from 2019 and onwards (but this will be at high investment cost). This means that there are zero GHG emissions in future years from the current (2015) emission of 397,000 metric tonnes of CO_2-equivalent emissions.

The potential avoided emission was calculated similar to the description in Sect. 8.6.1.1. With introduction of new solar PV in future years, the carbon dioxide avoided emission starts from 48,000 metric tonnes in 2019 to around 148,000 metric tonnes of avoided emissions by 2030 (see Fig. 8.10). For new biomass power plant, avoided CO_2 emissions starts from 47,000 metric tonnes in 2017 to 381,000 metric tonnes by 2030 while new wind farms have 103,000 metric tonnes of avoided

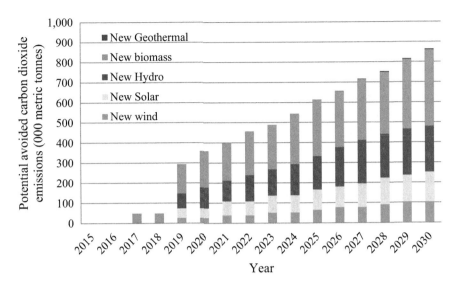

Fig. 8.10 Potential avoided carbon dioxide emissions from different renewable energy sources used for grid electricity generation

CO_2 emissions. New hydropower plants have 228,000 metric tonnes of avoided CO_2 emissions by 2030.

8.7   Discussion

EFL has planned for 5 MW solar power plant in Nadi, Fiji. This would require approximately 33,000 m² of land area and using Eq. 8.1, its generation potential is estimated to be around 9 GWh/annum. However, for diversifying Fiji's electricity supply sources, further capacity addition is needed for solar PV supported by wind and biomass. New hydropower plants are in EFL's plans as it is considered a reliable and firm generation. Section 8.6, estimates around 100 MW of new solar PV to generate around 170 GWh/year while avoiding around 150,000 tonnes of carbon dioxide emissions. With such a large capacity of solar PV coming on grid, battery storage or pumped hydro storage systems would be needed in times when solar PV and other existing generators are not able to meet the demand in a short span of time. Fiji is considering pumped hydro storage at Monasavu. EFL has requested for quotations from consultants for carrying out a feasibility study to explore the potential of building a pumped hydropower with the current Monasavu Hydropower Scheme (FEA 2017). However, to date there has not been any other reported plan for grid storage.

In terms of meeting its NDCs, Fiji has to make investments in not just new solar PV, but new biomass power plants, new wind farms, new geothermal power plants with new hydro. This would require significant amount of funding and government of Fiji alone will not be able to meet this financial commitment. Fiji will need financial support from donor agencies such as The World Bank (WB), Asian Development Bank (ADB), Global Environmental Facility (GEF), Korea International Cooperation Agency (KOICA), Japan International Cooperation Agency (JICA), Global Green Growth Institute (GGGI), The Deutsche Gesellschaft für Internationale Zusammenarbeit GmbH (GIZ), and others. Fiji can also access funding from Green Climate Fund (GCF) which has been set up to provide funding for developing countries to support projects related to climate change. In addition, Masdar funds and Leonardo DiCaprio funds are available for new solar powered projects to the population that are not connected to national grid and are using diesel generators to meet their electricity demand. Apart from external donor funds, government of Fiji is already providing tax incentives such as duty free importation of renewable energy equipment and 5-year tax holiday to any investor tax payer who invests in renewable energy projects. However, there can be more lucrative tax incentives devised so that the cost of renewable energy projects in Fiji to potential investors are further reduced.

There also needs to be a portfolio of policies to enable increased renewable energy share in electricity generation. These polices can include the development of a range of feed-in-tariffs based on cost of generation of electricity from various sources and technologies, setting up net-metering programs to upscale of roof-top solar systems, development of regulations which promote energy efficiency measures in existing industrial process, green building policies, forest logging and re-plantation policies, creation of new tax policies for new renewable energy projects, and many more. These policies would ensure transparency to potential investors and all stakeholders involved in electricity generation.

In addition, there needs to be regulations and legal framework in place to aid development of renewable energy for electricity generation. For grid electricity generation, independent body must be appointed for more independent power producers to come on board for grid electricity generation. The flat rate that is currently offered ($0.33/kWh) to IPPs may not be promising for some producers. There needs to be diverse range of tariffs offered by different sources of energy.

Moreover, capacity building and development programs must be enhanced to cater for growth in renewable energy generation. The programs offered by academic institutions must meet the needs for power companies. It must also be able to offer programs that are able to cater for future renewable energy technicians and managers.

8.8   Conclusions

Solar PV application for electricity generation is the cheapest for Fiji in order to reduce its emissions as the unit cost of solar PV module has significantly reduced in the recent past. With Fiji having average horizontal solar insolation of around 5.4 kWh/m^2/day and the capital cost of installation of solar PV ranging from

FJD3,100 to 3500/kW for rooftop systems, the solar PV generation potential was estimated using two methods. In method 1, different consumers of EFL are considered with monthly solar insolation data together with efficiency and area of a solar module used. It was found that 170 GWh of annual electricity generation is possible using solar PV roof-top installations. This results in 151,000 metric tonnes of avoided carbon dioxide emissions annually. In method 2, grid electricity demand and generation model was constructed in LEAP tool to model different future renewable energy generation technologies together with the existing generation technologies. It was found that around 167 GWh of solar PV annual generation is possible by 2030 with 99 MW of installed capacity. This led to an estimated carbon dioxide avoided emission of 148,000 metric tonnes/year by 2030.

Fiji is a small island developing state and its numerous geographically dispersed islands present unique challenges for 100% electrification. Solar PV can help establish distributed systems to provide electricity to un/underserved population. Government departments together with non-governmental agencies, private investors and development partners should work together to have more streamlined, efficient and transparent processes and regulations to achieve Fiji's NDC target.

Acknowledgements   Authors would like to sincerely thank Energy Fiji Limited (EFL) for providing 2016 hourly demand and generation data along with existing generators capacity. We are also grateful to Fiji Bureau of Statistics for providing other data related to this study.

References

Berno, T. (2017). *Sunergise Fiji – Investment case study*. https://cpb-ap-se2.wpmucdn.com/blogs. auckland.ac.nz/dist/0/269/files/2017/10/Sunergise-case-study-Final-281grp8.pdf. Accessed 23 Nov 2018.

Caruso, J. (2016). *UAE-pacific partnership fund: Fiji LaKaRo solar PV/diesel hybrid project case study*. https://www.slideshare.net/jameshamilton10/flinders-island-isolated-power-system-ips-connect-2016-j-caruso-masdar. Accessed 19 Dec 2018.

CBS. (2018). *Recent projects undertaken by CBS power solutions*. http://www.cbspowersolutions. com/projects/. Accessed 23 Nov 2018.

Datt, A. A., Singh, A., & Raturi, A. (2015). Performance of a solar EV charger in the Pacific Island Countries, Solar World Congress 2015, Daegu, South Korea.

Engerati. (2015). *Solar microgrids for Fiji*. https://www.engerati.com/article/solar-microgrids-fiji. Accessed 23 Nov 2018.

FBoS. (2010). *Fiji facts and figures as at July 2010*. Suva: Fiji Bureau of Statistics.

FBoS. (2018). *Household census survey 2018*. Personal Communication in June 2018 via email. Suva: Fiji Bureau of Statistics.

FDoE. (2018). *Solar energy system*, Personal Communication with Deepak in June 2018.

FEA. (2016). *Annual reports 2005–2016*. Suva: Fiji Electricity Authority.

FEA. (2017). *MR 323/2017 – Feasibility study of pumped storage: Hydropower at Monasavu – request for prices*. Fiji Electricity Authority. http://efl.com.fj/wp-content/uploads/2017/11/Tender-Specification-MR-323-2017.pdf. Accessed 16 Oct 2017.

FRA. (2018). *FRA's first solar streetlight project for Nailaga village in Ba*. Fiji Roads Authority. http://www.fijiroads.org/fras-first-solar-streetlight-project-nailaga-village-ba/

GoF. (2018). Hon. PM Bainimarama contribution as story-teller at the global climate action Summit Talanoa On "A rapid transition to a net-zero emissions society" – How do we get there? http://www.fiji.gov.fj/Media-Center/Speeches/HON%2D%2DPM-BAINIMARAMA-CONTRIBUTION-AS-STORY-TELLER-A.aspx. Accessed 22 Nov 2018.

GoF, GGGI. (2017). *Fiji's NDC energy sector implementation roadmap (2018–2030): Pathway to reaching national GHG mitigation targets in the energy sector.* Fiji: Government of Fiji and Global Green Growth Institute.

IPCC. (2006). *2006 IPCC guidelines for National Greenhouse Gas Inventories: Chapter 2 – Stationary combustion intergovernmental panel on climate change.* https://www.ipcc-nggip.iges.or.jp/public/2006gl/pdf/2_Volume2/V2_2_Ch2_Stationary_Combustion.pdf. Accessed 12 Sept 2018.

IRENA. (2017). *Solar supports village livelihoods and spurs business in Fiji.* https://www.irena.org/newsroom/articles/2017/Oct/Solar-Supports-Village-Livelihoods-and-Spurs-Business-in-Fiji. Accessed 28 Jan 2019.

ITP. (2014). *Review of existing subsidy and incentive schemes – Fiji. Final report.* IT Power for Fiji Department of Energy. www.fdoe.gov.fj. Accessed 29 Apr 2015.

ITP. (2015). Proposals for renewable energy support mechanisms Fiji: Final report 2015.

Katz, J., Denholm, P., & Pless, J. (2015). *Wind and solar on the power grid: Myths and misperceptions, greening the grid.* Golden: NREL National Renewable Energy Laboratory (NREL).

Lazar, J. (2016). *Teaching the "Duck" to Fly.* https://www.ice-energy.com/wp-content/uploads/2016/04/RAP_Lazar_TeachingTheDuck2_2016_Feb_2.pdf. Accessed 27 Nov 2018.

Masdar. (2015). *The UAE inaugurates three micro grid solar plants in Fiji.* Masdar: A Mubadala Company. https://masdar.ae/en/media/detail/the-uae-inaugurates-three-micro-grid-solar-plants-in-fiji. Accessed 15 June 2018.

Nand, R., & Raturi, A. (2015). *Rural electrification initiatives in Fiji – A case study of solar home systems, solar World Congress 2015.* Daegu: International Solar Energy Society.

NASA. (2017). *Surface meteorology and solar energy.* Atmospheric Science Data Center. https://eosweb.larc.nasa.gov/cgi-bin/sse/sse.cgi?skip@larc.nasa.gov. Accessed 7 Nov 2017.

Noguchi, R., Koyama, M., Ahamed, T., Genkawa, T., & Takigawa, T. (2013). Estimation of renewable energy potentials using geographic and climatic databases – A case study of the Tochigi Prefecture of Japan. *Agricultural Information Research, 22,* 71–83.

ParlimentFiji. (2018). *Standing committee on economic affairs: Report on Fiji electricity authority annual report 2016 – Parlimentry paper no. 103 of 2017.* Suva: Department of Legislature, Parliment House. http://www.parliament.gov.fj/wp-content/uploads/2018/03/FEA-2016.pdf. Accessed 4 Sept 2018.

PRDR. (2016). *Sunergise solar grid connected systems in Fiji.* Pacific Regional Data Repository for SE4ALL. http://prdrse4all.spc.int/node/4/content/sunergise-solar-grid-connected-systems-fiji. Accessed 6 July 2017.

SEANZ. (2017). *Powersmart solar – Vunabaka Resort Fiji.* Sustainable Energy Association New Zealand. https://www.seanz.org.nz/powersmart_solar_vunabaka

SMA. (2010). *Solar stand-alone power and backup power supply.* http://files.sma.de/dl/10040/INSELVERSOR-AEN101410.pdf. Accessed: 23 Nov 2018.

Stolz, E. (2012). $40K solar project launched at USP, Fiji Sun Online. Fiji Sun. http://fijisun.com.fj/2012/08/08/40k-solar-project-launched-at-usp/. Accessed 28 Apr 2017.

Sunergise. (2017). *Coca-Cola AMATIL launches energy efficient solar initiative.* https://www.sunergisegroup.com/news/2017/7/4/coca-cola-amatil-launches-energy-efficient-solar-initiative. Accessed: 15 Oct 2018.

Sunergise, ClayEnergy. (2017). *Relevant experience.* https://static.ptbl.co/static/attachments/137155/1473711772.pdf?1473711772. Accessed 23 Nov 2018.

SunnyPortal. (2018). USP/KOICA 45kW PV system profile. https://www.sunnyportal.com/Templates/PublicPageOverview.aspx?plant=f46fa170-f258-4618-bde4-63d53a65b45d&splang=en-US. Accessed 15 Oct 2018.

Syngellakis, K., Johnston, P., Hopkins, R., & Hyde, G. (2016). *Introduction to energy efficiency and renewable energy for hotels in Fiji – With applications to other Pacific Island countries.* Pacific Community (SPC) and Deutsche Gesellschaft für Internationale Zusammenarbeit (GIZ). https://www.pacificclimatechange.net/sites/default/files/documents/CCCPIR-Fiji_Energy%20 Efficiency%20and%20Renewable%20Energy%20for%20Hotels.pdf. Accessed 23 Nov 2018.

Valemei, J., 2014. Solar home systems, solar lights and solar water pumps. Personal Communication on 14/08/14, Rural Electrification Unit, Fiji Department of Energy.

WBG. (2016). *Global solar Atlas: Fiji.* The World Bank Group. http://globalsolaratlas.info/down-loads/fiji. Accessed 8 Sept 2017.

Chapter 9
A Life Cycle Analysis of the Potential Avoided Emissions from Coconut Oil-Based B5 Transportation Fuel in Fiji

Dhrishna Charan

Abstract Biodiesel B5 blend is being considered for the substitution of diesel fuel needs for the entire transportation sector in Fiji. The 2017 Nationally Determined Contributions Roadmap targets for all diesel fuel land transport in Fiji to begin using biodiesel fuel B5, which is made up of 95% diesel and 5% biodiesel, from the years 2021 to 2025 in an attempt to lower emissions.

Fiji is capable of producing enough coconut oil (CNO) to provide for the production of this B5 requirement. However, its copra production has been declining relentlessly since the peak production of some 33,000 tonnes per annum in 1977. The government has set up a Coconut Industry Task Force to reverse the decline. Based on a scenario that the government can increase its copra yield to 3.25 tonnes/ha on the currently available land of 15,000 ha, this chapter reveals that it is possible to replace the entire 450 million litres of imported diesel fuel per annum with B5.

The chapter also investigates the Greenhouse gas emissions of coconut methyl ester (CME) production in Fiji using a simple cradle to gate life cycle approach and compares it with the emissions arising from diesel use. Emission factors of 9.61 gCO_2eq emissions/MJ and 91.534 gCO_2eq emissions/MJ were calculated for CME and diesel respectively. An estimation of the avoided emissions without considering the emissions due to production of CME reveal avoided emissions of 59,000 tonnes of CO_2eq emissions per annum. The avoided emissions reduce to 47,500 tonnes when the emissions savings due to the CME production are taken into account. This is in excess of the savings of 37 kT CO_2eq/year targeted by Fiji's NDC Roadmap.

Keywords Life cycle assessments · Fiji · Biodiesel · Coconut methyl ester · Diesel · Green house gas emissions

D. Charan (✉)
School of Science and Technology, The University of Fiji, Lautoka, Fiji
e-mail: dhrishnac@unifiji.ac.fj

© Springer Nature Switzerland AG 2020
A. Singh (ed.), *Translating the Paris Agreement into Action in the Pacific*,
Advances in Global Change Research 68,
https://doi.org/10.1007/978-3-030-30211-5_9

9.1 Introduction

Biofuels have become the subject of intense interest lately as environmental issues such as climate change take on the global center stage. It is apparent that using biofuels presents a cleaner, more environmentally enduring and sustainable substitute to fossil fuels. Biodiesel from vegetable oil is a promising biofuel that has properties close to those of diesel. It is produced using a biochemical method known as transesterification which produces mono-alkyl esters of vegetable oil or animal fats (Wakil et al. 2012). There are many advantages of using biodiesel which include being of domestic origin, reduction of dependency on imported petroleum, biodegradability, high flash point and inherent lubricity (Knothe et al. 2006).

Biodiesel can be produced from coconut, soybean, sunflower, safflower, peanut, linseed, rapeseed and palm oil amongst others (Hossain et al. 2012). For the Pacific Islands, biodiesel from coconut oil is promising with coconut blends having the potential to replace diesel fuels under favorable conditions (Cloin 2005a). Coconut methyl ester (CME) is a type of fatty acid ester that is derived by transesterification of coconut oil with methanol. A study on the functional feasibility of CME in diesel engines showed that the technical specifications torque of biofuels is similar to that of diesel fuel (Chinnamma et al. 2015). A low carbon residue, minimal acidity and absence of sulphur elements make CME an eco-friendly fuel (Chinnamma et al. 2015). Many countries have formulated policies to promote biofuel production and use in the transportation sector, driven by goals which include reduction in greenhouse gas emissions, decarbonization of transport fuels and diversifying fuel supply (Acquaye et al. 2011).

Amid all these benefits, there are some reported drawbacks of biofuel use. The opponents of biofuels argue that biofuels compete with other food crops for land and water, do not deliver cost-effective carbon emission reductions and demand a disproportionate number of subsidies and incentives to become successful with the overall effect of negatively impacting biodiversity (Calumpang 2009). According to the Asian Development Bank (2013), no real greenhouse gas (GHG) emission reduction is achieved when emissions related to land use changes during biomass production is taken into account. Achten et al. (2010) find that the land occupation of palm oil triggers ecosystem quality loss of 30–45% when compared to natural vegetation. Using a worldwide agricultural model to estimate emissions from land use change, Searchinger et al. (2008) find that corn-based ethanol nearly doubles greenhouse gas emissions over 30 years and leads to increased greenhouse gas emissions for 167 years. In response to this, the European Union (EU) has defined a set of sustainability criteria called the EU Renewable Energy Directive for sustainable biofuels to ensure that the use of biofuels in transportation is done in a way that guarantees real carbon savings and protection of biodiversity. Some requirements of the directive include ensuring biofuels achieve GHG saving of 60% for new production plants in 2018 and that biofuels are not grown on land which have a high carbon stock or land with high biodiversity (European Commission 2018).

Apart from land use changes, the use of fertilizers during the plantation phase and the emissions from fossil fuel use during oil extraction and transportation also release notable greenhouse emissions which need to be considered to understand the true impact on the environment (Nanaki and Koroneos 2012). A life cycle assessment (LCA) is therefore required to evaluate the environmental implications of biodiesel production. It provides a systematic framework through which different stages of a product's life and their environmental impacts can be analyzed and the potential for impact reduction can be assessed (Simonen 2014). Typically, the boundaries for a LCA study include upstream activities including biomass cultivation and transport as well as processes involved in converting biomass to biofuel and the delivery of fuel to vehicles and consumption of fuel in the vehicle (Larson 2006).

Life cycle analysis is one of the most common techniques used to study and overcome the environmental challenges that come about due to changes in technological undertakings, especially at national levels. Pleanjai and Gheewala (2009) use a LCA approach to investigate the energy consumption of palm methyl ester (PME) production in Thailand. The energy analysis results for the study favor PME with a net energy balance (NEB) and net energy ratio (NER) of 100.84 GJ/ha and 3.58 respectively. A LCA of Jatropha diesel in India yielded similar results with NER of 1.7 and 1.4 for irrigated and rain-fed scenarios with GHG emission reduction values of 54% and 40% respectively (Kumar et al. 2012). The study finds that GHG emission in the individual stages of the life cycle assessment (LCA) is strongly dependent on co-product handling and irrigation. Pascual and Tan (2004) in their study of comparative LCA of coconut biodiesel and conventional diesel for Philippine's automotive transportation and industrial boiler application find that environmental impacts from emissions and energy consumption of coconut biodiesel were lower in all scenarios that were tested, especially when coconut residues were utilized for power cogeneration.

The government of Fiji is working towards using locally produced renewable energy to reduce their carbon footprint and to increase energy security. The SE4ALL Rapid Assessment and Gap Analysis and Fiji's National Energy Policy targets to increase the renewable energy share in electricity generation by 81% in 2020 and 99% by 2030 (Government of Fiji 2014). Fiji's 2017 Nationally Determined Contributions Implementation Roadmap (NDC-IR) is also aimed towards achieving 100% renewable energy by 2030 (Government of Fiji 2017). Given the high fossil fuel import bill the potential for producing transportation fuels made from biomass, especially from coconuts is attracting keen interest in Fiji. Apart from reducing the high import bills, use of biofuels improve the environment and increase the share of renewable energy in Fiji's total energy consumption.

The NDC-IR aims to reduce the total carbon dioxide emissions in Fiji's energy sector by around 30% by 2030 compared to the baseline year 2013, under the business as usual scenario. Specifically the NDC-IR targets for all diesel fuel used in land transport in Fiji to begin using biodiesel fuel "B5" with 95% diesel and 5% biodiesel between the years 2021 and 2025 with an expected greenhouse gas reduction of 35,000 tCO$_2$/year from 2021 through to 2030 (Government of Fiji 2017). The NDC implementation roadmap indicates importing biodiesel and providing it across all land transport sub-sectors (private, public, and commercial). However, it is worth exploring the potential of establishing a biofuels industry in Fiji so that B5 fuel can be sourced locally. Doing so will reduce Fiji's dependence on imported fuel.

According to the Biofuels International Website, biofuels derived from jatropha plant, pongamia plant, castor oil, waste vegetable oil and coconut oil have been investigated in Fiji (as cited in Singh 2012). Out of these, coconut oil is best placed to meet Fiji's biodiesel demand. However, the copra industry in Fiji has been declining in the past and long term policy for replanting and market promotion is needed. A review by the International Energy Agency (2011) find it essential to avoid large land-use changes which release huge amounts of greenhouse gas emissions if biofuels are to be used to provide any envisaged emission reduction in the transport sector. The greenhouse gas emission potential of biofuel production will depend on where and how the feed stocks are produced (Food and Agricultural Organisation 2008), the efficiency of biomass conversion to fuel and the efficiency of biofuel use (Larson 2006).

This research is focused on the evaluation of biodiesel production in Fiji using coconut oil (CNO) as feedstock. To ascertain whether there will be any net environmental benefit from such a proposed venture, a LCA with cradle to gate approach is carried out to support policy makers make informed decisions about the copra industry of the country and using locally sourced biodiesel. This chapter begins by providing an introduction on the status of the copra industry in Fiji and presents a picture of the existing coconut resources. A life cycle assessment of biodiesel is carried out to investigate the total emissions from the production of biodiesel and diesel. This is used to calculate the avoided emissions. The study explores whether it is environmentally feasible to use biodiesel B5 produced in Fiji to contribute to Fiji's Nationally Determined Contribution Implementation Roadmap targets. The prospect of reducing carbon emissions through use of coconut biodiesel provides a good incentive for Fiji to improve the feedstock production in the country and to invest more in the manufacture and use of biodiesel.

9.2 Methodology

This study is based on a scenario which considers future coconut nut yield improvements with an output of around 3.25 tonnes of copra per hectare with no significant changes in agricultural or energy technologies. This research explores the feasibility of using locally grown coconuts for meeting the NDC-IR target which aims to introduce biodiesel B5 blend across all diesel transport in Fiji. An important consideration in this research is that biodiesel from coconuts do not cause direct land use changes and therefore the increase in yield will come from existing coconut plantations. The production of coconut oil and CME is stipulated to be from two locations; a coconut mill and biodiesel plant in Suva which is located in Viti Levu and a coconut mill and biodiesel plant in Savusavu which is located in Vanua Levu. The locations are marked as red dots in Fig. 9.1. In both cases, the plant and the mill will be

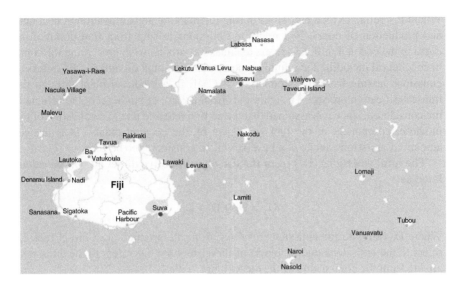

Fig. 9.1 Map of Fiji Islands (portion), showing the location of the proposed coconut mill and biodiesel plant. (Source: Google Maps)

close to the wharf to minimize the emissions resulting from the transportation of copra from ships to the coconut mill and from coconut oil mill to the biodiesel plant. The scenario assumes that relevant technology development and market shifts towards supporting the biodiesel industry will be made possible through relevant policy actions.

A preliminary (first order) estimate of the avoided emissions due to CME use in place of diesel in the transportation sector is made. This will be based the volume of CME produced. This is followed by a LCA for the CME system and petroleum diesel (ADO) system (reference system). The avoided emissions are recalculated based on the results of the LCA. The functional unit chosen for this study is greenhouse gas emissions per energy produced. The results are expressed as gCO_2eq emissions/MJ for a year. The emissions included in the study are carbon dioxide, methane and nitrous oxide. The environmental impact category considered is greenhouse gas effect. The Global Warming Potential (GWP) of methane and nitrous oxide is taken as 23 and 293. GWP measures the global warming impacts of a gas compared to carbon dioxide. The GWP for methane is 23; therefore, methane is around 23 time more potent as a heat-trapping gas compared to carbon dioxide. Other greenhouse gases such as hydrofluorocarbons, perfluorocarbons, and sulphur hexafluoride are not taken into account for this study as they are often insignificant in the bioenergy production chains (Wicke et al. 2008).

The total emissions (E_t) from CME production for a year will be calculated as:

$$E_T = E_{CP} + E_{TM} + E_{CM} + E_{BP}$$

Where E_{CP} refers to total emissions during the planting and cultivation of coconuts and production of copra, E_{TM} refers to emissions arising from transportation of copra to the coconut mill, E_{CM} refers to emissions arising from processing the copra to coconut oil including the refining process at the coconut oil mill and E_{BP} refers to emissions arising from the production of CME in a biodiesel plant through the transesterification process. The emissions from the use of biodiesel in the vehicle is not considered, since it is assumed that the carbon released due to combustion of the biodiesel component of the fuel was taken by the coconut plants. Thus the net amount of emissions due to combustion will be considered to be zero.

The total emissions (D_T) from the use of automotive diesel oil will be calculated as:

$$D_T = D_P + D_I + D_C$$

Where D_P refers to the emissions resulting from the production of diesel fuel, D_I refers to the emissions due to import of diesel fuel and D_C refers to the emissions arising from the combustion of the diesel fuel.

Since this study required original data for Fiji, local publications such as company reports and reports by NGO and related institutions are widely used. In the case where country specific data is not available, values found in reliable scientific studies are used. Furthermore, because the research done on LCA for CME was relatively less compared to PME (palm methyl ester), some input values from the production of PME was approximated to be the same for CME as the production process and the requirement inputs are similar for these two fuels. All data sources and assumptions made for this study are duly penned to maintain transparency. This study has been carried out according to the LCA methodology set by the International Organization for Standardization. The methodology used falls under the international standards series ISO 14040 and 14041 (ISO 14040 2006).

9.3   Potential of Local Biodiesel Production

A large proportion of households living in isolated dispersed islands and rural areas in Fiji still lack access to grid connected electricity which increases their dependency on imported expensive fossil fuels such as diesel and kerosene to meet their energy needs. One of the main factors that add to the high cost of petroleum products in Fiji Islands, like other Pacific Islands, is the nature of the fuel supply chain. Small markets have resulted in a lack of economies of scale that limits the potential for competition between fuel suppliers and ensures that fuel prices continue to remain high (Morris 2006). The monopoly position of multi-national oil companies in many countries has allowed these companies to earn returns on investments which are comparable in nature to large rapidly growing economies (Morris 2006).

The government through its Biofuel unit at the Department of Energy has introduced several incentives to develop biofuel industry in Fiji. The government initiative is timely considering the near demise of the copra industry which has been producing well below its potential for several years now. Government subsidies for coconut oil production from copra are an alternative social security system for communities especially in times of low copra prices on the world market (Cloin 2005b). Supporting a local industry that cuts down on fuel imports can profit Fiji's fragile rural island communities through enhancements in balance of payments and job creation (Nair 2014).

Biodiesel from the transesterification of coconut oil is one of the more promising alternative fuels for transportation in Fiji. In 2013, the total annual coconut production was 270 million nuts and the copra production was 45,000 metric tonnes (Food and Agriculture Organization of the United Nations 2013). Fiji's National Agricultural Census of 2009 revealed that there was 17,757 ha of coconut plantations around the country, with 14,270.06 ha classified as bearing (Department of Agriculture, Economic Planning and Statistics Division 2009). This was a significant reduction from the 49,812 ha that existed in 1991. Table 9.1 contains data on the area, production and sales of coconut (copra) by province in the year 2009. The data for Table 9.1 has been extracted from the 2009 Agricultural census.

According to Fiji's Ministry of Agriculture website, around 19,918 metric tonnes of copra was produced in the year 2015 (Fiji Ministry of Agriculture 2015). A recent estimate for 2018, places the whole nut production at 41 million nuts per year, out of which 35% was used for copra (Vula 2018). Since six nuts produce 1 kg of copra (Food and Agriculture Organization of the United Nations 2013), this represents a total production of 2392 metric tonnes of copra. Currently there is an estimated ten million scattered and planted coconut trees in an area of approximately 65,000 ha, mostly under a mixed cropping and livestock farming systems (Tuilevuka 2018). Most of the copra, almost 80%, is from small-holder producers.

A research by Singh (2012), found Fiji to be well placed to meet all its diesel engine needs in transportation using biodiesel blend B5 through small improvements in the efficiency of CNO production in Fiji in the year 2012. However, the situation in Fiji has been changing. The coconut industry has been experiencing a decline in the past few years, which was exacerbated by tropical cyclone Winston. This has led to a huge reduction in coconut production. The situation has been compounded due to a surge in the number of vehicles in Fiji over the past few years, largely aided by the Fijian government providing duty relief on the importation of used motor vehicles into the country (Das 2018) with new hybrid vehicles enjoying zero duty (Consumer Council of Fiji 2016). This increase in vehicles has led to an increase in the volume of automotive diesel oil (ADO) imported.

A good place to start in order to understand the current potential of the coconut industry to meet the country's biodiesel need (B5) is to estimate the amount of biodiesel that can be produced and compare it with the amount of diesel fuel used in vehicles today. As shown in Table 9.2, the ADO import in 2011 was 393 million litres (Pacific Community 2013). The import of diesel has been increasing over the years. The automatic diesel oil import in the year 2017 was 442.23 million litres (Pacific regional data repository 2018). We will work around a round figure of 450 million litres of ADO import per year in this study. This will mean that around 23 million

Table 9.1 Production and sales of coconut (copra) by province per year in the 2009 agricultural census

Province	Farms with crop	% farms	Planted area (HA)	% planted area (HA)	Bearing area (HA)	Total production (KG)	Total production sold (KG)	Sold locally (KG)	Total value (FJ$)
Naitasiri	17	0.5	17.79	1.8	2.96	66,359	502	0	150.66
Rewa	136	21.6	56.48	49.9	53.91	191,102	112,926	112,721	65,714.7
Serua	11	1.5	5.27	2.8	5.27	10,723	9973	9973	797.87
Tailevu	65	1.9	114.2	13.5	113.32	106,179	93,507	93,507	38,009.85
CENTRAL	229	2.5	193.73	7.7	174.46	374,363	216,909	216,202	104,673.09
Ba	121	5.4	162.39	14.7	137.86	127,736	57,935	57,935	28,008.2
Nadroga	34	1.3	22.27	2.3	21.9	26,141	4061	4061	2030.64
Ra	206	4.6	145.52	8.9	143.46	250,971	224,029	224,029	12,160.73
WESTERN	361	3.9	330.8	8.9	303.21	404,848	286,025	286,025	151,099.57
Bua	517	16.5	3195.54	75.1	2606.75	939,832	915,288	617,587	259,434.67
Cakaudrove	740	10.6	9117.05	73.5	9.43.55	3,726,768	3,637,125	3,637,125	1,826,116.78
Macuata	201	11.4	394.28	32.9	382.33	182,426	148,059	141,000	206,099.51
NORTHERN	1457	12.3	12,706.87	71.1	12,032.63	4,849,026	4,700,472	4,395,712	2,291,650.96
Lau	590	42.6	1624.65	92.4	1621.77	4,270,209	3,025,320	2,973,672	356,078.15
Lomaiviti	71	2.7	59.66	9.7	46.51	367,637	361,617	345,055	81,590.04
Rotuma	47	25.8	94.04	83.6	90.48	368,112	355,052	354,192	59,184.85
EASTERN	708	11	1778.35	56.2	1758.77	5,005,958	3,741,988	3,672,919	696,853.04
FIJI	2755	7.5	15,009.04	55.1	14,270.06	10,634,196	8,945,394	8,570,858	3,244,276.66

Source: Department of Agriculture, Economic Planning and Statistics Division (2009)

Table 9.2 Oil import data, Fiji from 2009 to 2011

Type of oil	2009 (litres)	2010 (litres)	2011 (litres)
Lubricating oil and mineral turpentine	5,861,901	4,825,808	5,075,597
Grease	205,519	187,512	150,216
Other lubricating oils, min Turps and grease	275,099	443,433	188,114
Automotive diesel	112,757,762	391,697,149	393,037,274
Industrial diesel	172,192,600	139,447,009	94,034,776

Source: Pacific Community (2013)

litres of CME will be needed to meet the NDC goal of B5 blend for all land transport in Fiji using diesel. One tonne of copra can produce 600 l of oil (Singh 2012). By which account, the entire copra production of 2392 metric tonnes in 2018 would have produced 1.44 million litres of coconut oil. Using an extraction rate of 0.8 from coconut oil to CME through transesterification, the resulting CME production would be around 1.15 million litres. This is very less compared to the 23 million litres needed. Therefore, a radical increase in copra production compared to 2018 levels is needed to meet the national B5 target through locally produced CNO biodiesel.

The Government together with other organizations such as the Pacific Community (SPC) and the European Union (EU) is working towards coconut rehabilitation programs which include initiatives such as the one million coconut tree planting/replanting program, coconut-based cropping, livestock diversification program, promotion of value-added coconut, amongst others to develop the coconut industry (Food and Agriculture Organization 2013). The government has merged the Coconut Industry Development Authority (CIDA) under the Ministry of Agriculture and developed a coconut industry taskforce in 2010 to allow for a more coordinated effort. Currently targeted efforts concentrate on replacing both fallen trees due to TC Winston and senile trees with around 40,000 seed nuts raised and supplied to farmers to regenerate farmer's interest back into the coconut industry (Vula 2018). The Government working with Copra Millers of Fiji Ltd., is targeting to plant around 30,000 coconut trees per year for the next few years (Tuilevuka 2018). Considering that it takes 6 years for a coconut tree to start producing nuts; that 50 coconuts are produced in a year by each new tree that is planted; and that six nuts gives 1 kg copra in Fiji (Food and Agriculture Organization 2013), we can expect around an additional 2000 tonnes of copra per year in future. If all this new production is used for biofuel production and added to the 2018 yield of 2392 tonnes, we can expect around 2.6 million litres of CME in future. This still falls noticeably short of the 23 million litres required for the B5 production.

Investigating Fiji's 2020 agriculture sector policy agenda, we get a sense of the Ministry of Agriculture's plans for the coconut industry. The agenda aims to bring the coconut production to the 1977 levels with a production of 33,000 metric tonnes of copra, which will require 615,323 trees to be planted each year for 5 years (Bacolod 2014). If the 1977 levels were achieved, we can extract around 15.8 million litres of biodiesel, which is around 3.5% of the ADO import. This path is a little closer to achieving the 23 million litres target. However, a major rehabilitation of the copra industry is needed for B5 production in Fiji.

9.4   Scenario Development for a Future Copra Industry

For the purpose of this study, we will consider a scenario in which all government initiatives are successful and there is an increased interest amongst the farming communities in the coconut industry. Since there is no recent data on the area of coconut plantations around the country since the 2009 National Agricultural Census, we will assume that the area of bearing coconut plantations has remained mostly steady and is around 15,000 ha with the current production of copra of around 0.16 tonnes/hectare. This is based on the 2018 coconut production of 41 million nuts, out of which 35% was used for copra. This is well below the estimated national potential of 3–10 tonnes of copra production per hectare (Singh 2012). The only possible way for Fiji to be in a position to meet its biodiesel demands through local production, is to significantly increase coconut yield through replanting and refurbishment of old plantations.

We will therefore consider a scenario for the future where the coconut yield from existing coconut plantations increases, with a subsequent increase in the percentage of coconut harvested for copra, bringing the average production to somewhat around 3.25 tonnes of copra/hectare. This will produce around 48,750 tonnes of copra. In this scenario, Fiji will be in a position to produce 29.3 million litres of coconut oil per annum. This will not induce any land use changes as the production will be from the 15,000 ha of plantation which already exists. Using an extraction rate of 0.8, transesterification of 29.3 million litres of coconut oil will produce 23.4 million litres of CME. This represents almost 5.2% of the automotive diesel imports of around 450 million litres (estimated), and is sufficient to meet Fiji NDC Roadmap target of using B5 biodiesel in all diesel vehicles.

The following sections on LCA of biodiesel production in Fiji is based on an upward trend in coconut production with significant increases in replanting and new planting in the 15,000 ha of area available for coconut farming. A potential production of 3.25 tonnes of copra per hectare is assumed. Figure 9.2 tracks the assumptions made for the scenario developed for this study.

Fig. 9.2   Scenario development for the case of the LCA

9.5   System Boundary for the LCA Study

In this study, cradle to gate LCA is carried out which considers the processes of cultivation of coconut plants, collection and handing of coconuts, transportation of dried copra to the coconut mill and the production of coconut methyl ester (CME) in a biodiesel plant. The distribution of CME and its subsequent combustion is not included. Figure 9.3 outlines the system boundary used for this study. Input with regards to infrastructure including the construction of the coconut mill and the biodiesel production plant is not included. The production of greenhouse gas emissions from the production of machinery and equipment, construction of buildings, and production are minor compared to overall emissions in the system (Wicke et al. 2008). There is always a possibility of utilizing coconut residues (shell and husk) for energy production, as this adds to emission reduction. However, in the case of Fiji Islands, a sufficient proportion of the residues are used by households for handicrafts, firewood etc. In view of this, the utilization of coconut residues is not considered for this study. CME production will be from first generation production technology based on conventional techniques of production using a

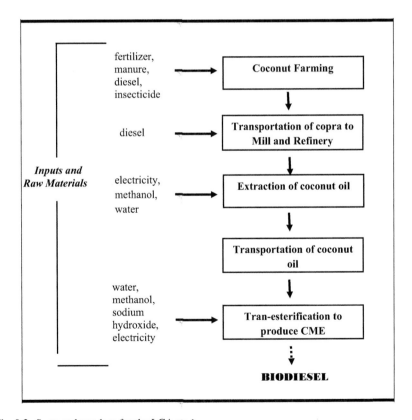

Fig. 9.3 Systems boundary for the LCA study

process called transesterification. The system boundary for the study includes four processes; coconut plantation, coconut oil extraction, CME production and all the transportation at these stages. The net emissions from CME use in vehicle engines is zero as the emissions resulting from the use of CME is equal to the carbon that is removed from the atmosphere via photosynthesis as the coconut plants grow. Comprehensive LCA take into account various co-products, by-products and side effects of the activities associated with each stage of the biofuel life cycle. However, for this research a simple LCA will be carried out where these products and side effects will not be considered. According to Pehnelt and Vietze (2013), the more co-products, allocation, distribution effects and environmental effects a LCA considers, the more complex the whole process becomes, carrying with it assumptions, which increases the uncertainty and usually blurs the results obtained.

9.6   Estimation of Inputs and Results

9.6.1   Coconut Plantation

A coconut tree has a life span of around 80 years with peak coconut nut production between 15 and 30 years of age (Foale 2003). The tree begins to fruit around 6 years after planting. The coconut palm is able to grow well in many soil types, but thrives in well drained sandy loam soil (Rangaswami and Mahadevan 1998). For the coconut plantation stage, several farm processes such as land ploughing and setup, application of fertilizer, picking of coconut nuts and copra production is normally considered for a LCA study. Since Fiji currently has sufficient land to produce the required volumes of CME and since coconut trees have a life span of 60–80 years, no new land will be cleared for cultivation. Instead of clearing land, this study assumes an upward trend in coconut production with significant increases in replanting and planting of new trees in the 15,000 ha of area already available for coconut planting. Land use changes for biofuel feedstock planting especially from forest and grasslands cease the prevailing carbon sequestration which is equivalent of additional emissions (Searchinger et al. 2008). Since the scenario in this study does not involve land use changes, these extra emissions are avoided.

Therefore, the first source of emissions released in coconut plantations considered in this study is from the use of tractors for ploughing the land and preparing it for planting. A study on some common tractor models used for ploughing operations in sandy loam soil find an average consumption of 23.5 l of diesel fuel per hectare (Adewoyin and Ajav 2013). This consumption rate is approximated to be the same for this study and used to compute the greenhouse gases resulting from ploughing the 15,000 ha of land using tractors. Table 9.3 shows the inputs related to use of the tractor for this study.

In terms of fertilizer use, Index Mundi (2011) used FAO's electronic files and website to compute that 29.99 kg of fertilizer was used per hectare in Fiji in the year 2014. Nitrogenous, potash, and phosphate fertilizers are commonly used in Fiji.

Table 9.3 Tractor usage with related emissions and energy factors for an area of 15,000 ha

Fuel consumption (L/ha)	23.5[a]
Total fuel requirement (L)	352,500[b]
Emissions per litre (CO$_2$eq/L)	2.7[c]
Gross energy of diesel (MJ/litre)	38.6

[a]Estimated from Adewoyin and Ajav (2013)
[b]Based on 15,000 ha of land
[c]Estimated from Pacific Community (2004)

Table 9.4 Fertilizer consumption of common crop commodities in Fiji in the year 1994

Fiji commodity (FAO)	Area (000Ha)	N	P$_2$O$_5$	K$_2$0	N	P$_2$O$_5$	K$_2$0	Total
Year 1994		Average (kg/ha)			Consumption (000Mt)			
Cocoa	2.0	70.0	60.0	120.0	0.1	0.1	0.2	0.4
Coconut	60.0	20.0	0.0	70.0	1.2	0.0	4.2	5.4
Rice	15.0	60.0	30.0	70.0	0.9	0.5	1.1	2.5
Sugar cane	75.0	100.0	40.0	81.0	7.5	3.0	61	16.6
Total	152.0				9.7	3.6	11.6	24.9
1994 overall consumption estimate (FAO)					9.0	3.0	6.0	18.0

Source: Food and Agriculture Organization (2004)

According to the Food and Agriculture Organisation (2004), consumption estimates for coconuts plantations in Fiji were 20 kg/Ha and 70 kg/Ha for N and K$_2$0 (potash) fertilizers respectively in 1994. The fertilizer uses in common crops in 1994 is shown in Table 9.4.

In Fiji, there has been a radical shift in the ownership of coconut plantations from a few large plantations in the 1990s to a large number of small holder plantations in the recent years. Fertilizer application in smallholder coconut plantations is not practiced in Fiji with most managed plantations using green manure (Zieroth et al. 2007). Ministry of Agriculture, Fisheries & Forest and Provincial Development has also recently indicated that it is not viable to apply fertilizer in the coconut plantations since most of the plantations are small-holder plantations and the excessive price of agro inputs would be impracticable due to the low returns from coconut (Food and Agriculture Organization 2013). Instead of fertilizers, coconut diversification through intercropping and livestock integration approaches is recommended to be more useful in enhancing soil fertility. Therefore, the emissions from fertilizer use will be considered to be negligible for this study. Additionally, it is assumed that rainfall will provide sufficient water for good plant productivity and coconut palms will be mainly rain fed.

The collection of the coconuts is carried out by climbing the trees and picking the nuts. The nuts are also allowed to fall down naturally, after which it can be collected. Coconut oil can be extracted directly from the fresh nut or from dried kernel (copra). Copra, through sun drying, is preferred in Fiji. The process from harvesting of coconuts to the production of copra is mainly through manual labour and contributes zero emissions. The copra is transported and sold to large processing mills where the oil is extracted. The transportation of copra to the coconut mills has fossil fuel inputs and is discussed in the next section.

9.6.2   Transportation of Copra from Plantation to Mill

Copra is transported from the plantations to coconut mills by using trucks and by shipping from remote areas. Referring back to Table 9.1, we can get a good estimate of the percentage of coconut production from each province. The total production of copra in 2009 was 10,634 tonnes. The total production of copra needed to achieve B5 blends for all automotives using diesel in Fiji is 48,780 tonnes, which will produce the 23.4 million litres of CME being considered under the scenario. Therefore, an increase in copra production by around 4.6 folds is needed compared to the 2009 production. The major assumption under production of future copra is that there will be a uniform 4.6 fold increase across all the 13 copra producing provinces. The estimated increase in copra production from each province is given in Table 9.5.

The copra is transported to two coconut mills, one located in Suva and the other in Savusavu. An average value of distance between the provinces and their nearest mill is estimated using satellite images. For the Central, Western and Northern Division, transportation will be through land transport; specifically using 13 tonne trucks running on automotive diesel fuel. The number of trips made annually to a province is based on the total potential production in the province divided by the carrying capacity of the truck, which in this case is a 13 ton truck. The total distance travelled is therefore calculated as the distance travelled (per round trip) to a province multiplied by the number of trips made. Transportation from the three provinces in the Eastern Division will be through ships using marine diesel oil. Table 9.6 contains the distance dimensions for the land transportation and Table 9.7 contains those for the sea transportation considered for this study.

Table 9.5 Production rate of copra in 2009 and the total future potential based on the scenario of a 4.6 fold increase

Division	Province	Total production 2009 (kg)[a]	Total potential production after 4.6 fold increase (kg)[b]	Division	Province	Total production 2009 (kg)[a]	Total potential production after 4.6 fold increase (kg)[b]
Central	Naitasiri	6.64E+04	3.04E+05	Northern	Bua	9.40E+05	4.31E+06
	Rewa	1.91E+05	8.77E+05		Cakaudrove	3.73E+06	1.71E+07
	Serua	1.07E+04	4.92E+04		Macuata	1.82E+05	8.37E+05
	Tailevu	1.06E+05	4.87E+05				
Western	Ba	1.28E+05	5.86E+05	Eastern	Lau	4.27E+06	1.96E+07
	Nadroga	2.61E+04	1.20E+05		Lomaiviti	3.68E+05	1.69E+06
	Ra	2.51E+05	1.15E+06		Rotuma	3.68E+05	1.69E+06

[a]Extracted from the 2009 National Agricultural Census Fiji
[b]Estimation based on a 4.6 fold increase in production compared to the 2009 Census

Table 9.6 Copra transportation though land transport to the coconut mill: distance dimensions and related energy usage and emissions for the scenario considered

Division	Province	Total potential production (tonnes)	Average distance nearest mill (km)[a]	No. of trips annually	Distance travelled annually (km)	Litres of diesel consumed (l)[b]	CO_2 Emissions (kg)[c]	$CH4$ Emissions (kg)[d]	N_2O Emissions (kg)[d]
Central	Naitasiri	3.04E+02	30	23	1.40E+03	7.02E+03	1.90E+04	4.45E-03	4.19E-03
	Rewa	8.77E+02	16.7	67	2.25E+03	1.13E+04	3.04E+04	7.14E-03	6.71E-03
	Serua	4.92E+01	55.9	4	4.23E+02	2.12E+03	5.71E+03	1.34E-03	1.26E-03
	Tailevu	4.87E+02	49.6	37	3.72E+03	1.86E+04	5.02E+04	1.18E-02	1.11E-02
Western	Ba	5.86E+02	226	45	2.04E+04	1.02E+05	2.75E+05	6.46E-02	6.07E-02
	Nadroga	1.20E+02	168	9	3.10E+03	1.55E+04	4.18E+04	9.82E-03	9.24E-03
	Ra	1.15E+03	123	89	2.18E+04	1.09E+05	2.94E+05	6.90E-02	6.49E-02
Northern	Bua	4.31E+03	102.47[e]	332	6.18E+04	3.09E+05	8.34E+05	1.96E-01	1.84E-01
	Cakaudrove	1.71E+04	57.44[f]	1315	1.41E+05	7.05E+05	1.90E+06	4.47E-01	4.20E-01
	Macuata	8.37E+02	96.1	64	1.24E+04	6.19E+04	1.67E+05	3.92E-02	3.69E-02

[a]Estimated from satellite images
[b]United Nations Environmental Programme (2013)
[c]2.7 kgCO_2/litre, based the Pacific Community (2004)
[d]CH_4 and N_2O emissions is taken for Diesel Medium and Heavy Duty Vehicle from (Source Environmental Protection Agency 2014). An emission factor of 23 is used for CH_4 and 293 for N_2O
[e]9.35 km is on horseback and the analysis has been adjusted likewise
[f]3.8 km is on horseback and the analysis has been adjusted likewise

Table 9.7 Copra transportation though shipping to the coconut mill for the Eastern Division: distance dimensions and related energy usage and emissions

Province	Total potential production (tonnes)	Average distance to nearest wharf (km)	Landing wharf	How many trips annually[a]	Total average distance (km)/year	Number of days at sea per trip	MDO consumed per year (Mg)[b]	Total energy (MJ)[c]	CO_2 emissions (kg)[d]	N_2O emissions (kg)[e]	CO_2 eq emissions (kg)[f]
Lau	1.96E+04	241	Suva	24	5784	1	8.60E+02	3.79E+08	1.81E+06	1.74E–02	1.81E+06
Lomaiviti	1.69E+03	85	Suva	24	5784	0.25	2.15E+02	9.46E+07	1.56E+05	1.74E–02	1.56E+05
Rotuma	1.69E+03	534	Savusavu	24	5784	1.5	1.29E+03	5.68E+08	1.56E+05	1.74E–02	1.56E+05

[a]Based on the case where ships visit remote islands twice monthly

[b]Estimated from Norwegian University of Science and Technology (n.d.) and using the equation 16.263 + 0.001*tons Mg/day for all ships (Jun et al. 2002) to calculate the consumption of MDO

[c]Estimated using the caloric value for diesel as 44 kJ/g

[d]Based on average emission of 16 g of CO_2/ton-km for short – sea shipping from (Cefic, E. C. T. A 2011)

[e]N_2O emissions are assumed to be the same as those for heavy-duty diesel engine as specified by Jun et al. (2002), which according to the IPCC (2006a) emissions guideline is 3 mg/km (Table 3.2.3). CH_4 emissions assumed to be negligible (Jun et al. 2002)

[f]GWP of 23 is used for CH_4 and 293 for N_2O

9.6.3 Palm Oil Production in Coconut Mill

Copra Millers Fiji Limited is Fiji's main producer of coconut oil, which produced around 4.5 million litres of coconut oil in 2007 with a capacity to produce double that amount (United Nations Environmental Programme 2013). There are ten coconut oil mills with installed capacity of 51 MT (shown in Table 9.8). This capacity will have to increase significantly to accommodate the processing of 23.4 million litres of coconut oil. Two large scale coconut oil processing plant will need to be constructed to handle the increased influx of copra under the scenario considered for this study: one in Suva and one in Savusavu. There are two steps considered for milling in this study. The initial step is the production of coconut oil, followed by the refining of coconut oil. Both these processes require electricity and water inputs. This analysis will use an extraction rate of 600 l of oil production per tonne of copra (Singh 2012). The greenhouse gas emissions from the coconut mill and the transesterification process are based on data from literature.

Initially the coconut oil is extracted from dried copra. The main steps for the coconut oil production is cleaning the copra; crushing and cutting the copra to fine particle sizes; cooking the crushed copra; oil extraction; screening of coconut oil; and the filtration of coconut oil to remove any solids (Raghavan 2010). The coconut oil is then refined through physical refining whereby a weak solution of phosphoric acid is added, followed by neutralization during which sodium hydroxide is added to convert free fatty acids in oil to soap stock, which is then removed. The processes of bleaching and deodorization of oil follows to remove excess moisture (Raghavan 2010).

Sumiani and Sune (2007), studying the life cycle assessment of crude oil in Malaysia find that around 17 kWh of electricity is required to process 1 tonne of FFB (fresh fruit brunch). Based on this, an average of 17 kWh of electricity is assumed to be required to process 1 tonne of copra into coconut oil. The processed coconut oil will pass through a refinery with diesel requirements of 200 MJ/t and electrical energy requirements of 23.4 kWh/t (Wicke et al. 2008). All electricity consumed during these processes will be purchased from the grid. Energy Fiji

Table 9.8 Existing coconut oil mills in Fiji with their related capacity

Name	Capacity
Copra Millers Fiji Ltd	24 MT
Ocean Soap	8 MT
Biodiesel Group Fiji Ltd	5 MT
Niu Industries	2 MT
Koro Biofuel	2 MT
Cicia Biofuel	2 MT
Rotuma Biofuel	2 MT
3 other mills (Vanua Balavu, Gau, Lakeba)	6 MT
Total	51 MT

Source: Food and Agriculture Organization (2013)

Limited (EFL), a government owned company solely responsible for supplying power throughout the Fiji Islands, generated 923,628 MWh of electricity in 2016. Fifty four percent of this was generated from renewable sources such as hydro, wind and biomass and 46% was from thermal generation from diesel generators (Energy Fiji Limited 2016). UNDP's project on exploring the Fiji's electrical energy generation for the future, projected a cumulative generation of 18,015 GWh of energy from 2010 to 2025 with an overall emission of 5747.4 ktons of CO_2, under the business as usual scenario (United Nations Development Programme 2010). Baseline Scenario conditions and analysis from this study is shown in Table 9.9. Using the information from Table 9.9, the emission factor for electricity from the grid is calculated to be 0.32 kg CO_2/kWh for the near future. The average efficiency of a large modern diesel engine (>1000 kW output) for the Pacific region is 0.284 litres/kWh (Pacific Community 2004). Default IPCC emission factors for stationary combustion in manufacturing industries of 74,100 kg/TJ for CO_2, 3 kg/TJ for CH_4 and 0.6 kg/TJ for N_2O is used for analysis (IPCC 2006b). Table 9.10 provides a summary of the inputs considered for the coconut milling stage. The water requirements for milling and biodiesel production factory will be met through rainwater harvesting.

Table 9.9 Future power generation and GHG emissions in Fiji based on a baseline scenario analysis by UNDP

Year	2010	2015	2020	2025
Total power demand, GWh	942.0	1096.0	1278.0	1435.0
Total power generation, GWh	857.0	1127.4	1215.0	1280.0
RE electricity, GWh	494.0	723.4	765.0	785.0
Fossil fuel generated electricity, GWh	363.0	404.0	450.0	495.0
%RE electricity	57.6%	64.2%	63.0%	61.3%
Total energy usage for power generation, ktoe	422.2	754.2	908.6	991.5
RE consumption, ktoe	309.3	628.3	768.3	837.5
Non-RE consumption, ktoe	112.9	125.9	140.3	154.3
%RE in power generation mix	73.3%	83.3%	84.6%	84.4%
GHG emissions in power generation, ktons CO_2	316.4	353.0	393.6	432.7
Summary				
Overall power generation (2010–2025), GWh	18,015.0			
Overall RE electricity production (2010–2025), GWh	10,697.3			
Overall energy consumption (2010–2025), ktoe	11,416.4			
Overall RE consumption (2010–2025), ktoe	9366.4			
Overall GHG emissions (2010–2025), ktons CO_2	5747.4			

Source: United Nations Development Programme (2010)

Table 9.10 Input, emissions and energy consumption from the production of coconut oil

Land Area (ha)	15,000
Copra production (t/ha)	3.25
Total copra production (t)	48,750
Milling	
Electrical energy requirement/tonne of copra (kWh/t)	17[a]
Total electrical energy consumed (kWh)	828,750
CO_2 emissions factor for electrical grid energy (kg/kWh)	0.32[b]
Refinery	
Electrical energy requirement/tonne of copra (kWh/t)	23.4[c]
Total electrical energy consumed (kWh)	1,140,750
Diesel requirements (MJ/t)	200[c]
Average efficiency of large diesel engine (L/kWh)*	0.284[d]
Total diesel energy required (kWh)	2,708,333
Diesel requirement (L)	769,166.7
Emissions CO_2 (kg)	722,475[e]
Emissions CH_4 (kg)	29.25[e]
Emission N_2O (kg)	5.85[e]

[a]The electricity requirement is sourced out from Sumiani and Sune (2007)
[b]Calculated using information contained in United Nations Development Programme (2010)
[c]Based on values from Wicke et al. (2008)
[d]Carbon dioxide emissions taken from the Pacific Community (2004)
[e]Default IPCC emission factors for stationary combustion in manufacturing industries (IPCC 2006b)

9.6.4 Coconut Methyl Ester (CME) Production in Biodiesel Plant

The transesterification of the coconut oil would be done using methanol in the presence of the catalyst sodium hydroxide. Methanol is not produced in Fiji and will need to be imported. The resulting coconut oil methyl ester (biodiesel) will then be blended with petroleum diesel to form blend B5. This study does not consider the by-products from the biodiesel production. We will assume that the biodiesel production will take place in a plant in close proximity to the coconut oil mills so that the emissions associated with transportation of coconut oil to the plant is negligible.

Using an extraction rate of 0.8 for transesterification, an input of 1.25 l of coconut oil will be needed to produce 1 l of biodiesel together with inputs of methanol and sodium hydroxide. Since no data on transesterification of coconut oil in Fiji was available, the data used is based on Wicke et al. (2008) study on Palm oil diesel in Malaysia with the assumption that the GHG emissions in the production of PME and CME is approximately same due to the almost identical transesterification process and conversion efficiency. The transesterification reaction for converting triglycerides into methyl esters is given in Fig. 9.4.

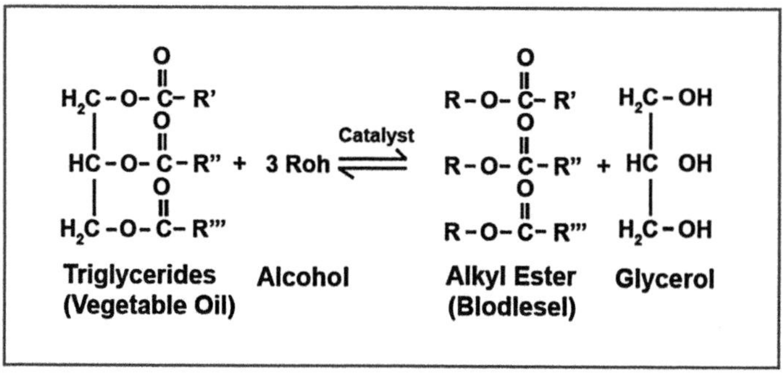

The chemical reaction diagram shows:

Triglycerides (Vegetable Oil) + 3 Roh (Alcohol) → (Catalyst) → Alkyl Ester (Biodlesel) + Glycerol

Fig. 9.4 Transesterification reaction for converting triglycerides into methyl esters (Velasquez et al. 2009)

The amount of methanol and NaOH required is 100 kg per tonne of coconut oil and 6 kg per tonne of coconut oil respectively. Density of methanol, sodium hydroxide and coconut oil used is 0.792 kg/l, 2.13 kg/l and 0.903 kg/l respectively. Based on the scenario under consideration, some 29.3 million litres of oil is to be produced, requiring 4.102 million litres (3704.1 tonnes) of methanol together with 158.52 tonnes of NaOH.

9.6.4.1 Methanol Production and Import

There are two stages to consider under this – Methanol production and importing methanol. Both these stages release emissions.

Methanol Production

Petrochemicals such as methanol need to be addressed in detail because their global production volume and associated greenhouse gas emissions are relatively large (IPCC 2006a). Methanol is made by way of steam reforming of natural gas which produces methanol and by-products CO_2, CO, and H_2 from synthesis gas. Methanol would need to be imported into the country for the transesterification process. Since New Zealand is a close neighbor of Fiji, this study considers methanol that is imported from New Zealand in order to minimize emissions from transportation. About 2.4 million tonnes of methanol is produced annually in New Zealand, out of which 95% is exported to the Asia Pacific region (Methanex 2018). Referring to the New Zealand Greenhouse Gas Inventory from 1990 to 2016, the emission factor for methanol production from mixed feed in New Zealand is reported to be 63.44 tCO_2/TJ (0.06244 $kgCO_2$/MJ) (New Zealand Government 2016). According to the New Zealand Ministry for the Environment website, 125.28 $ktCO_2$ eq emissions were emitted in 2016 due to methanol production. The annual report of the company Methanex (the sole producer of methanol in New Zealand) reports that around 2.2

Table 9.11 Inputs, emissions and energy consumption factors associated with methanol production and import in the transesterification process

Methanol shipping	
Methanol requirement (L/L CNO)	0.14[a]
Methanol consumed for transesterification (ML)	4.102
Methanol consumed for transesterification (t)	3704.106
Emissions from methanol shipping (CO_2kg/t-km)	0.017[b]
Distance between New Zealand and Fiji Port (km)	2889.12[c]
Diesel use - methanol shipping (kg/t-km)	0.0055[c]
Total Diesel consumed in diesel shipping (kg)	117717.7
MDO calorific value (MJ/kg)	44
Methanol production	
Methanol emissions per production ($kgCO_2$eq/t methanol produced)	57.9
Methanol imported (t)	3704.106[d]

[a]Based on Wicke et al. (2008)
[b]Emissions based on Fitzgerald et al. (2011)
[c]Distance between New Zealand and Fiji is calulated from PortWorld Distances (2018)
[d]The amount of methanol that needs to be imported is based on the method specified in Wicke et al. (2008)

million tonnes of methanol was produced in 2016 (Methanex 2016). Using these figures, an emissions factor of 0.0579 tCO_2eq emissions for every tonne of methanol produced is obtained. We can make reasonable estimates on the emissions attributable to Fiji's use of methanol based on the amount of methanol it will need to import for biodiesel transesterification. Details of this analysis can be found in Table 9.11.

Methanol Import

A study by Fitzgerald et al. (2011) on New Zealand's greenhouse gas emissions from international maritime transport find a mean rate of fuel consumption of 5.5 g per t-km and CO_2 emissions factor of 17 g per t-km. Using these values, the emissions arising from methanol import can be computed. The fuel used for shipping methanol is taken to be marine diesel fuel with a calorific value of 44 kJ/g. The average distance between a major port in New Zealand and Fiji has been conservatively calculated to be 2889.12 km (PortWorld Distances 2018). Other inputs with regards to methanol use can be found in Table 9.11.

9.6.4.2   Sodium Hydroxide Use

Sodium hydroxide will be used as the catalyst for the transesterification process. Sodium hydroxide is widely used for agrochemical and cleaning purposes and is readily available in local chemical agencies in Fiji (Singh 2008). The production of

Table 9.12 Inputs, emissions and energy consumption associated with NAOH use in the transesterification process

NaOH consumed (L/L CNO)	2.54E−03[a]
Total NaOH used in transesterification (ML)	0.0744
NaOH Density (kg/l)	2.13
Total NaOH used in transesterification (t)	1.59
NaOH emission (CO_2eq/kg)	1.2[a]
Energy used in NaOH production (MJ/kg)	18.25[b]

[a]Based on Wicke et al. (2008)
[b]According to the Switzerland Federal Office for the Environment, Forest and Agriculture (as cited in Pleanjai, Gheewala 2009).

1 kg sodium hydroxide produces 1.2 kg CO_2 eq emissions (Wicke et al. 2008). According to the Switzerland's Federal Office for the Environment, Forest and Agriculture, 18.25 MJ of energy is consumed in production of 1 kg of NaOH production (as cited in Pleanjai and Gheewala 2009). These inputs are detailed in Table 9.12. The amount of NaOH used in the transesterification process is much less than methanol and therefore emissions arising from the import of NaOH will be considered to be negligible in comparison to the emissions arising from the other processes involved in transesterification stage.

9.6.4.3 Energy Requirements

According to Smeets, Junginger & Faaij (2005), the electrical energy requirement for general vegetable oil transesterification is 250 kWh/m^3 (as cited in Wicke et al. 2008). As in the case of coconut oil mill, CO_2 emission factor of as 0.32 kg CO_2/kWh is used for the electricity consumption in the transesterification process. Table 9.13 lists the inputs associated this stage.

9.6.5 Reference System: Automotive Diesel Oil

This study involves investigating the feasibility of producing and using biodiesel in Fiji compared to a reference fossil fuel. The fuel for the reference system is automotive diesel oil (ADO). About 450 million litres of ADO is imported in Fiji today (estimated). Considering volume by volume displacement, the LCA is based on the 5% of the diesel that will be displaced by the CME, which is around 23.4 million litres. An attempt is made to estimate the life cycle emissions of diesel use in Fiji by investigating the three stages of diesel use which include emissions released during the production of diesel, emissions due to importing diesel to Fiji and emissions

Table 9.13 Inputs, emissions and energy consumption associated with electricity use in the transesterification process

Energy requirement (kWh/l CNO)	0.25[a]
Energy requirement (MJ/l CNO)	0.9
Litres of coconut oil to be produced (Ml)	29.3
Emissions related to energy use (kgCO$_2$/kWh)	0.32

[a]The energy requirement is based on research by Smeets, Junginger & Faaij (2005) which is cited in Wicke et al. (2008)

Table 9.14 Automotive diesel oil energy conversions, CO$_2$ emissions and measurements

Fuel	Unit	Typical density kg/ litre	Typical density l/ ton	Gross energy MJ/kg	Gross energy MJ/litre	Oil equiv. toe/unit (net)	kg CO$_2$ equivalent Per GJ	Per litre
Automotive diesel (ADO)	Ton	0.84	1190	46	38.6	1.08	70.4	2.7

Source: Pacific Community (2004).

released from the use of diesel. Standard emissions from the combustion of diesel are taken to be 2.7 kgCO$_2$/l (Pacific Community 2004). Table 9.15 summaries the inputs required at each phase together with the emissions and the energy consumption indexes. Table 9.14 contains a list of energy conversion units used for this study.

9.6.5.1   Diesel Production Phase

In order to estimate the emissions arising from the production of diesel used in Fiji, the emission factors reported by countries from where Fiji imports diesel is considered. According to the World Integrated Trade Solution (a World Bank Group), the top partner countries and regions from where Fiji imported its fuel in 2016 included Singapore, Korea, Rep., Australia, Swaziland and Switzerland (World Integrated Trade solution 2018). 81.53% of all fuel imports were from Singapore. The analysis used in this paper is based on the ADO imports from Singapore. A study by Tan et al. (2010) investigating the cradle to gate LCI of fuels and electricity generation in Singapore find that 0.94 kg of CO$_2$ emissions and 5.55×10^{-4} kg of N$_2$O emissions are emitted per each kg of diesel manufactured. The energy consumption per kg of diesel for solely the production stage was not available for Singapore. This can be due to the fact that Singapore itself depends greatly on foreign crude oil production and imports (Chua et al. 2010). For this reason, an international emission factor is considered for diesel fuel. Eriksson and Ahlgren (2013) in their well to tank life cycle analysis for diesel, report an emission factor of 14–17 gCO$_2$eq/MJ. A value of 17 gCO$_2$eq/MJ is used in the analysis.

Table 9.15 Emissions and energy consumption from the production, import and use of diesel fuel in Fiji

From consumption	
Total imports (l)	23,400,000.00
Total energy (MJ)	856,000,000[a]
Total emissions (CO_2e kg)	63,200,000[a]
From production	
CO_2 emissions (kg/kg diesel)	0.94[b]
N_2O emissions (kg/kg diesel)	0.000555[b]
CO_2 eq emissions (kg/kg diesel)	1.10
From shipping	
CO2 emissions (kg/t-km)	0.003[c]
Average distance between Singapore and Fiji Ports (km)	8319.00
Typical density of diesel (kg/l)	0.84
Total tonnes of diesel carried as freight (t)	19,700.00
CO_2 emissions (kg)	981000.00
HFO consumption (kg/tkm)	0.00
Total HFO used as fuel (Kg)	605000.00
HFO calorific value (MJ/kg)	44.00
Total energy released in HFO combustion (MJ)	56,600,000.00

[a]Computed from the values from Table 9.14
[b]Based on study by Tan et al. (2010)
[c]Based on study by Freese (2017)

9.6.5.2 Diesel Shipping Phase

There are emissions associated with the transportation of diesel fuel through shipping. In the absence of a national value for Singapore, an average value of 3 g CO_2 emitted per metric ton of freight and per km of transportation from modern ships is used to quantify the emissions related to transportation from Singapore to Fiji (Freese 2017). IPCC default water borne navigation CH_4 and N_2O emission factors of 7 kg/TJ and 2 kg/TJ (IPCC 2006a) is low compared to CO_2 emissions and will be considered to be negligible for this study. Fuel consumption of 0.0037 kg/tkm by a container ship using HFO carrying bulk goods is used (Schmied and Knörr 2012), with the HFO calorific value of 41 kJ/g.

9.6.5.3 Diesel Combustion Phase

The final phase under consideration is the combustion of diesel which produces around 2.7 kg CO_2eq emissions per litre. Table 9.15 provides more information on this.

9.7   Estimation of Avoided Emissions

The interest in using biodiesel in Fiji arises due to its potential to cut transport emissions to meet its climate protection targets. An attempt is made in this section to show how much GHG emissions will be avoided according to the scenario considered in this study. This is then compared to the target set by the NDC-IR. Most studies have found that substituting biodiesel for diesel reduces greenhouse gases due to the sequestration of carbon through growth of the feedstock. Calculating avoided emissions presents an approach for evaluating the emissions avoided by using B5 blend in place of diesel for a year. First order estimate of avoided emissions is found for two cases. Case A considers, the no LCA approach, whereby the emissions avoided does not consider the emissions due to the production and use of CME. Therefore for case A, the emissions from CME production is kept at zero and emissions avoided is from diesel not burned due to CME use. While for case B, the emissions due to the production and use of biodiesel is considered for a refined value of avoided emissions: one which considers all the fossil fuel inputs.

9.7.1   Avoided Emissions for Case A: Emissions Considering CME Production Without LCA

The avoided emissions for case A is based on the energy content of diesel and biodiesel B5. The energy content in 1 l of diesel is more than the energy content of 1 l of biodiesel, which consequently means that in terms of volume, a smaller amount of diesel will have the same energy content as a larger amount of biodiesel. In order to find how much energy diesel has compared to biodiesel, the heating values of these fuels are considered. Heating values indicate the energy density of fuels. The heating value of diesel is taken as 45.71 MJ/kg and the heating value of CME is taken as 40.37 MJ/kg (Hossain et al. 2012). By using the volumetric mass density of diesel as 840 kg/m^3 (Pacific Community 2004) and 885 kg/m^3 for CME (Barabás and Todoruț 2011), it is found that 0.93 l of diesel produces around as much energy as 1 l of CME. Burning automotive diesel releases 2.7 kg CO_2eq emissions per litre (Pacific Community 2004). By which account, approximately 58,750 tonnes of CO_2 emissions could be avoided when B5 is used instead of diesel. This reduction is greater than the 35,000 tCO_2/year targeted under the NDC roadmap. However, this estimation does not consider the emissions released in the production of CME. There are various stages in CME production that have fossil fuel inputs and these are considered in the next section.

9.7.2 Avoided Emissions for Case B: Considering Emissions Due to CME Production and Use

Armed with data from the LCA analysis, new estimates for avoided emission is made: one which is inclusive of the emissions released during the production of CME. Table 9.16, shows the emissions generated at each stage of the CME production and use under the scenario considered in this study. A total of 11,200 tonnes of CO_2 eq emissions are released from CME production each year. Considering that 59,000 tonnes is released from diesel use under case A, the avoided emissions is calculated as the difference between emissions that would have resulted from diesel use minus the emissions due to CME production and use. This results in an emissions avoided of 47,500 tonnes per year. The emissions avoided after LCA is smaller compared to the emissions without the LCA, since the production of CME has fossil fuel inputs which release emissions into the atmosphere. It is clearly important to consider the emissions from CME production to gauge the actual emissions saving. The avoided emissions still go beyond the NDC-IR expected target of 35,000 tonnes of emission reduction, which makes biodiesel an environmentally sustainable option for reducing GHG emissions.

Table 9.16 Greenhouse gas emissions released at different stages of CME production for the scenario considered for a year

Stage	Emissions (kgCO$_2$eq)
Palm plantation stage	
Tractor	948,000.00
Transportation of copra to coconut mill	
Land transport	3,620,000.00
Sea transport	2,130,000.00
Coconut mill	
Oil extraction process – electricity	265,000.00
Oil refining process	
Electricity	365,000.00
Diesel usage	725,000.00
Transesterification stage	
NAOH production	190,000.00
Electricity usage during transesterification	2,340,000.00
Methanol	
Production	214,000.00
Shipping	364,000.00
Total emissions	11,200,000.00

9.8   Discussion

The inputs covered in the sections above are used to calculate the GHG emissions and the energy input associated at each stage of the biodiesel production. Table 9.17 shows the GHG emission factors for the four stages of production of CME namely coconut plantation; oil mill and refinery; biodiesel plant; and transportation from the plantation to oil mill. The diesel emission factors for the production, import and combustion of diesel is also shown in Table 9.17.

The total emissions associated with CME and diesel is calculated as 9.61 gCO$_2$eq/MJ and 91.534 gCO$_2$eq/MJ respectively. The emissions factor for biodiesel is lower than those reported in other biodiesel studies. This is primarily due to the scenario condition used in this study of no land changes for the production of CME. Major source of emissions in many studies were due to large scale land changes. For this study, land use change is assumed to be zero since it is based on existing plantations.

Table 9.17 Emission factor based on GHG emissions per energy output for different stages of CME and diesel production and use for a year under the scenario considered

Stage	Emissions factor (gCO$_2$ emissions/MJ CME)
Biodiesel	
Palm plantation stage	
Tractor	0.98
Transportation to coconut mill	
Land transport	0.99
Sea transport	0.03
Coconut milling	
Oil extraction process – electricity	1.25
Oil refining process	
Electricity	1.25
Diesel usage	1.05
Transesterification stage	
NAOH production	0.93
Electricity usage during transesterification	1.25
Methanol	
Production	0.88
Shipping	0.99
Grand total eissions per mj cme	9.61
Diesel	
Reference system – automotive diesel oil	Emissions factor (gCO$_2$ emissions/MJ Diesel)
Diesel production	17.00
Diesel shipping	0.73
Diesel combustion	73.80
Grand total emissions per mj for the reference system	91.53

In addition, other factors such as no fertilizer use, close proximity of biodiesel plant and the coconut mill and placement of mill and plant close to the wharf further reduce the emissions due to production of CME and subsequently result in a lower emission factor.

The GHG emissions in the plantation stage for biodiesel, accounted for 10% of the emissions. The plantation stage is usually the largest contributor towards GHG emissions in biofuel production, especially for cases where degraded, marginal or forested areas are converted to agricultural land for biofuel planting. Since we did not cover land use changes in our calculation, the emissions associated with the plantation stage is marginal, contributing to around only 10% of overall emissions. This is largely from the use of tractors for preparing the land.

The GHG emissions associated with the transportation of dried copra to the oil mill is almost as same as the plantation stage. However, the transportation of copra may increase due to shortage of supply at times, in which case a greater number of trips will have to be made to collect the required amount of copra. An average distance is taken from a province to the coconut mill station, which likely adds uncertainties since some coconut plantations are located in remote locations. While it is important to consider the location and capacity of each farm for supply allocation, such detailed study is beyond the scope of this paper. There are possibilities to reduce the emissions at this stage in future as trucks used for transporting copra begin using B5. Contribution from sea transport was small compared to the overall emissions from transportation with an emission factor of 0.0000288 gCO_2eq/MJ.

The coconut oil mill stage, with the extraction and refining of coconut oil contributed to around 24% of emission factor. There is possibility to reduce the emissions at this stage if the fiber residue from coconuts is used as fuel to generate the electricity for the oil mill processes.

The GHG emissions were largest in the transesterification process (around 40%), half of which is due to the production and import of methanol. Because of the higher GHG emissions in the scenario, the potential for GHG saving in future is greatest at the transesterification process. To reduce the emissions due to methanol use, locally produced sugar-based ethanol can be considered as a substitute.

In the case of diesel fuel, the emission factor was calculated as 91.534 gCO_2eq/MJ. The emission factor for diesel found in this study correlates with the values reported in other works. The European Union (EU) (2009) set the reference value for GHG emissions from fossil fuel at 83.8 g CO_2eq/MJ. The default emission factor according to the European Commission for diesel specifically is taken as 88.7 g CO_2eq/MJ (Edwards et al. 2012). Martínez-González et al. (2011) provides a comparative analysis of high sulfur diesel (DS3000) and low sulfur diesel (DS500) using a LCA approach and find a value for 107.4 gCO_2eq/MJ in DS500 and 103.9 gCO_2eq/MJ in DS3000 fuels. The largest source of emissions in this study is due to combustion of diesel fuel which releases large amounts of carbon dioxide.

The lower total emissions and emission factor calculated for biodiesel indicates that it is an environmentally sustainable fuel. The NDC-IR target for B5 fuel is capable of reducing emissions greater than the 35,000 tonnes reported in the plan, given that the set-up of a biodiesel industry in future is similar to the scenario developed for this study.

9.9 Conclusion

The study compared the GHG emissions factors due to use of biodiesel from CME and diesel and explored the possibility of locally producing biodiesel to meet the NDC Implementation roadmap targets. An evaluation of the current coconut oil production reveal that a significant rehabilitation of the coconut industry is needed to achieve the national B5 fuel target through locally produced biodiesel. By exploring a scenario for biofuel production in Fiji, this study compares the LCA of biodiesel production with automotive diesel oil. The analysis provides results in favor of CME production in Fiji. The emission factor of biodiesel was lower compared to diesel which led to emission saving through use of B5 in place of diesel. The emissions avoided surpass the targeted reduction in the NDC-IR. B5 fuel from coconut oil is shown to have potential as an alternative fuel in place of diesel fuel in Fiji in terms of the environment. A full chain energy analysis and economic impacts should be further studied for overall assessment.

References

Achten, W. M., Vandenbempt, P., Almeida, J., Mathijs, E., & Muys, B. (2010). Life cycle assessment of a palm oil system with simultaneous production of biodiesel and cooking oil in Cameroon. *Environmental Science & Technology, 44*(12), 4809–4815.

Acquaye, A. A., Wiedmann, T., Feng, K., Crawford, R. H., Barrett, J., Kuylenstierna, J., Duffy, P., Koh, S., & McQueen-Mason, S. (2011). Identification of 'carbon hot-spots' and quantification of GHG intensities in the biodiesel supply chain using hybrid LCA and structural path analysis. *Environmental Science & Technology, 45*(6), 2471–2478.

Adewoyin, A. O., & Ajav, E. A. (2013). Fuel consumption of some tractor models for ploughing operations in the sandy-loam soil of Nigeria at various speeds and ploughing depths. *Agricultural Engineering International: CIGR Journal, 15*(3), 67–74.

Asian Development Bank. (2013). *Energy outlook for Asia and the Pacific: October 2013.* Retrieved from https://www.adb.org/sites/default/files/publication/30429/energy-outlook.pdf

Bacolod, E. D. (2014). *Fiji 2020 agriculture sector policy agenda.* Suva, Fiji: Ministry of Agriculture.

Barabás, I., & Todoruț, I. A. (2011). *Biodiesel quality, standards and properties.* In Biodiesel emissions and by-products. InTech, Available from: http://www.intechopen.com/books/biodiesel-quality-emissions-and-by-products/biodiesel-qualitystandards-and-properties

Calumpang, L. M. (2009). *Biofuels in Asia: An analysis of sustainability options.* Policy Brief Series, 2009, Southeast Asian Regional Center for Graduate Study and Research in Agriculture (SEARCA), vol. 2009, pages 1–2.

Cefic E. C. T. A. (2011). Guidelines for measuring and managing CO_2 emission from freight transport operations. *Cefic Report, 1*, 1–18.

Chinnamma, M., Bhasker, S., Madhav, H., Devasia, R. M., Shashidharan, A., Pillai, B. C., & Thevannoor, P. (2015). Production of coconut methyl ester (CME) and glycerol from coconut (*Cocos nucifera*) oil and the functional feasibility of CME as biofuel in diesel engine. *Fuel, 140*, 4–9.

Chua, C. B. H., Lee, H. M., & Low, J. S. C. (2010). Life cycle emissions and energy study of biodiesel derived from waste cooking oil and diesel in Singapore. *The International Journal of Life Cycle Assessment, 15*(4), 417–423.

Cloin, J. (2005a). Biofuels in the Pacific: Coconut oil as a biofuel in Pacific islands. *Refocus, 6*(4), 45–48.

Cloin, J. (2005b). *Coconut oil as a biofuel in Pacific Islands: Challenges and opportunities.* Suva: South Pacific Applied Geoscience Commission.

Consumer Council of Fiji. (2016). *Consumer watch – 2016–2017 National budget.* Retrieved on April 4, 2018, from http://www.consumersfiji.org/upload/Consumer%20Watch/June%20 2016%20Consumer%20Watch.pdf

Das, V. (2018, December 8). Importation of used or reconditioned motor vehicles into Fiji and the issuance of import licenses. *Fiji Sun.* Retrieved from https://fijisun.com.fj/2018/12/08/impor-tation-of-used-or-reconditioned-motor-vehicles-into-fiji-and-the-issuance-of-import-licences/

Department of Agriculture, Economic Planning and Statistics Division. (2009). *Report on the Fiji National Agricultural Census.* Suva: Department of Agriculture, Economic Planning and Statistics Division.

Edwards, R., Mulligan, D., Giuntoli, J., Agostini, A., Boulamanti, A., Koeble, R., Marelli, L., Moro, A., & Padella, M. (2012). Assessing GHG default emissions from biofuels in EU legisla-tion – Review of input database to calculate "Default GHG emissions", following expert con-sultation 22–23 November 2011, Ispra (Italy), Retrieved from http://publications.jrc.ec.europa. eu/repository/bitstream/JRC76057/reqno_jrc76057_default_values_report__online_version1. pdf

Energy Fiji Limited. (2016). *EFL annual report 2016.* Suva: Energy Fiji Limited.

Environmental Protection Agency. (2014). *Emission factors for greenhouse gas inventories.* Retrieved on the December 15, 2018, from https://www.epa.gov/sites/production/files/2015-07/documents/emission-factors_2014.pdf

Eriksson, M., & Ahlgren, S. (2013). LCAs of petrol and diesel: A literature review. *Department of Energy and Technology, Swedish University of Agricultural Sciences, SLU (1654-9406).* Retrieved from https://pub.epsilon.slu.se/10424/17/ahlgren_s_and_ eriksson_m_130529.pdf

European Commission. (2018). *Sustainability criteria: Proposal for updated sustainability crite-ria for biofuels, bioliquids and biomass fuels.* Retrieved from https://ec.europa.eu/energy/en/ topics/renewable-energy/biofuels/sustainability-criteria

European Union (EU). (2009). Directive 2009/28/EC of the European Parliament and of the coun-cil of 23 April 2009 on the promotion of the use of energy from renewable sources and amend-ing and subsequently repealing directives 2001/77/EC and 2003/30/EC. *Official Journal of the European Union, L140,* 16–61.

Fiji Ministry of Agriculture. (2015). *Crop and livestock production performance.* Retrieved from www.agriculture.gov.fj

Fitzgerald, W. B., Howitt, O. J., & Smith, I. J. (2011). Greenhouse gas emissions from the inter-national maritime transport of New Zealand's imports and exports. *Energy Policy, 39*(3), 1521–1531.

Foale, M. (2003). *Coconut Odyssey: The Bounteous possibilities of the tree of life.* Retrieved from the Australian Centre for International Agricultural Research website: https://www.aciar.gov. au/node/8251

Food and Agriculture Organization. (2004). *Fertilizer use by crop.* Retrieved on the December 31, 2018, from http://www.fao.org/fileadmin/templates/ess/ess_test_folder/Publications/ Agrienvironmental/FUBC5thEditioncomplete.pdf

Food and Agriculture Organization. (2008). *The state of food and agriculture 2008: Biofuels: Prospects, risks and opportunities.* Retrieved on the December 31, 2018 from http://www.fao. org/3/a-i0100e.pdf

Food and Agriculture Organization. (2013). *Report of the FAO high level expert consultation on coconut sector development in Asia and the Pacific region.* Retrieved from http://www.fao.org/ fileadmin/templates/rap/files/meetings/2013/Coconut.pdf

Freese N. (2017). *CO$_2$ emissions from international maritime shipping – Regulations, challenges and possibilities* (Working Paper Series 2017: 4). Accessed on 12 Dec 2018 from http://www. unepdtu.org

Government of Fiji. (2014). *Sustainable Energy for All (SE4All): Rapid assessment and gap analy-sis.* Retrieved from prdrse4all.spc.int/system/files/fiji_raga.pdf

Government of Fiji. (2017). Fiji NDC implementation roadmap 2017–2030 – Setting a pathway for emissions reduction target under the Paris Agreement, UNFCCC Accessed on the 15th of May 2018 from http://prdrse4all.spc.int/sites/default/files/1._fijis_energy_sector_outlook_ndc_roadmap.pdf

Hossain, M. A., Chowdhury, S. M., Rekhu, Y., Faraz, K. S., & Islam, M. U. (2012). Biodiesel from coconut oil: a renewable alternative fuel for diesel engine. World Academy of Science. *Engineering and Technology, 68*, 1289–1293.

Index Mundi. (2011). *Fiji – Fertilizer consumption (kilograms per hectare of arable land).* Available on-line at: https://www.indexmundi.com/facts/fiji/indicator/AG.CON.FERT.ZS

International Energy Agency. (2011). *Technology roadmap: Biofuels for transport.* Retrieved from https://www.iea.org/publications/freepublications/publication/biofuels_roadmap_web.pdf

IPCC. (2006a). *IPCC guidelines for national greenhouse gas inventories – Volume 2: Chapter 3.* National Greenhouse Gas Inventories Programme. (IGES, 2006).

IPCC. (2006b). *IPCC guidelines for national greenhouse gas inventories – Volume 2: Chapter 2.* National Greenhouse Gas Inventories Programme. (IGES, 2006).

ISO 14040. (2006). *Life cycle assessment: Principles and framework in environmental management.* Geneva: International Organization for Standards.

Jun, P., Gillenwater, M., & Barbour, W. (2002). *CO_2, CH_4, and N_2O emissions from transportation-water-borne-navigation* [Background paper]. Good practice guidance and uncertainty management in national greenhouse gas inventories. Retrieved from http://www.ipcc-nggip.iges.or.jp/public/gp/bgp/2_4_Water-borne_Navigation.pdf

Knothe, G., Sharp, C. A., & Ryan, T. W. (2006). Exhaust emissions of biodiesel, petrodiesel, neat methyl esters, and alkanes in a new technology engine. *Energy & Fuels, 20*(1), 403–408.

Kumar, S., Singh, J., Nanoti, S. M., & Garg, M. O. (2012). A comprehensive life cycle assessment (LCA) of Jatropha biodiesel production in India. *Bioresource Technology, 110*, 723–729.

Larson, E. D. (2006). A review of life-cycle analysis studies on liquid biofuel systems for the transport sector. *Energy for Sustainable Development, 10*(2), 109–126.

Martínez-González, A., Casas-Leuro, O. M., Acero-Reyes, J. R., & Castillo-Monroy, E. F. (2011). Comparison of potential environmental impacts on the production and use of high and low sulfur regular diesel by life cycle assessment. *CT&F - Ciencia, Tecnología y Futuro, 4*(4), 123–138.

Methanex. (2016). Company annual report 2016. Retrieved on December 10, 2018 from https://www.methanex.com/sites/default/files/investor/annual-reports/Full%20AR%20Methanex.pdf

Methanex. (2018). *Methanol production.* Retrieved on December 10, 2018, from https://www.methanex.com

Morris, J. (2006). *Small Island states bulk fuel procurement of petroleum products: Feasibility study.* Suva: In Pacific Islands Forum Secretariat.

Nair, K. (2014). Prospects for biofuel industry in Fiji; engine performance of CNO blended biodiesel. *American Journal of Energy Research, 2*(4), 99–104.

Nanaki, E. A., & Koroneos, C. J. (2012). Comparative LCA of the use of biodiesel, diesel and gasoline for transportation. *Journal of Cleaner Production, 20*(1), 14–19.

New Zealand Government. (2016). *New Zealand's greenhouse gas inventory.* Retrieved on December 15, 2018, from http://www.cobop.govt.nz/vdb/document/616

Norwegian University of Science and Technology. (n.d.). *Fuels and fuel convertors* [lecture notes]. Retrieved December 13, 2018, from https://www.ntnu.edu/documents/20587845/1266707380/01_Fuels.pdf/1073c862-2354-4ccf-9732-0906380f601e

Pacific Community. (2004). *Pacific Regional energy assessment 2004 – An assessment of the key energy issues, barriers to the development of renewable energy to mitigate climate change, and capacity development needs for removing the barriers.* Retrieved November 20, 2018, from https://www.sprep.org

Pacific Community. (2013). *Consultancy for In-Country waste oil audit for Fiji.* Retrieved November 28, 2018, from https://www.sprep.org

Pacific regional data repository. (2018). *Fiji petroleum imports and re-exports 2015–2017.* Retrieved on April 25, 2019, from http://prdrse4all.spc.int/data/fiji-petroleum-imports-and-re-exports-2015-2017

Pascual, L. M., & Tan, R. R. (2004). Comparative life cycle assessment of coconut biodiesel and conventional diesel for Philippine automotive transportation and industrial boiler application. *Technological Forecasting and Social Change, 58*, 83–103.

Pehnelt, G., & Vietze, C. (2013). Recalculating GHG emissions saving of palm oil biodiesel. *Environment, Development and Sustainability, 15*(2), 429–479.

Pleanjai, S., & Gheewala, S. H. (2009). Full chain energy analysis of biodiesel production from palm oil in Thailand. *Applied Energy, 86*, S209–S214.

PortWorld Distances. (2018). Accessed on http://www.portworld.com

Raghavan, K. (2010). *Biofuels from coconuts*. Wageningen: Fuels from Agriculture in Communal Technology (FACT) Foundation.

Rangaswami, G., & Mahadevan, A. (1998). *Diseases of crop plants in India* (4th ed.). New Delhi, PHI Learning Pvt Ltd.

Schmied, M., & Knörr, W. (2012). *Calculating GHG emissions for freight forwarding and logistics services in accordance with EN 16258*. Bern: European Association for Forwarding, Transport, Logistics and Customs Services (CLECAT).

Searchinger, T., Heimlich, R., Houghton, R. A., Dong, F., Elobeid, A., Fabiosa, J., Tokgoz, S., Hayes, D., & Yu, T. H. (2008). Use of US croplands for biofuels increases greenhouse gases through emissions from land-use change. *Science, 319*(5867), 1238–1240.

Simonen, K. (2014). *Life cycle assessment* (1st ed.). Abingdon, UK: Routledge.

Singh, R. (2008). *Investigating the potential of biodiesel production in Fiji: Facilitating sustainable production in the Fiji Region*. Master's thesis, University of the South Pacific. Retrieved from http://digilib.library.usp.ac.fj/gsdl/collect/usplibr1/index/assoc/HASH09df.dir/doc.pdf

Singh, A. (2012). Biofuels and Fiji's roadmap to energy self-sufficiency. *Biofuels, 3*(3), 269–284.

Sumiani, Y., & Sune, B. H. (2007). Feasibility study of performing an life cycle assessment on crude palm oil production in Malaysia. *International Journal of Life Cycle Assessment, 12*(1), 50–58.

Tan, R. B., Wijaya, D., & Khoo, H. H. (2010). LCI (Life cycle inventory) analysis of fuels and electricity generation in Singapore. *Energy, 35*(12), 4910–4916.

Tuilevuka, N. (2018, August 12). Progamme to develop coconut industry launched. *Fiji Sun*. Retrieved from http://fijisun.com.fj/2018/08/12/progamme-to-develop-coconut-industry-launched/

United Nations Development Programme. (2010). *Fiji renewable energy power project (FREPP) report*. Retrieved on the December 12, 2018, from http://www.fj.undp.org/content/dam/fiji/docs/ProDocs/Fij_FREPP_00076656.pdf

United Nations Environmental Programme. (2013). *Emissions reduction profile: Fiji*. Roskilde, Denmark.

Velasquez, J. M., Jhun, K. S., Bugay, B., Razon, L. F., & Tan, R. R. (2009). Investigation of process yield in the transesterification of coconut oil with heterogeneous calcium oxide catalyst. *Journal-The Institution of Engineers, Malaysia, 70*(4).

Vula, M. (2018, March 22). Coconut industry's future bright: Pillay. *Fiji Sun*. Retrieved from http://fijisun.com.fj/2018/03/23/coconut-industrys-future-bright-pillay/

Wakil, M. A., Ahmed, Z. U., Rahman, M. H., & Arifuzzaman, M. D. (2012). Study on fuel properties of various vegetable oil available in Bangladesh and biodiesel production. *International Journal of Mechanical Engineering, 2*(5), 10–17.

Wicke, B., Dornburg, V., Junginger, M., & Faaij, A. (2008). Different palm oil production systems for energy purposes and their greenhouse gas implications. *Biomass and Bioenergy, 32*(12), 1322–1337.

World Integrated Trade Solution. (2018). *Fiji fuel imports by country and region 2016*. Retrieved on December 15, 2018, from https://wits.worldbank.org/CountryProfile/en/Country/FJI/Year/2012/TradeFlow/Import/Partner/All/Product/Fuels

Zieroth, G., Gaunavinaka, L., & Forstreuter, W. (2007). *Biofuel from coconut resources in Rotuma: A feasibility study on the establishment of an electrification scheme using local energy resources*. Fiji: PIEPSAP, SOPAC.

Chapter 10
Pongamia Biodiesel Production Potential in Vanua Levu: A Full LCA of Emissions Reduction

Salvin Sanjesh Prasad

Abstract Biofuels are becoming increasingly popular as alternative fuels for transport due to rising oil prices and the need for energy security in the Pacific Island Countries (PICs). Indigenous production of biodiesel from second generation feedstock has the potential to reduce the dependence on costly diesel fuel imports and minimize Greenhouse Gas (GHG) emissions while avoiding the food versus fuel controversy associated with first generation biodiesel. Biodiesel production from Pongamia oil has been receiving increasing attention recently, as the oil is inedible and the trees have the ability to survive on many types of soils, including marginal lands.

Vanua Levu is the second largest island of Fiji and has significant land resources available for agriculture, from which approximately 58,897 ha of marginal land can be made available for establishing Pongamia plantations. Using the entire available land resource, the production of 488,834,780.40 L of Pongamia oil per annum has been projected.

While Pongamia oil can be produced domestically and converted to a nominally carbon-neutral Pongamia biodiesel, the life cycle production of such fuels entails emissions to the environment due to the use of fossil fuels and other GHG-producing agents. In this work the environment impacts of these GHG emissions were assessed via Life Cycle Assessment (LCA) in terms of Global Warming Potentials (GWPs), Acidification Potentials (APs) and Eutrophication Potentials (EPs) in air. The Pongamia biodiesel system shows a net Carbon Dioxide (CO_2) emission of 18.32 g CO_2eq/MJ, which is five times lower in comparison with diesel production system that shows a net CO_2 emission of 98.03 g CO_2eq/MJ. The blends of B5, B10, B15, B20 and B100 show 4.07%, 8.13%, 12.20%, 16.26% and 81.31% reduction in CO_2 emissions, respectively, in comparison with diesel. Moreover, the net Sulphur Dioxide (SO_2) emission in Pongamia biodiesel system is 0.056 g SO_2eq/MJ, which

S. S. Prasad (✉)
School of Pure Sciences, Fiji National University, Labasa, Fiji
e-mail: salvin.prasad@fnu.ac.fj

© Springer Nature Switzerland AG 2020   233
A. Singh (ed.), *Translating the Paris Agreement into Action in the Pacific*,
Advances in Global Change Research 68,
https://doi.org/10.1007/978-3-030-30211-5_10

makes relatively low contributions towards the Acidifation Potential (AP) and Eutrophication Potential (EP) of the air. The indigenous production of Pongamia biodiesel shows total possible avoided emissions of approximately 3,202,733.49 kg CO_2eq if diesel fuel is replaced by Pongamia biodiesel produced from the 154 ha of existing Pongamia farms. The reduced emissions of such GHGs indicates that Pongamia biodiesel is a suitable alternative for diesel fuel in outer and remote islands of developing countries in the PICs for operating inter-island shipping vessels, fishing boats and small diesel power plants for household electrification.

Keywords Land resource · Pongamia oil · Biodiesel · Ethanol · Diesel · LCA · GHG emissions · Global warming potential (GWP) · Acidification potential (AP) · Eutrophication potential (EP)

10.1 Introduction

With the reduction in mineral fossil fuel reserves through indiscriminate extraction with lavish consumption and the increase in environmental concerns due to climate change there is an urgent need to investigative alternative fuels which are clean and viable (Keles 2011; Murugesan et al. 2009). With the exception of hydroelectricity and nuclear energy, the majority of the world's energy needs are supplied through petro-chemical sources (Machacon et al. 2001). The impact of this scenario has become a major problem for the developing countries like the Pacific Island Countries (PICs), mainly due to their substantial dependence on imported fossil fuels. An escalating rise in demand for this non-renewable source of energy, due to rising population and forecasted shortage of fossil fuel reserves, has contributed towards significant rise in its market price. The absence of indigenous fossil fuel resource leaves the PICs (including Fiji), with no option but to import these fuels at high costs.

The Fiji National Energy Policy (NEP) indicates that the transportation sector has been the main user of imported fossil fuel, accounting for over 60% of Fiji's total petroleum consumption in recent years (Department of Energy 2013). The NEP and Fiji Bureau of Statistics indicate that the largest imported fuel in Fiji was diesel, which was over 50% of the total fuel imports in the years 2008–2011 (Department of Energy 2013; Whiteside 2013). The dependence on fossil fuels can be minimized by giving priority to developments in alternative, indigenous and cheaper sources of fuel supply. Biofuels are becoming one of the optional substitutes for fossil fuel, as they are more reliable and sustainable form of fuel energy and more environmentally friendly with low carbon emissions (Arpornpong et al. 2014).

The second generation biofuels are currently gaining attention as alternative fuels, primarily for remote and outer islands of PICs to run diesel powered fishing boats, inter-island shipping vessels and smaller diesel power plants for household electricity with no or minimum engine modifications. Unlike the first generation biofuels, the second generation biofuels avoid the food versus fuel controversy, are not cost competitive, and benefit from the ability of growing their feedstocks on marginal lands. They also show reduced emissions when used as fuels. One biofuel feedstock that has great potential to produce a biodiesel substitute for neat diesel in compression ignition engines is Pongamia oil (Kazakoff et al. 2011). Pongamia oil is inedible and contains some toxic components that are not suitable for human food (Atabani et al. 2013; Rahman et al. 2011). The use of Pongamia oil for biofuel production would resolve the issue of food versus fuel, which currently exists in the case (for instance) of coconut oil.

Pongamia seeds have great potential for oil production. The oil content of the kernel is 40% (w/w) (Karmee and Chadha 2005). The Pongamia tree grows fast and begins fruiting after 4–7 years. It has a productive lifespan of at least 60 years (Csurhes and Hankamer 2010). Currently, a private company (Biofuels International Ltd) has commenced Pongamia research and small scale farming in approximately 154 ha of land in a preliminary trial phase in Vanua Levu. The island has a total of approximately 580, 938 ha land resource available for all agricultural activity (Department of Environment 2010), including some possible sites for establishing Pongamia plantation. Pongamia farming further contributes towards intensified afforestation, thus capturing more carbon in form of CO_2 and increasing carbon shares as per the Clean Development Mechanism (CDM) (Lau et al. 2009).

Identification of possible sites for establishing Pongamia farms is one of the foremost important aspects of biodiesel production, as most of the agricultural lands are utilized for crops and vegetables in most developing countries. On the other hand, if the land resource for establishing Pongamia plantations is made available, it is also vital to investigate GHG emissions for the production cycle of Pongamia biodiesel via a full Life Cycle Assessment (LCA) (Hou et al. 2011). The production cycle of Pongamia biodiesel must show reduced emissions of GHGs over the diesel cycle to qualify as an alternative fuel for diesel engines.

This chapter investigates the physical feasibility and some of the important environmental impacts of producing Pongamia biofuels on the island of Vanua Levu in Fiji. It discusses the production suitability, identifies possible sites for establishing Pongamia plantations and projects the production potential of Pongamia oil from the total available land. In addition, the chapter analyses the emissions in the production cycle of Pongamia biodiesel via a full LCA in terms of global warming potentials (GWPs), acidification potentials (APs) and eutrophication potentials (EPs) to compare emissions from the biodiesel production cycle with the diesel production cycle. In addition to providing and accurate determination of the avoided emissions, these results are needed to assess the viability of selecting Pongamia biodiesel as a substitute for diesel in the remote islands of PICs for running inter-island shipping vessels, fishing boats and rural electrification.

10.2 Assessment of Climatic Suitability

Vanua Levu is located 64 km to the north of Viti Levu in Fiji. Vanua Levu enjoys a tropical maritime climate without great extremes of heat or cold. The average monthly maximum and minimum temperature data indicates that the temperature in Vanua Levu has been fluctuating between 10.0 and 36.5 °C (Fiji Meteorological Service 2010, 2011). This monthly temperature range is within the literature tolerance range of −1 to 50 °C for suitability of Pongamia (Meher et al. 2004; Wani et al. 2006).

The annual mean total rainfall data indicates that Vanua Levu receives annual mean rainfall of between 2277 and 2484 mm per annum (Fiji Meteorological Service 2010, 2011). The rainfall data of Vanua Levu is also nearly within the literature rainfall tolerance range of 500–2500 mm per annum, which is suitable for Pongamia (Meher et al. 2004; Wani et al. 2006).

10.3 Soil Assessment for Pongamia Farming

Pongamia can grow on most soils, as this plant is very tolerant to saline and alkaline conditions. Rasul et al. (2012) have mentioned that Pongamia can survive on soils with pH range from 4 to 10 and according to Sangwan et al. (2010), this plant will have optimum growth on soils with pH from 6.5 to 8.5. The soil type in Vanua Levu is diverse at different locations, ranging from mostly acidic to alkaline (ProAnd Associations 2013). The acidity and alkalinity assessments of soil type in Vanua Levu is classified in the range of strongly acidic (with pH less than 5.2), moderate acidic (with pH from 5.3 to 5.9), slightly acidic (with pH from 6.0 to 6.5), neutral (with pH from 6.6 to 7.0), slightly alkaline (with pH from 7.1 to 7.5) and moderate to more alkaline (with pH greater than 7.6). These soil types have been described on the Vanua Levu map in Fig. 10.1 using appropriate legends.

According to Fig. 10.1, most of the lands in Vanua Levu are acidic. According to Rasul et al. (2012), Pongamia has the capacity to survive on acidic soils as well and this indicates that such plant is suitable for most of the lands in Vanua Levu.

10.4 Classes of Land Area in Vanua Levu for Agriculture and Forestry

The total available land area in Vanua Levu is divided in four major classes (Department of Environment 2010). The sort of agricultural activity and forestry is highly dependent on the class of land, as summarized in Table 10.1.

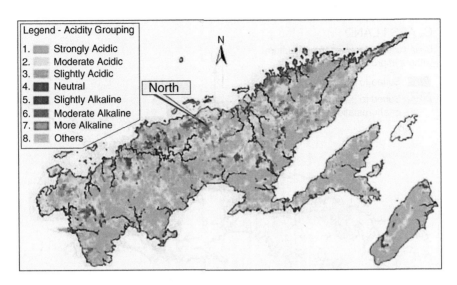

Fig. 10.1 Map of Vanua Levu showing different types of soils (ProAnd Associations 2013)

Table 10.1 Classes of land area in Vanua Levu (Department of Environment 2010)

Land Category	Land Area	Descriptions
Class I	85,143 ha	Land which may be used without improvement, other than by the occasional addition of fertilizers. This is the land that can be used immediately without serious danger of soil erosion or depletion, with the techniques now commonly applied in Vanua Levu in arable farming. Class I soil is commonly flat land.
Class II	73,225 ha	Lands which may be used if more advanced techniques are employed. In general, the techniques of regular fertilizer application or minor drainage or soil conservation measures are needed if class II land is to be used safely. The combined area of Class II land and Class I may be considered to be the area which can be used immediately with the level of technology available in Vanua Levu. (Class II soil is undulating and hilly land).
Class III	242,121 ha	Land that can be fully utilized after major improvements, such as regular and heavy application of fertilizers, major drainage schemes or major soil conservation measures are carried out. These improvements require a large capital investment per unit area and may be considered outside the range of available technology at least until more of the land of better class is fully utilized. However, class III lands are more suitable for forestry. (Class III soils are mainly steep mountainous land or wet land).
Class IV	180,449 ha	Land that is largely unsuited to permanent agriculture, although only some part of it might be used for forestry. (Class IV soils are mainly steep mountainous land).

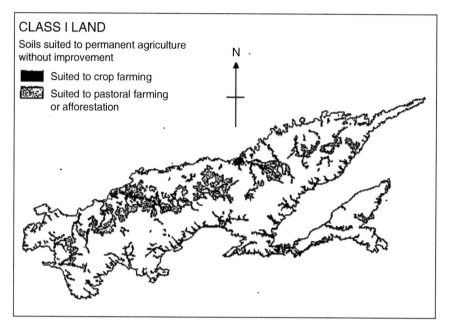

Fig. 10.2 Class I Land allocation in Vanua Levu (Department of Environment 2010)

The classes of land allocations suitable for agricultural activity and forestry shown in Table 10.1 have been identified on Vanua Levu maps using appropriate legends in Figs. 10.2, 10.3 and 10.4.

According to the information given in Table 10.1, maps shown in Figs. 10.2, 10.3 and 10.4, and (Gaunavinaka 2015) most of Class I, II and III soils occupy sugarcane, rice, coconut plantations, crops/vegetable farming, pine plantations and grass land. Class I and II lands are arable lands, which are highly suitable for crops and vegetables. Commercial scale cattle farming has not been reported in Vanua Levu. Some Class III and IV soils are used for softwood-pine plantation and hard wood-mahogany plantation. Other class III and IV soils are mostly covered with scattered, medium and dense forest. Moreover, most of class IV land is very steep and mountainous, due to which, only some parts of these lands can be utilized for forestry.

10.5   Categories of Land Ownership in Vanua Levu

Vanua Levu has three categories of land ownership in Fiji, namely Native Land, Free Hold Land and State Land. The land area covered under Native Land, State Land and Freehold Land are approximately 544,592.17 ha, 8822.96 ha and 80,831.87 ha, respectively (Department of Environment 2010). The availability of unutilized lands or marginal lands as possible sites for Pongamia farming depends entirely upon the land owners. The Native Lands and State lands are mostly leased to farmers.

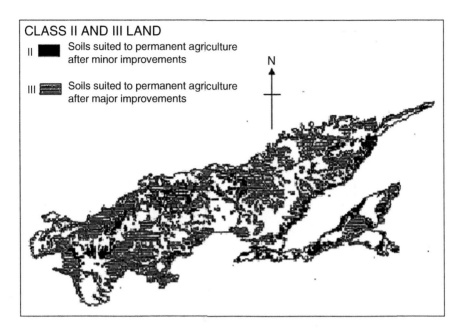

Fig. 10.3 Class II and III Land allocation in Vanua Levu (Department of Environment 2010)

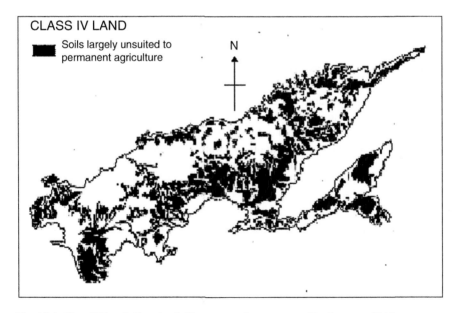

Fig. 10.4 Class IV Land allocation in Vanua Levu (Department of Environment 2010)

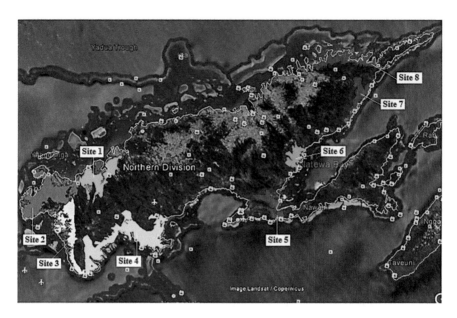

Fig. 10.5 Model of the eight possible Pongamia plantation sites in Vanua Levu

10.6   Site Availability for Pongamia Farming

The analysis of the soil type, classes of land area and categories of land ownership, indicate that the best possible sites for establishing Pongamia farms are most of the class III lands with a minority of some unoccupied class I, II and IV lands and other areas occupying grassland and forest. Google Earth Pro software has been utilized to model the most suitable and available sites for Pongamia farming. The majority of the possible Pongamia plantation sites selected are Native Lands, with the minority being provided by Freehold and State Lands. Figure 10.5 shows eight suitable sites modeled using Google Earth pro software for Pongamia plantations.

The actual land areas for the eight selected sites have been also determined using Google Earth pro software. However, the available land areas have been approximated by eliminating the land areas occupied by drains, creeks, rivers, some farm houses, livestock farming, rocky surface, some boggy areas, very steep/undulating land topography, etc. from the actual land area. The land characteristics of all the eight possible sites are shown in Table 10.2.

Table 10.2 shows the total available land area identified for Pongamia farming. The total land area available for Pongamia farming including existing 154 ha Pongamia farms is 58,897 ha.

Table 10.2 Land characteristics of possible available sites

Possible Sites	Major Region or District Names	Available Land Area of Site (ha)
Site 1	Lekutu, Delainaroga, Nasarowaqa, Dreketi	6,704
Site 2	Bua, Koroinasolo, Navunievu, Navakasiga	13,270
Site 3	Dama, Naruwai, Nabouwalu, Vuya	7,245
Site 4	Nabou, Solevu, Nasolo, Nasavu, Sawani, Daria, Kubulau,	16,833
Site 5	Navatu, Nacavanadi, Bucalevu	2,891
Site 6	Korotasere, Vuinadi, Wavu, Vatukuca	3,395
Site 7	Sese, Naboutini, Lakeba	2,930
Site 8	Namuka, Qelewara, Korokalo, Qaranivai, Lagi	5,475
Existing Pongamia plantation at Lekutu and Nasarowaqa farms		154
Total available land area of Pongamia plantations (with existing farms) = 58,897 ha		
Total available land area of Pongamia plantations to be established = 58,743 ha		

Fig. 10.6 Image of Pongamia pods and tree on the existing Pongamia plantation of Biofuels International Limited

10.7   Assessment of Seed Requirement to Establish Pongamia Farms on Total Available Land

For optimum utilization of land resource, 500 seedlings have been considered per hectare of land at a planting spacing of 4 m within rows by 5 m between rows (Kesari et al. 2010). Considering the seed germination rate of 80% (Wani et al. 2006) and seedling survival rate of 95% at normal climatic conditions (Syamsuwida et al. 2015), the total number of seeds required for total available land is nearly 36,714,382 seeds.

The existing 154 ha of Pongamia farms have been assessed for supplying the seeds to establish plantations on total available land. Figure 10.6 shows a picture taken from one of the existing Pongamia plantations of Biofuels International Limited.

The plant spacing of the existing Pongamia farms, as shown in Fig. 10.6, is approximately an average of 4.5 m within rows and 4.5 m between rows. It is to be noted that Pongamia bears seeds once per year. Considering the yearly yield of seed production to be an average of 9–90 kg seeds/tree (Karmee and Chadha 2005) and the average weight of one Pongamia seed to be 3.18×10^{-3} kg at normal climatic conditions (Raut et al. 2011a), the total number of seed production from 154 ha of farms is approximately 1,181,929,245 seeds, which successfully meets the seed demand to establish farms on total available land.

10.8 Production Potential of Pongamia Oil from Total Available Land Area and from Existing 154 ha of Pongamia Farms

The production potential of Pongamia oil from total available land has been considered at full harvest. A planting spacing of 500 trees/ha (Kesari et al. 2010) and seed production yield with an average of 9–90 kg seeds/tree (Karmee and Chadha 2005) has been considered. Pongamia has up to 40% (w/w) of oil per seed (Karmee and Chadha 2005; Sangwan et al. 2010), from which approximately 31% (w/w) of the oil can be extracted using oil expelling machinery (Nabi et al. 2009). The annual yield of Pongamia oil produced from 58,897 ha of Pongamia plantations is approximately 450,705,667.50 kg. By considering the density of Pongamia oil as 0.922 g/cm^3 (through laboratory tests using a pycnometer), the annual yield of Pongamia oil from total available land is 488,834,780.40 L.

The existing 154 ha of Pongamia farms raised by Biofuels International Limited have started fruiting. At full harvest, the annual yield of Pongamia oil is approximately 1,278,173.02 L.

10.9 Indigenous Ethanol Production as a Feedstock for Biodiesel

Ethanol is a vital input component of biodiesel (ethyl ester) production through the process of transesterification (Baiju et al. 2009). Apart from Pongamia oil production, local production of ethanol would further reduce the cost of ethanol import as a feedstock for biodiesel. On the same note, the current decrease in market price of sugar produced in Fiji provides opportunities for Fiji Sugar Corporation (FSC) to begin local production of ethanol for its success (Oxfam International 2005).

10.10 Life Cycle Assessment (LCA) of Pongamia Biodiesel Production

Pongamia biodiesel has very high scope of production in Vanua Levu, yet it entails emissions of GHG's to the environment, which are GWP's, AP's and EP's in air. The impact of these environmental emissions from Pongamia Biodiesel system has been assessed using LCA. The analysis involves setting the goal and scope of work with appropriate system boundary, gathering the life cycle data inventory for all inputs and outputs of Pongamia biodiesel system, evaluating the impact assessment and interpreting the data.

10.10.1 Goal and Scope of Work

The goal of carrying out LCA is to assess the GWP's and AP's. The system boundary of the production cycle has been scoped based on cradle-to-grave for Pongamia biodiesel. The system boundary has been divided into three major parts, which are inclusive of Pongamia cultivation (plantation, nurturing and harvest), Pongamia oil production (crude Pongamia oil extraction and refining) and transesterification into biodiesel (reaction process and biodiesel purification). The LCA has been carried out for establishing Pongamia farms on total available land and the functional unit has been defined as 1 MJ of energy available in Pongamia biodiesel (B100). The primary and secondary data sources have been used to compile the inventory analysis. The system boundary and processes are presented in Fig. 10.7.

10.10.2 Life Cycle Inventory

Cultivation
- The weight of seeds required to establish Pongamia farms is estimated at 500 seedlings per hectare at a planting spacing of 4 m within rows by 5 m between rows (Kesari et al. 2010). The seed germination rate of 80% (Wani et al. 2006) and seedling survival rate of 95% has been considered at normal climatic conditions (Syamsuwida et al. 2015). The average weight of one Pongamia seed is approximated at 3.18×10^{-3} kg (Raut et al. 2011a).
- Polybags requirement for raising Pongamia seedlings is estimated by considering the weight of 330 polybags at 1.65 kg (Chandrashekar et al. 2012).
- Pongamia has the capability to replenish nitrogen itself in the soil once the lateral roots develop root nodules. Organic fertilizer is utilized only during plantation phase at approximately 2 kg per pit (Chandrashekar et al. 2012).

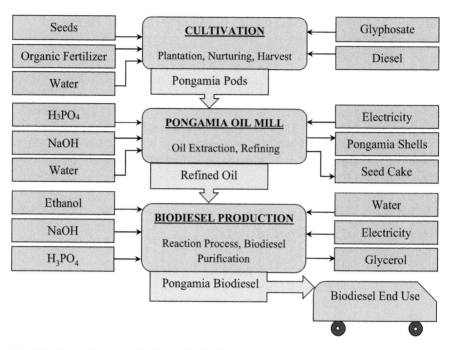

Fig. 10.7 System boundary for Pongamia biodiesel production

- The diesel requirement for irrigating seedlings during dry seasons is estimated at an average total head measurements of 19.29 m and using the pump data (CottonInfo 2015).
- Glyphosate is selected for weed control. Its requirement is estimated at 100 ml per knapsack spray (16 L) and a spray delivery of ten knapsack sprays/ha (Bajera 2014).

The annual yield of Pongamia seed production from 58,897 ha of Pongamia plantations is estimated at 500 trees/ha (Kesari et al. 2010) and seed production yield with an average of 9–90 kg seeds/tree (Karmee and Chadha 2005).

Pongamia Oil Mill: Crude Pongamia Oil Extraction and Refining
- The electricity consumption for decorticating Pongamia seeds (removal of shells from seeds) is estimated using a decorticator with processing capacity of 4 tons/h and input power of 29 HP. The electricity consumption for extracting oil using oil press machine is estimated at a processing capacity of 100 tons/day and input power of 22 kW. Other energy requirement during oil extraction and refining are estimated using the data of oil extraction and refining from Samoa at an oil production capacity of 1176 L/day (Demafelis et al. 2009). This includes daily energy consumption of conveyor, pump, filter press, lights and other miscellaneous, which are 8 kWh, 12 kWh, 22.92 kWh, 4 kWh and 2 kWh, respectively.
- The estimated weight of shell of one Pongamia pod is determined by taking the difference between the pod weight (inclusive of seed) to be average of

2.80×10^{-3} kg to 7.64×10^{-3} kg and the average seed weight of 3.18×10^{-3} kg (Raut et al. 2011b).

- Crude Pongamia oil extraction from Pongamia seeds is estimated at 31% (w/w) using oil press machine (Nabi et al. 2009).
- Seed cake produced from Pongamia seeds is estimated at 69% (w/w) using oil expelling machinery (Nabi et al. 2009).
- Top degumming method is selected for degumming or refining Pongamia oil to remove impurities, such as phospholipids (Kulkarni et al. 2014; Zufarov et al. 2008). According to this method, phosphoric acid (14% – weight) is utilized in amount of 0.1% by weight of crude Pongamia oil for degumming. Crude Pongamia oil is initially added with water at 5% by volume and heated to 80 °C. NaOH (20% water solution) is utilized in amount of 0.1% by weight of crude Pongamia oil for partially neutralizing phosphoric acid.
- Using top degumming method, 10.020 g of crude oil yields 9.6993 g of degummed or refined Pongamia oil (Kulkarni et al. 2014; Zufarov et al. 2008). The density of refined Pongamia oil is considered as 0.920 g/cm^3 (through laboratory tests using a pycnometer).

Biodiesel Production: Transesterification into Ethyl Esters
- The input of ethanol is estimated at an ethanol to refined oil ratio of 3:5 (v:v) (Sales 2011). It has to be noted that sugarcane ethanol from FSC has lots of prospects to cater for ethanol requirement in biodiesel production.
- The requirement for NaOH catalyst is estimated by considering 1.5 g of NaOH at 120 ml of ethanol and 200 ml of refined oil (Sales 2011).
- H_3PO_4 requirement for neutralizing the catalyst is estimated at 0.00342 kg per L of biodiesel (Kittithammavong et al. 2014).
- Water requirement for biodiesel purification is estimated at 1000 L per 19891.49 L of biodiesel (Sampattagul et al. 2011).
- Electricity requirement for transesterification and biodiesel purification is estimated at 34 kWh per 900 kg of Pongamia pod (equivalent to 126 kg of Pongamia biodiesel) (Chandrashekar et al. 2012). The density of Pongamia biodiesel is considered as 0.90 g/cm^3 (Rengasamy et al. 2014).
- The yield of Pongamia biodiesel from refined oil in a plant is estimated at 96.5% (w/w) (Sales 2011). The density of Pongamia biodiesel has been considered as 0.90 g/cm^3 (Rengasamy et al. 2014).
- The energy content of Pongamia biodiesel is 36.5 MJ/kg (Chandrashekar et al. 2012).
- The yield of glycerol output is estimated at 2000 tons per 20,000 tons of biodiesel (Sales 2011).

Transportation at All Stages
Eight ton trucks are selected to cater for all transportation needs for the production cycle and 20,000 L oil tanker trucks are selected for transporting Pongamia biodiesel for end use. All transportation distances are return trips and are determined

using Google Earth Pro software. The numbers of return trips are also taken into account. The transportation details are as follows:

- Pongamia seeds are transported from 154 ha of existing Pongamia farms in Lekutu to all eight sites for raising seedlings in nurseries. One return trip is estimated to cart 8 tons of seeds.
- Pongamia seedlings are transported from nurseries at the eight sites to farms within the site for plantation. One return trip is estimated to cart 7052 seedlings using built-in stacks on the truck tray.
- Organic fertilizer is prepared at site 1 and transported to farms at sites 1, 2, 3 and 4. In sites 5, 6, 7 and 8, organic fertilizer is prepared within the site and transported to farms. One return trip is estimated to cart 8 tons of fertilizer.
- Glyphosate is transported from Nabouwalu (proposed site for biodiesel factory) at site 3 to all the eight farm sites. One return trip is estimated to cart 2000 L of glyphosate.
- Pongamia pods are transported from all the eight farm sites to oil extraction and biodiesel production site, proposed to be located at site 3 in Nabouwalu, which is also the main port of entry in Vanua Levu. One return trip is estimated to cart 8 tons of Pongamia pods.
- Pongamia biodiesel are transported for end use from site 3 in Nabouwalu to Labasa. One return trip is estimated to cart 20,000 L of Pongamia biodiesel.

The one way transportation distance for entire transportation needs during the production cycle is summarized in Table 10.3.

The life cycle inventory data for inputs and outputs of Pongamia biodiesel production are presented in Table 10.4. The inputs and outputs are determined per L of biodiesel (/L B100) and per MJ of energy from Pongamia biodiesel (/MJ). The functional unit is selected as /MJ so that the GHG emissions from biodiesel could be compared with the reference system of diesel emissions.

The life cycle inventory data presented in Table 10.4 is utilized as an input data to carry out life cycle impact assessment.

Table 10.3 Data for transportation at all stages from production to end use

Transportation Distance (1 Way)	Site 1	Site 2	Site 3	Site 4	Site 5	Site 6	Site 7	Site 8
Seed Transport (km)	Site of seed distribution	36.6	42.1	69.9	89.2	116.0	157.0	151.0
Seedling Transport (km)	7.67	10.17	8.66	13.94	3.40	4.42	3.64	5.29
Organic Fertilizer Transport (km)	7.67	36.6	42.1	69.9	3.40	4.42	3.64	5.29
Glyphosate Transport (km)	42.1	25.5	8.66	39.8	90.3	158.1	199.1	193.1
Pongamia Pod Transport (km)	42.1	25.5	8.66	39.8	90.3	158.1	199.1	193.1
Pongamia Biodiesel Transport (km)	134 (Transportation distance from Nabouwalu to Labasa)							

Table 10.4 Inventory data for calculations

Stages	Amount	
	/L B100	/MJ
Pongamia Cultivation		
Input		
Seeds (kg)	0.00024	7.31E-06
Polybags (kg)	0.00039	1.19E-05
Organic fertilizer (kg)	0.15697	4.78E-03
Diesel (water pump - irrigating seedlings) (L)	8.0E-07	2.44E-08
Glyphosate (L)	0.00050	1.52E-05
Diesel (truck round trip - seed transport) (km)	5.9E-06	1.80E-07
Diesel (truck round trip - seedling transport) (km)	0.00016	4.87E-06
Diesel (truck round trip - fertilizer transport) (km)	0.00110	3.35E-05
Diesel (truck round trip - glyphosate transport) (km)	0.00003	9.13E-07
Diesel (truck round trip - Pongamia pod transport) (km)	0.08121	2.47E-03
Output		
Harvested seeds (kg)	3.10800	9.46E-02
Pongamia Oil Extraction and Refining		
Input		
Electricity (kWh)	0.07440	2.26E-03
H_3PO_4 (oil refining) (kg)	0.00196	5.97E-05
NaOH (acid neutralizing) (kg)	0.00289	8.80E-05
Water (L)	0.05225	1.59E-03
Output		
Pongamia shells (kg)	1.99381	6.07E-02
Pongamia seedcake (kg)	2.14452	6.53E-02
Crude Pongamia oil (L)	1.04499	3.18E-02
Refined Pongamia oil (L)	1.01374	3.09E-02
Biodiesel Production		
Input		
Ethanol (L)	0.60825	1.85E-02
NaOH (catalyst) (kg)	0.00760	2.31E-04
H_3PO_4 (biodiesel purification) (kg)	0.00342	1.04E-04
Electricity (kWh)	0.24286	7.39E-03
Water (L)	1.01097	3.08E-02
Diesel (oil tanker round trip - biodiesel transport) (km)	0.01340	4.08E-04
Output		
Pongamia biodiesel (B100) (L)	1.00000	3.04E-02
Glycerol (kg)	0.09000	2.74E-03

10.10.3   Life Cycle Impact Assessment

The life cycle impact assessment is aimed at evaluating the contributions of GHG's towards global warming, acidification and eutrophication in air. The basic approach to determine GHG emissions is given in Eq. 10.1.

$$GHG\,\text{Emissions} = \text{Activity Data} \times \text{Emissions Factor} \quad (10.1)$$

Where:

Activity data is the quantity of input in the production process

Emissions Factor is the quantity of GHG emission by respective input per functional unit

Source: (Kittithammavong et al. 2014)

The associated emissions of GHGs are given in terms of emissions factor and these are collected from the literature.

10.10.3.1   Global Warming Potential (GWP)

The GHG's for GWP are CO_2 (Carbon Dioxide), CO (Carbon Monoxide) and N_2O (Nitrous Oxide), which are assessed in grams of CO_2 equivalent per MJ of energy in biodiesel (g CO_2eq/MJ). The conversions are carried out by considering the emission of 1 kg of N_2O to be equivalent to 298 kg of CO_2eq (Hull 2009) and the emission of 1 g of CO to be equivalent to 1.57 g of CO_2eq (Lower 2004). The emissions of GHG's as GWPs during the stages of Pongamia cultivation, oil extraction/refining, biodiesel production and biodiesel end use (through tailpipe emissions) are presented in Table 10.5.

According to Table 10.5, the production cycle of Pongamia biodiesel system, shows total emission of 93.81 g CO_2eq/MJ. However, the tailpipe emissions of CO_2 come from the biomass carbon which is present in the oil used as feedstock. This CO_2 is subtracted from the diesel engine emissions as part of the biological recycle of carbon (Sheehan et al. 2000). Thus, the net CO_2 is calculated by setting the biodiesel end use (i.e. CO_2 emissions from the tailpipe) in Table 10.5 to zero. This gives a net CO_2 emission of 18.32 g CO_2eq/MJ in the Pongamia biodiesel system. Biomass derived fuels, such as Pongamia biodiesel, reduces the net atmospheric carbon by

Table 10.5 GHG emissions – CO_2eq

Items	Emissions factor: (literature values)	Emissions factor: g CO_2eq/Item	g CO_2eq/MJ
Cultivation Phase			
Polybags production and discharge (kg)	5.5 kg CO_2/kg polybag (Chandrashekar et al., 2012)	5500.0 g CO_2eq/kg polybag	0.06545
Organic fertilizer (kg) (N_2O)	0.01 kg N_2O-N/kg N × 44/28 (kg N_2O/ha) (FAO UN, 2015; IPCC-GNGGI, 2006)	3.8 g CO_2eq/kg fertilizer	0.01816
Diesel fuel – irrigation (L)	15.6 kg CO/m^3 diesel (Commonwealth of Australia, 2008)	24.0 g CO_2eq/L diesel	5.86E-07
Glyphosate (L)	16 kg CO_2eq/kg glyphosate (Silalertruksa & Kawasaki, 2015)	16000.0 g CO_2eq/kg glyphosate	0.24320
Diesel fuel – transportation for cultivation phase (km)	774 g CO_2/km (Samaras & Zierock, 1999)	774.0 g CO_2eq/km diesel	1.94232
Sub Total			2.26913
Pongamia Oil Extraction and Refining			
Electricity (kWh)	0.5095 ton CO_2eq/MWh (UNEP RISO, 2013)	509.5 g CO_2eq/kWh	1.15147
Phosphoric acid (kg)	3124.7 g CO_2eq/kg H_3PO_4 (JRC Science for Policy Report, 2017)	3124.7 g CO_2eq/kg H_3PO_4	0.18654
Sodium hydroxide (kg)	529.7 g CO_2eq/kg NaOH (JRC Science for Policy Report, 2017)	529.7 g CO_2eq/kg NaOH	0.04661
Sub Total			1.38462
Biodiesel Production			
Ethanol Production (L)	0.548 kg CO_2eq/L ethanol (Numjuncharoen, Papong, Malakul, & Mungcharoen, 2015)	548.0 g CO_2eq/L ethanol	10.13800
Sodium hydroxide (kg)	529.7 g CO_2eq/kg NaOH (JRC Science for Policy Report, 2017)	529.7 g CO_2eq/kg NaOH	0.12236
Phosphoric acid (kg)	3124.7 g CO_2eq/kg H_3PO_4 (JRC Science for Policy Report, 2017)	3124.7 g CO_2eq/kg H_3PO_4	0.32497
Electricity (kWh)	0.5095 ton CO_2eq/MWh (UNEP RISO, 2013)	509.5 g CO_2eq/kWh	3.76521
Diesel fuel – biodiesel transportation (km)	774 g CO_2/km (Samaras & Zierock, 1999)	774.0 g CO_2eq/km diesel	0.31580
Sub Total			14.66634
Biodiesel End Use *(CO2 emissions from tailpipe)*			
Biodiesel combustion in engine (B100) (L)	2.48 ton CO_2/m^3 B100 (Coronado, de Carvalho Jr, & Silveira, 2009)	2480.0 g CO_2eq/L B100	75.49467
Sub Total			75.49467
Total			**93.81476**

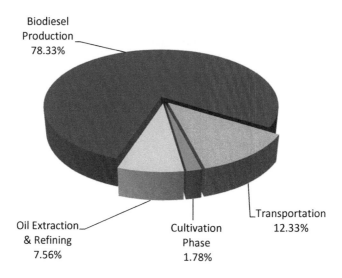

Fig. 10.8 CO_2 emissions in Pongamia biodiesel system

participating in the relatively rapid biological cycling of carbon to the atmosphere (via engine tailpipe emissions) and from the atmosphere (via photosynthesis). The net result is that such fuels partially displace fossil fuel emissions of the diesel fuel they displace. It is also of interest that 1 ha of Pongamia plantation effectively sequesters the total CO_2 released (1.5 ton/ha) during the life cycle (Chandrashekar et al. 2012).

The CO_2 emissions at different stages of life cycle Pongamia biodiesel production is shown in Fig. 10.8.

According to Fig. 10.8, the biodiesel production stage is the biggest contributor for CO_2 emissions in the system (78.33%). Ethanol production is the major contributor of increased emissions during biodiesel production stage, yet the magnitude of its emissions is not so high. The emissions data considered for ethanol production has been estimated using a well to tank analysis using sugarcane ethanol, where the input production process energy is supplemented by bagasse as biomass (CRC Report 2013). Such analysis shows that if indigenous sugarcane ethanol is produced by FSC as biodiesel feedstock, the net CO_2 emissions from Pongamia biodiesel system will be low and revolve around 18.32 g CO_2eq/MJ.

The CO_2 emissions from the reference system of diesel production have been considered using a well to wheel approach, i.e. from crude oil extraction, up to the end use through combustion in diesel engine. The net CO_2 emissions of diesel system has been estimated to revolve around 98.03 g CO_2eq/MJ of energy from diesel (CRC Report 2013). The net CO_2 emissions from Pongamia biodiesel system indicates that it is a suitable alternative to substitute diesel fuel with reduced emissions. However, if Pongamia biodiesel is blended with diesel to avoid any engine modifications, the blends will also show reduced emissions. According to Fiji NEP, biofuel

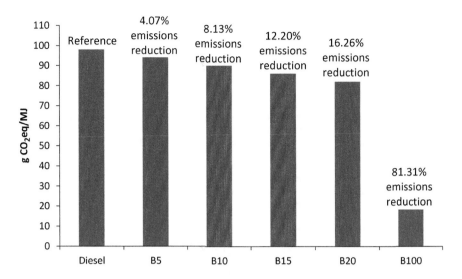

Fig. 10.9 Comparison of net CO_2 life cycle emissions for diesel and Pongamia biodiesel blends

standards for B5 (blend of 5% vegetable oil ester and 95% diesel) has already been approved. The comparisons of CO_2 emissions of B5, B10, B15, B20 and B100 with diesel (reference) are shown in Fig. 10.9.

According to Fig. 10.9, Pongamia biodiesel system with B100 shows a significant 81.31% reduction in CO_2 emissions in comparison with diesel. The GWP reduces five times in comparison with diesel. Moreover, the blends of B5, B10, B15 and B20 shows 4.07%, 8.13%, 12.20% and 16.26% reduction in CO_2 emissions, respectively.

The existing 154 ha of Pongamia farms have the potential to produce 1,278,173.02 L of Pongamia oil annually at full harvest. Such yield could produce nearly 1,223,145.27 L of biodiesel (B100). Considering the CO_2 emissions from diesel at 98.03 g CO_2eq/MJ (CRC Report 2013) and from Pongamia biodiesel at 18.32 g CO_2eq/MJ, the total possible avoided emissions by replacing diesel with 1,223,145.27 L of Pongamia biodiesel is approximately 3,202,733.49 kg CO_2eq.

10.10.3.2   Acidification Potential (AP) and Eutrophication Potential (EP) in Air

The GHG's for acidification and eutrophication in air are SO_2 (Sulphur Dioxide), NO_x (Oxides of Nitrogen) and NH_3 (Ammonia), which are assessed in grams of SO_2 equivalent per MJ of energy in biodiesel (g SO_2eq/MJ). The conversions are carried out by considering the emission of 1 kg of NO_x to be equivalent to 0.70 kg of SO_2eq and the emission of 1 kg of NH_3 to be equivalent to 1.88 kg of SO_2eq (Benders et al. 2012). The emissions of GHGs as APs and EPs in air during the stages of Pongamia

Table 10.6 GHG emissions – SO$_2$eq

Items	Emissions factor: (Literature)	Emissions factor: g SO$_2$eq/Item	g SO$_2$eq /MJ
Cultivation Phase			
Polybags production and discharge (kg) (NO$_x$)	0.0714 kg SO$_2$/kg polybag (Chandrashekar et al., 2012)	71.4 g SO$_2$eq/kg polybag	0.00085
Polybags production and discharge (kg) (NO$_x$)	0.126 kg SO$_2$eq/kg polybag (Chandrashekar et al., 2012)	126.0 g SO$_2$eq/kg polybag	0.00150
N Volatilization (kg) (NH$_3$)	0.1× 0.01 kg N$_2$O-N/kg N × 44/28 (kg N$_2$O/ha) (FAO UN, 2015; IPCC-GNGGI, 2006)	2.37E-03 g SO$_2$eq/kg fertilizer	0.00001
Diesel fuel - irrigation (L) (NO$_x$)	0.2 × 72.5 kg NO$_x$/m^3 diesel (Commonwealth of Australia, 2008)	10.2 g SO$_2$eq/L diesel	2.49E-07
Diesel fuel – transportation for cultivation phase (km) (NO$_x$)	10.4 g NO$_x$/km (Samaras & Zierock, 1999)	7.3 g SO$_2$eq/km diesel	0.01832
Sub Total			0.02068
Pongamia Oil Extraction and Refining			
Electricity (kWh) (NO$_x$)	0.28 g NO$_x$/kWh (Kawashima et al., 2015)	0.2 g SO$_2$eq/kWh	0.00045
Sub Total			0.00045
Biodiesel Production			
Electricity (kWh) (NO$_x$)	0.28 g NO$_x$/kWh (Kawashima et al., 2015)	0.2 g SO$_2$eq/kWh	0.00148
Diesel fuel – biodiesel transportation (km) (NO$_x$)	10.4 g NO$_x$/km (Samaras & Zierock, 1999)	7.3 g SO$_2$eq/km diesel	0.00298
Sub Total			0.00446
Biodiesel End Use			
Biodiesel combustion in engine (B100) (L)	0.15 g SO$_2$eq/ha (Chandrashekar et al., 2012)	1.0 g SO$_2$eq/L 100	0.03044
Sub Total			0.03044
Total			**0.05603**

cultivation, oil extraction/refining, biodiesel production and biodiesel end use (tail-pipe emissions) are presented in Table 10.6.

According to Table 10.5, the production cycle of Pongamia biodiesel system, shows total emission of APs and EPs of 0.056 g SO$_2$eq/MJ. The emission of APs and EPs in the life cycle production of Pongamia biodiesel is relatively low. Thus, its contributions towards acidification and eutrophication in air are insignificant.

10.11   Conclusions

The total available land area for establishing Pongamia plantations in Vanua Levu is 58,897 ha. Approximately 488,834,780.40 L of Pongamia oil can be produced from the entire available land. The LCA of Pongamia biodiesel system shows a net CO$_2$

emission of 18.32 g CO_2eq/MJ, which is very low in comparison with diesel production system that shows a net CO_2 emission of 98.03 g CO_2eq/MJ. The GWP reduces five times in comparison with diesel. The blends of B5, B10, B15, B20 and B100 show 4.07%, 8.13%, 12.20%, 16.26% and 81.31% reduction in CO_2 emissions, respectively, in comparison with diesel. The life cycle emission of Pongamia biodiesel system shows a net SO_2 emission of 0.056 g SO_2eq/MJ, which has relatively low AP and EP in air. The production of Pongamia biodiesel from existing 154 ha of Pongamia farms and replacement of diesel with the Pongamia biodiesel produced gives total avoided emissions of approximately 3,202,733.49 kg CO_2eq. The analysis of GHG emissions indicate that Pongamia biodiesel is a potential substitute for diesel due to its lower emissions level, which could be utilized for operating inter-island shipping vessels, fishing boats and small diesel power plants for household electrification in remote and outer islands of PICs.

Acknowledgement The author would like to sincerely thank Biofuels International Limited and Department of Lands for providing data to identify suitable sites for establishing Pongamia farms.

References

Arpornpong, N., Attaphong, C., Charoensaeng, A., Sabatini, D. A., & Khaodhiar, S. (2014). Ethanol-in-palm oil/diesel microemulsion-based biofuel: Phase behavior, viscosity, and droplet size. *Fuel, 132*, 101–106.

Atabani, A., Silitonga, A., Ong, H., Mahlia, T., Masjuki, H., Badruddin, I. A., & Fayaz, H. (2013). Non-edible vegetable oils: A critical evaluation of oil extraction, fatty acid compositions, biodiesel production, characteristics, engine performance and emissions production. *Renewable and Sustainable Energy Reviews, 18*, 211–245.

Baiju, B., Naik, M., & Das, L. (2009). A comparative evaluation of compression ignition engine characteristics using methyl and ethyl esters of Karanja oil. *Renewable Energy, 34*(6), 1616–1621.

Bajera, B. G. (2014). *Corn farmers' practices in herbicide spraying under no-till farming.* Retrieved from www.cropsreview.com/corn-farming.html

Benders, R. M., Moll, H. C., & Nijdam, D. S. (2012). From energy to environmental analysis: Improving the resolution of the environmental impact of Dutch private consumption with hybrid analysis. *Journal of Industrial Ecology, 16*(2), 163–175.

Chandrashekar, L. A., Mahesh, N., Gowda, B., & Hall, W. (2012). Life cycle assessment of biodiesel production from pongamia oil in rural Karnataka. *Agricultural Engineering International: CIGR Journal, 14*(3), 67–77.

Commonwealth of Australia. (2008). *National pollutant inventory – emissions estimation techniques for combustion engines* (pp. 1–89). Retrieved from http://www.npi.gov.au/system/files/resources/afa15a7a-2554-c0d4-7d0e-d466b2fb5ead/files/combustion-engines.pdf

Coronado, C. R., de Carvalho, J. A., Jr., & Silveira, J. L. (2009). Biodiesel CO_2 emissions: A comparison with the main fuels in the Brazilian market. *Fuel Processing Technology, 90*(2), 204–211.

CottonInfo. (2015). *Fundamentals of energy use in water pumping.* Retrieved from http://www.cottoninfo.com.au/sites/default/files/documents/Fundamentals%20EnergyFS_A_3a.pdf

CRC Report. (2013). *Transortation fuel life cycle assessment: Validation and uncertainty of well-to-wheel GHG estimates.* Retrieved from Canada: https://crcao.org/reports/recentstudies2013/E-102/CRC%20E%20102%20Final%20Report.pdf

Csurhes, S., & Hankamer, C. (2010). *Invasive plant risk assessment: Pongamia (Millettia pinnata syn. Pongamia pinnata)* (pp. 11–12). Queensland, Australia: Department of Agriculture and Fisheries Biosecurity.

Demafelis, R., Dominigo, A., & FAO Consultants. (2009). *Samoa biofuel study report*. Retrieved from http://www.fao.org/docrep/013/am012e/am012e00.pdf

Department of Energy. (2013). *Fiji national energy policy*. Retrieved from http://www.fijiroads.org/wp-content/uploads/2016/08/Final20DRAFT20Fiji20National20Energy20Policy20Nov202013.pdf

Department of Environment. (2010). *Fiji's fourth national report to the United Nations convention on biological diversity*. Retrieved from https://www.cbd.int/doc/world/fj/fj-nr-04-en.pdf

FAO UN. (2015). Global database of GHG emissions related to feed crops. Global database of GHG emissions related to feed crops Retrieved 8/10/2018, from Food and Agriculture Organization of the United Nations and Livestock Environmental Assessment and Performance Partnership http://www.fao.org/fileadmin/user_upload/benchmarking/docs/leap_user_guide_version1_2015.pdf

Fiji Meteorological Service. (2010). *Climatological summary for Nabouwalu, Fiji Islands*. Retrieved from http://www.met.gov.fj/climate_services.php

Fiji Meteorological Service. (2011). *Climatological summary for Labasa Airfield, Fiji*. Retrieved from http://www.met.gov.fj/climate_services.php

Gaunavinaka, L. (2015). *National land use development plan web – GIS and National Land Register Project*. Retrieved from http://gisconference.gsd.spc.int/presentations_2015/Day2/Session3/PacGIS_RS_2015_Development_plan_web-GIS.pdf

Hou, J., Zhang, P., Yuan, X., & Zheng, Y. (2011). Life cycle assessment of biodiesel from soybean, jatropha and microalgae in China conditions. *Renewable and Sustainable Energy Reviews, 15*(9), 5081–5091.

Hull, C. (2009). *GHG lifetimes and GWPs*. Retrieved from https://climatechangeconnection.org/wp-content/uploads/2014/08/GWP_AR4.pdf

IPCC-GNGGI. (2006). N2O emissions from managed soils, and CO_2 emissions from lime and urea applications *2006 IPCC Guidelines for National Greenhouse Gas Inventories* (Vol. Volume 4, pp. 1–54). USA.

JRC Science for Policy Report. (2017). *Definition of input data to assess GHG default emissions from biofuels in EU legislation*. Retrieved from Luxembourg: https://ec.europa.eu/energy/sites/ener/files/documents/default_values_biofuels_main_reportl_online.pdf

Karmee, S. K., & Chadha, A. (2005). Preparation of biodiesel from crude oil of Pongamia pinnata. *Bioresource Technology, 96*(13), 1425–1429.

Kawashima, A. B., de Morais, M. V. B., Martins, L. D., Urbina, V., Rafee, S. A. A., Capucim, M. N., & Martins, J. A. (2015). Estimates and spatial distribution of emissions from sugar cane bagasse fired thermal power plants in Brazil. Journal of Geoscience and Environment Protection, 3(06), 72.

Kazakoff, S. H., Gresshoff, P. M., & Scott, P. T. (2011). Pongamia pinnata, a sustainable feedstock for biodiesel production. *Energy Crops, 4*, 233–258.

Keles, S. (2011). Fossil energy sources, climate change, and alternative solutions. *Energy Sources, Part A: Recovery, Utilization, and Environmental Effects, 33*(12), 1184–1195.

Kesari, V., Das, A., & Rangan, L. (2010). Physico-chemical characterization and antimicrobial activity from seed oil of Pongamia pinnata, a potential biofuel crop. *Biomass and Bioenergy, 34*(1), 108–115.

Kittithammavong, V., Arpornpong, N., Charoensaeng, A., & Khaodhiar, S. (2014). *Environmental life cycle assessment of palm oil-based biofuel production from transesterification: Greenhouse gas, energy and water balances*. Paper presented at the A Presentation at International Conference on Advances in Engineering and Technology (ICAET'2014), March.

Kulkarni, V., Jain, S., Khatri, F., & Thulasi, V. (2014). Degumming of Pongamia Pinnata by acid and water degumming methods. *International Journal of ChemTech Research, 6*, 1–11.

Lau, L. C., Tan, K. T., Lee, K. T., & Mohamed, A. R. (2009). A comparative study on the energy policies in Japan and Malaysia in fulfilling their nations obligations towards the Kyoto protocol. *Energy Policy, 37*(11), 4771–4778.

Lower, S. (2004). *Chemical equations and calculations – Basic chemical arithmetic and stoichiometry*. Retrieved from http://www.chem1.com/acad/webtext/intro/int-4.html

Machacon, H. T. C., Matsumoto, Y., Ohkawara, C., Shiga, S., Karasawa, T., & Nakamura, H. (2001). The effect of coconut oil and diesel fuel blends on diesel engine performance and exhaust emissions. *JSAE Review, 22*(3), 349–355.

Meher, L., Naik, S., & Das, L. (2004). Methanolysis of Pongamia pinnata (karanja) oil for production of biodiesel. *Journal of Scientific and Industrial Research, 63*(2004), 913–918.

Murugesan, A., Umarani, C., Subramanian, R., & Nedunchezhian, N. (2009). Bio-diesel as an alternative fuel for diesel engines – a review. *Renewable and Sustainable Energy Reviews, 13*(3), 653–662.

Nabi, M. N., Hoque, S. N., & Akhter, M. S. (2009). Karanja (Pongamia Pinnata) biodiesel production in Bangladesh, characterization of karanja biodiesel and its effect on diesel emissions. *Fuel Processing Technology, 90*(9), 1080–1086.

Numjuncharoen, T., Papong, S., Malakul, P., & Mungcharoen, T. (2015). Life-cycle GHG emissions of cassava-based bioethanol production. *Energy Procedia, 79*, 265–271.

Oxfam International. (2005). *The Fiji Sugar Industry*. Retrieved from https://www.oxfam.org.nz/sites/default/files/oldimgs/fijian%20sugar%20industry.pdf

ProAnd Associations. (2013). *A glime for Fiji; Prepared for the Fiji Market Development Facility-Fiji Islands* (pp. 1–48). Cardno, Australian Aid.

Rahman, M., Islam, M., Rouf, M., Jalil, M., & Haque, M. (2011). Extraction of alkaloids and oil from Karanja (Pongamia pinnata) seed. *Journal of Scientific Research, 3*(3), 669–675.

Rasul, A., Amalraj, E., Kumar, G., Grover, M., & Venkateshwarlu, B. (2012). Characterization of rhizobial isolates nodulating *Millettia Pinnata* in India. Retrieved from Hyderabad, India: https://onlinelibrary.wiley.com/doi/pdf/10.1111/1574-6968.12001

Raut, S., Narkhede, S., Rane, A., & Gunaga, R. (2011a). Seed and fruit variability in Pongamia pinnata (L.) Pierre from Konkan region of Maharashtra. *Journal of Biodiversity, 2*(1), 27–30.

Raut, S., Narkhede, S., Rane, A., & Gunaga, R. (2011b). Seed and fruit variability in Pongamia pinnata (L.) Pierre from Konkan region of Maharashtra. *Journal of Biodiversity, 2*(1), 27–30.

Rengasamy, M., Anbalagan, K., Mohanraj, S., & Pugalenthi, V. (2014). Biodiesel production from pongamia pinnata oil using synthesized iron nanocatalyst. *International Journal of Chem Tech Research, 6*, 4511–4516.

Sales, A. (2011). *Production of biodiesel from sunflower oil and ethanol by base catalyzed transesterification* (Dissertation). Retrieved from http://urn.kb.se/resolve?urn=urn:nbn:se:kth:diva-41158.

Samaras, Z., & Zierock, K. (1999). *Emissions inventory guidebook – Road transport* (pp. 1–100). Retrieved from https://www.eea.europa.eu/ds_resolveuid/NB1XRT5JWI

Sampattagul, S., Nutongkaew, P., & Kiatsiriroat, T. (2011). Life cycle assessment of palm oil biodiesel production in Thailand. *Journal of Renewable Energy and Smart Grid Technology, 6*(1), 1–14.

Sangwan, S., Rao, D., & Sharma, R. (2010). A review on Pongamia Pinnata (L.) Pierre: A great versatile leguminous plant. *Nature and Science, 8*(11), 130–139.

Sheehan, J., Camobreco, V., Duffield, J., Shapouri, H., Graboski, M., & Tyson, K. (2000). *An overview of biodiesel and petroleum diesel life cycles*. Retrieved from https://www.nrel.gov/docs/legosti/fy98/24772.pdf

Silalertruksa, T., & Kawasaki, J. (2015). Guideline for greenhouse gas emissions calculation of bioenergy feedstock production and land use change (LUC): A case study of Khon Kaen Province, Thailand.

Syamsuwida, D., Putri, K. P., Kurniaty, R., & Aminah, A. (2015). Seeds and seedlings production of bioenergy tree species Malapari (Pongamia pinnata (L.) Pierre). *Energy Procedia, 65*, 67–75.

UNEP RISO. (2013). *Emissions reductio profile – Fiji*. Retrieved from Denmark: https://www.google.com/search?ei=Eja5W-epOM2ItQWqoaS4Bg&q=EMISSIONS+REDU CTION+PROFILE+fiji&oq=EMISSIONS+REDUCTION+PROFILE+fiji&gs_l=psy-ab.3...5544.6885.0.7281.5.5.0.0.0.0.377.1063.3-3.3.0....0...1c.1.64.psy-ab..2.1.372...33i22i29i 30k1.0.7Qco3FrT1Y8

Wani, S. P., Osman, M., D'Silva, E., & Sreedevi, T. (2006). Improved livelihoods and environmental protection through biodiesel plantations in Asia. *Asian Biotechnology Development Review, 8*(2), 11–29.

Whiteside, B. (2013). *National energy forum – Energy: Fiji macroeconomic perspective*. Retrieved from Fiji: http://www.rbf.gov.fj/docs2/Governor%20Reserve%20Bank%20of%20 Fiji%20Barry%20Whiteside%E2%80%99s%20presentation%20at%20the%20National%20 Energy%20Forum%20on%203%20April%20at%20the%20Holiday%20Inn%20Suva.pdf

Zufarov, O., Schmidt, S., & Sekretár, S. (2008). Degumming of rapeseed and sunflower oils. *Acta Chimica Slovaca, 1*(1), 321–328.

Chapter 11
Economic Viability of Jatropha Biodiesel Production on Available Land in the Island of Viti Levu

Praneet Anand Reddy

Abstract With declining petroleum reserves and global environmental concerns, most countries of the world today are looking at plant-derived biofuels as alternative fuels for diesel engines. This GIS-based study estimates the production capacity of jatropha biodiesel on the island of Viti Levu in Fiji to be approximately 73,500 tonnes per annum from the whole of the 49,000 ha of available farm land distributed over ten possible jatropha sites.

The capital cost of this project is approximately F$170,000 while the estimated cost of establishing 49,000 Ha of jatropha plantation is F$17,280.89 million over its 30 year commercial lifetime. The comparison of three case scenarios of jatropha-based biofuel production reveals a profitable industry is possible only when straight Jatropha oil is considered as the basis for the fuel industry. This fuel source shows a much reduced cost of establishment of F$6201.67 million and revenue generation of $F6215.8 million as compared to the biodiesel option, yielding a net profit of F$14.13 million over the commercial lifetime of the project of 30 years.

The net avoided emissions from diesel substitute have been determined to be 8.791 tonnes of carbon, established from the net difference between the CO_2 emission of 79.71 g CO_2eq/MJ for jatropha biodiesel and 98.03 g CO_2eq/MJ for diesel in combusting 89.7×10^{15} J of fuel.

Keywords Residual land · Geographical Information System (GIS) · Jatropha biodiesel · Economic viability · Net avoided emissions

P. A. Reddy (⊠)
Fiji Airways, Air Pacific Limited, Nasoso, Nadi, Fiji

© Springer Nature Switzerland AG 2020   257
A. Singh (ed.), *Translating the Paris Agreement into Action in the Pacific*,
Advances in Global Change Research 68,
https://doi.org/10.1007/978-3-030-30211-5_11

11.1 Introduction

The utility of biodiesel as a transportation fuel has recently been demonstrated by McDonald's of UAE, who have used 100% biodiesel produced by recycling their used cooking oil on their entire fleet of trucks in the past 4 years as of November, 2015 and covered over five million kilometres reducing the carbon footprint by at least 4000 tonnes (Graves 2015). Likewise, a variety of feedstock from crops can be used to produce biofuels. They include edible oils, such as rapeseeds, soybean, mustard, palm and coconut oil, animal fats, such as lard and tallow, and non-edible oils such as castor, pongamia, neem and Jatropha.

A demonstration of Jatropha oil as a pure plant oil substitute for diesel fuel was provided during World War II when Indonesian and Taiwanese military were forced by circumstances to use Jatropha oil as source of fuel. In more recent times, the demand for biodiesel as a substitute for petroleum diesel fuel has risen, due both to declining petroleum reserves and global commitments to mitigate climate change.. Fiji and the Pacific as a whole face tremendous energy challenges due to the lack of indigenous fossil fuel sources and high import bills for their transportation fuels. Fuel imports in 2016 were 16% of Fiji's National import bill which amounted to US$ 346 million (Fiji NDC Implementation Roadmap 2017–2030, 2017).

The International Energy Agency has set a goal for biofuels to meet more than a quarter of the world's demand for transportation fuel by 2050 (Secretariat 2011). In the Pacific, Fiji's NDC Implementation Roadmap 2017–2030 was launched by the country's Attorney General and Minister for Climate Change on sixth November, 2017 as part of the nation's commitment to its Nationally Determined Contributions (NDCs) as specified under the UNFCCC's Paris Agreement.

Fiji's NDC is aimed at reducing CO_2 emissions from the energy sector by as much as 30% by the year 2030. This reduction is targeted from both the supply and demand side energy use including grid improvements and energy efficiency, together with transformations in the transportation sector. Through such strategies, the energy sector expects to achieve a reduction of 627,000 tonnes of CO_2/year by 2030 at an estimated cost of US$ 3.089 billion (Fiji NDC Implementation Roadmap 2017–2030, 2017).

The biomass and waste to energy (WTE) sector is estimated to be the largest GHG emission mitigating contributor with 212,000 tonnes of CO_2/year at an estimated US$ 237 million cost of establishment and development (Fiji NDC Implementation Roadmap 2017–2030, 2017).

Jatropha has been in the global race of biofuel production since the twentieth century. From being diesel substitute in Madagascar, Benin and Cape Verde during World War II to the high flying engines of B747–300 Japanese Airliner, jatropha oil has been a remarkable alternative to fossil fuels. Jatropha fuel is an advanced biofuel processed from inedible source of oil and thus known as the second generation biomass fuel technology. It is one of the top contenders towards carbon emission reductions and a good candidate for Fiji's Nationally Determined Contributions. It emissions reduction potential is expected to be not dissimilar to that of Pongamia biodiesel considered in the previous chapter, which shows a net Carbon Dioxide (CO_2) emission of 18.32 g CO_2eq/MJ, which is five times lower in comparison with the diesel production system..

Jatropha oil burns smoke free and has high saponification value (SV). The oil has good oxidation stability when compared to soybean oil, lower viscosity as compared to castor oil, better freezing point as compared to coconut/palm and lower manufacturing cost than ethanol, and is potentially economically viable (Kantar 2008). It also has a higher acetone number (51) than other vegetable oils and petrol-diesel (46–50) (Kureel 2006).

Jatropha grows in marginal/poor soil conditions and is resistant to drought. It has also been found to survive almost anywhere in rugged conditions, deserts and even on gravel, stony/sandy and saline soils (Kumar et al. 2008). Typically it grows up to 6–7 m in height and on average, 10–20 male flowers to each female flower are produced on the same inflorescence. The fruits (three bi-valved) are produced in winter or dry seasons under sufficiently high temperatures (Biswas et al. 2006). Each inflorescence yields a bunch of approximately ten or more of void fruits containing two or three large black seeds. These seeds come with an average oil content of around 37%. The table below summarizes the different properties of jatropha oil and jatropha biodiesel as compared to fossil diesel (Table 11.1).

Generally, good fuels have the characteristics of low viscosity, high energy content, high saponification value, high cetane number (CN) and low iodine value. The energy content is the energy produced by the fuel upon complete combustion. The viscosity of the fuel is related to its number of carbon atoms in a molecule. The fuel gets oilier, waxier and finally solid as the number of carbon atoms increases. The measure of the average length of the carbon chains is termed as the saponification value (SV). Lower average lengths of the carbon chains indicate a higher SV, hence reducing the ignition delay. The iodine value (IV) is the measure

Table 11.1 Fuel Properties of Jatropha Oil, Jatropha biodiesel compared to fossil diesel (Parawira 2010)

Property	Jatropha oil	Jatropha biodiesel	Palm biodiesel	Diesel	Biodiesel standards	
					AST D 6751–02	DIN EN 14214
Density (15 °C, kgm^{-3})	940	880	855	850	–	860–900
Viscosity ($mm^2 s^{-1}$)	24.5	4.8	4.5	2.6	1.9–6.0	3.5–5.0
Flash point (°C)	225	135	174	68	>130	>120
Pour point (°C)	4	2	16	−20	–	–
Water content (%)	1.4	0.025		0.02	<0.03	<0.05
Ash content (%)	0.8	0.012		0.01	<0.02	<0.02
Carbon residue (%)	1.0	0.20	0.02	0.17	–	<0.30
Acid value ($mgKOHg^{-1}$)	28.0	0.40		–	<0.80	<0.50
Calorific value ($MJkg^{-1}$)	38.65	39.23	41.3	42	–	–

of the unsaturated bonds in the hydrocarbon chain (Divakara et al. 2009). Lower IV indicates a good fuel characteristics and thus low deposits of gums and cokes. Cetane number is the time delay between the ignition and fuel combustion (fuel ignition delay).

The project aims to achieve its goals with the available land lying idle or those presently contributing nothing to individuals or to the economy for that matter. Given the survival nature of the crop, Jatropha can withstand medium and flat terrain and not to mention the idle flat valleys and non-vegetative fields. The suitability of residual land on Viti Levu is determined from the environmental factors (Sect. 11.4) which lie on the extreme ends of the agricultural needs or trivial in nature to farmers but at the same time within the agronomical traits of jatropha cultivation (Sect. 11.2). These critical environmental factors is further categorised as suitable and unsuitable conditions for jatropha propagation and cultivation. The process of land selection is facilitated by the use of GIS framework.

With geographical base, GIS framework is computer based tool utilised in data management and analysis with visual outputs such as maps and 3D images. Data is initially stored digitally in the database and as required, is mapped by manipulation onto base map. The map serves as a geographical container which holds data for analysis by using spatial location to obtain desired outcomes. Spatial (data is referenced to locations on earth) analysis allows for integrating over multiple layers of data or manipulating single set of data and representing evaluated results in form of imageries or scenes. Basically, GIS creates geographical data, is reorganized and/or restructured over time, analyzed as considered necessary and displayed as visual geographic information usually on map.

Thus, the research focuses on determination of the total residual land available on Viti Levu suitable for Jatropha cultivation and crude oil production. Additionally, the capital and maintenance costs, and economic viability of setting-up such biodiesel venture is estimated for engrossed investors. It further looks into the total volume of diesel displaced by Jatropha biodiesel in terms of equivalent energy content. Finally, the paper reports the net avoided emissions by operating such biodiesel industry successfully.

11.2 Jatropha Agronomics

11.2.1 Profiling and Propagation

Jatropha curcas is a Mexican and Central American native plant belonging to the Euphorbiaceae family (Janick and Robert 2008). It is most commonly found and suited for tropical and sub-tropical regions for its growth. The plant is cultivated within 30° limits of the equator, known as the Jatropha cultivation belt (Wahl et al. 2009). Lower altitudes of 0–500 m above sea-level are preferred for growth with 1000–1500 mm of rainfall for optimum yield (Abugre 2011). It is known that

Jatropha plants can survive bare minimum rainfall which may be as low as 250–300 mm. The deep tap roots of the crop suits the dry living conditions for its endurance. The most favourable temperature range for the Jatropha plants is between 20 and 28 °C with well aerated, drained soils. The plants prefer semi-arid sandy and/or loam soils (Gour 2006) with a pH in the range 6–8.5 (FACT 2007). However, it is believed that these rigid plants are tolerant to water-logged saline or irrigated alkaline soils.

Planting of Jatropha crops have been reported to be successful by both seedlings and cutting transplanting methods. While seedlings develop tap roots for deep penetration and survival in dry conditions, cuttings establish best yielding clones through cross breeding techniques. The seedlings and/or cuttings are planted with a spacing of 2.5/3 m within rows by 3 m in between rows with alternating plants on each row to avoid self-shading and allowing for 5 m alley after every fourth row (Brittaine and Lutaladio 2010). The spacing intervals of larger dimensions (6 × 6 m) have also been reported due to insufficient soil nutrients, reduced amount of sunlight hours, and poor growing conditions. Consequently, Jatropha plant density ranges from 1100 to 2500 plants per hectare.

Jatropha is known to be a miracle plant for its ability to survive on marginal lands with poor soil and under dry conditions. While it is true that Jatropha can withstand these harsh conditions, is itself a pest repellent; the plantation does require manure/fertiliser, irrigation and pesticides for first 2–3 months of its growth. High survival attributes does not necessary mean higher productivity under those extreme agricultural environment for commercial purposes. At least 600 mm of rainfall is needed to flower and set fruits and fruiting maybe frequent in wetter conditions. If a good harvest is expected, Jatropha plant needs nutrients and water just like any other crop (McKenna 2009) as potrayed by the applied controlled treatment below (Fig. 11.1).

	TREATMENTS (GRAMS PER PLANT)				
	T1	T2	T3	T4	T5
Pods/plant	97.1	90.1	131.4	45.9	53.6
Pod weight (g)	350.9	248.7	390.8	130.7	148.9
Seeds/plant	247	210	341	131	133
Seed weight/plant (g)	168	143	233	83	87
Threshing%	48	57.4	59.5	63.4	58.3
100 seed weight (g)	68	67.8	68	63.1	65.2

T1=50 g Urea + 38 g SSP; T2= 50 g Urea + 76 g SSP; T3= 100 g Urea +38 g SSP; T4= 100 g Urea + 76 g SSP and T5 = Control

Fig. 11.1 Jatropha yield under conditional treatment (Brittaine and Lutaladio 2010)

11.2.2 Harvesting and Harnessing

Seeds are ready for harvesting between 2 and 3 months after flowering and the yield increases with plant maturity. Individual trees produce around 0–2 kg of seeds (Francis et al. 2005) while the plantations over a hectare have been found to be yielding 0.1–8 tonnes of seeds (Openshaw 2000; Heller 1996). High potential seed yield is expected under optimum conditions with good soil, higher rainfall and ideal management practices. In the third year, about 2–3.5 tonnes of seeds maybe harvested, which increases sharply to about 5–12 tonnes/ha from sixth year onwards (Achten et al. 2008; (FACT 2007; Jongschaap et al. 2007). The miracle bean has been generally found to have 18–45% (on average 37%) oil content and one of the top contenders to substitute *dirty* diesel. Jatropha plants have a lifespan of around 50 years (Brittaine and Lutaladio 2010) but its economic life is been estimated to be between 30 and 40 years which can be commercially marketed for its productivity.

The harvesting takes place manually by beating the branches with a stick and picking fallen fruits from the ground. Traditional oil extraction methods are highly labour intensive, requiring some 12 h to produce 1 l of oil. The process requires manual threshing by beating picked fruits contained in the sack with wood, sorting out and roasting the seed kernels, pounding them to a paste, adding water and boiling, and then separating the oil by skimming and filtering (Fig. 11.2).

Commercially, the fruit after being picked is then taken to the processing mill where the shells are removed using mechanical decorticator running at a capacity of 50 kg per day. The seeds are then pressed (with or without treatment) by engine

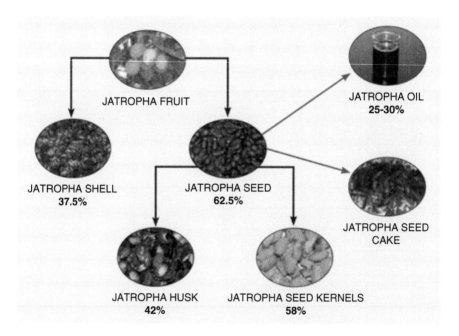

Fig. 11.2 Jatropha fruit composition (Abreu 2008)

Table 11.2   Basic technical data for Jatropha Curcas plantation

Particulars	Data
Spacing	2×2.5 m
Number of plants/ha	2000
Yield commences	3rd year
Fresh fruit to dry seed	60% (at least)
Yield per ha	2 tonnes (3rd – fifth year); 5 tonnes (sixth year onwards)
Oil recovery	30% (at least)
Crude oil production	1.5 tonnes at full maturity per ha (minimum)
Productive life span of Jatropha	33 years
Lifetime oil production/ha	45 tonnes (minimum)

driven expellers with an extraction rate of 55 l oil per hour (Henning 2008). The oil extraction of such industrial plants are 75–80%, producing about 1–15 l of oil from 4 to 75 kg seeds (Achten, et al. 2008). The recovery of oil may be enhanced through treatment of seeds such as pre-heating, solvent extraction and repeated pressings.

11.2.3   Yield

The best method for cultivating Jatropha is by transplanting pre-cultivated plants in poly bags. This includes sowing of Jatropha seeds in seedbeds (50% survival), then transplanting into poly bags (80% survival) and later transplanted in the fields (Heller 1996). Propagation of Jatropha can also be done by cuttings but these do not develop tap roots.

The table below shows the basic technical data for jatropha cultivation and production (Table 11.2).

The oil content of jatropha seed can range from 18.4% to 42.3% (Heller 1996) but generally lies around 37%. The oil is almost all stored in the seed kernel, which has an oil content of around 50–55% (Jongschaap et al. 2007).This compares well to groundnut kernel (42%), rape seed (37%), soybean seed (14%) and sunflower seed (32%) (Brittaine and Lutaladio 2010).

11.3   Jatropha Biodiesel (Jatropha Methyl Ester: JME)

11.3.1   JME Production

The production of Jatropha Methyl Ester (jatropha biodiesel) may commence through standalone or an integrated system by the transesterification process.

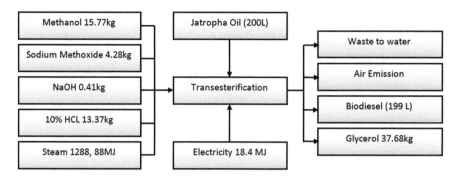

Fig. 11.3 Biodiesel production from Jatropha Oil (Source: Pruecksakorn and Gheewala 2008)

Standalone Biodiesel systems are those that have the vegetable oil as their inputs for biodiesel production and include equipment such as refining and transesterification plants.

Integrated Biodiesel systems are those that have the mature fruits/seeds as their inputs for biodiesel production and include equipment such as expellers and filter presses together with refining and transesterification plants.

The flowchart as given below shows a typical process to produce biodiesel from Jatropha oil, using the transesterification process (Prueksakorn and Gheewala 2008) (Fig. 11.3).

Transesterification is the reaction of oil (ester containing three hydrocarbon chains) with an alcohol in the presence of a catalyst. The main product of this reaction is biodiesel, which is an ester containing one hydrocarbon chain and is thus lighter and less viscous, and has other properties that make it a better fuel for diesel engines than the vegetable oil.

11.4   The Potential for Jatropha Production on Viti Levu

11.4.1   Geography

The archipelago of Fiji lies between 15 and 22° south of equator and longitudes 177° west to 174° east. The group consists of more than 300 islands with Viti Levu and Vanua Levu as the two largest islands covering over 85% of land area. The land tenure system is mostly dominated by iTaukei (Native Land) at 87% of the titles, followed by 7% freehold land and 6% state land. These figures have been shifting since the conversion of state land into Native as shown below (Table 11.3).

Viti Levu, the largest island of the country covers a land area of approximately 10,429 km² of the total 18,333 km². The land use capability (LUC) classifications system of Fiji ranks eight classes of land in order of their agricultural versatility. Class 1 is considered the best land with little or no limitations for agriculture while Class 8 is not suited with severe limitations (Land Use Planning Section). The land

Table 11.3 Land tenure in Fiji (Tuilau 2012)

Land tenure system	Proportion of total land area (%)		
	2000[a]	1990[b]	1970[c]
Freehold land	7	8.17	7
State land	6	9.45	10
iTaukei land	87	82.38	83

Sources:
[a]After lease expiries and tenure conversion, 2002
[b]Ward, 1995
[c]Nayacakalou, 1971

Table 11.4 Rainfall zoning in Fiji

Zone symbol	Description	LUC class
A	Very high: Over 4000 mm (400 cm)	II–VIII
B	High: 3000–4000 mm (300–400 cm)	I–VIII
C	Moderate: 2000–3000 mm (200–300 cm)	I–VIII
D	Low: 1500–2000 mm (150–200 cm)	II–VIII
E	Very low: Less than 1500 mm (150 cm)	IV, VI–VIII

uses as a percentage of total land area are as follows: agricultural land (23.26%), arable land (9.03%), permanent crops (4.65%), permanent meadows and pastures (8.6%) and forest coverage (55.68%) (Trading Economics 2019).

The land availability for jatropha plantation on Viti Levu was determined on the combination of the factors that were most suited for jatropha agronomy and at the same time marginal with respect to the country's agricultural and/or economic needs. The critical environmental factors considered are rainfall, slope, temperature, altitude, and drainage.

With the aid of Geographical Information System (GIS) software, most of Viti Levu's geographical factors listed above (Slope, Altitude and Drainage) were colour coded on the satellite map for a more digitized comprehensive analysis.

11.4.2   Rainfall

Fiji enjoys a tropical climate with only two seasons annually. Over the year the nation is subject to tropical cyclones and occasional flooding along the low lying areas. Rainfall is a factor that affects time of planting, harvesting, pest and disease controls, erosion, crop choices and other productivity factors of the land (Table 11.4).

It is known that Jatropha plants can survive bare minimum rainfall which may be as low as 250–300 mm. However, 1000–1500 mm of rainfall are preferred for growth with optimum yield. The deep tap roots of the crop suits the dry living conditions for its endurance. Group "C" as shown above is considered the ideal range in Fiji. The average rainfall over Viti Levu is shown below (Fig. 11.4):

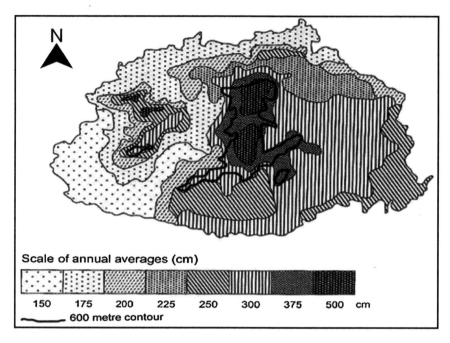

Fig. 11.4 Rainfall map of Viti Levu (Prasad 2010)

The north-western side of the island has drier and rigid conditions consequentially due to lower rainfall compared to the south-eastern region and a possible site for jatropha cultivation.

11.4.3   Slope

The land relief or slope describes the relief and shape of the land and is one of the critical factors in determining land's agricultural capabilities. Overall, a third of the country is too steep or rocky to be farmed. Steeper slopes lead to farm run-aways and measures such as contouring is taken to conserve soil and water (Fig. 11.5).

The figure above shows the group symbol (A–G) used to classify land pertaining to its degree of slope from 0 to 35° and beyond. Intensive farming is suitable for lands with slope 0–7°, and 8–20° is also suited for farming or agricultural use after minor improvements, while slope above 20° need major improvements for any use or is rather unsuited for any agricultural purposes (but compatible with livestock or pastoral) and is mostly forest covers.

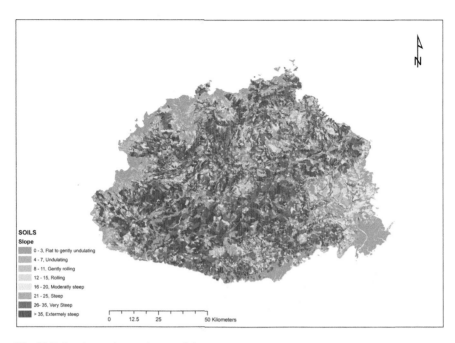

Fig. 11.5 Land grouping on degree of slope

11.4.4 Temperature

All plant life has upper, lower and optimum temperature range in which they thrive. Temperature range is as important as other factors in plant growth and reproduction. Seed germination, photosynthesis, flowering, pollination and fruiting are some of the conditional onsets of temperature. The most favourable temperature range for the Jatropha plants is between 20 and 28 °C with a pH in the range 6–8.5.The mean annual temperature of Fiji is about 26 °C with 22 °C and 30 °C as the average national minimum and maximum temperatures.

The temperature over Viti Levu has been assumed to be the national average of 26 °C which comfortably falls within the range of 20–28 °C of Jatropha's preference.

11.4.5 Altitude

The height above sea level is essential to plant as it matters to humans. Water table, wind temperature and humidity are some of the factors affected by the altitude. The growth and production of crops may fall behind schedule if not cultivated at recommended altitudes (Table 11.5).

Table 11.5 Land category from height above sea level (A.S.L)

Group symbol	Description	LUC class
A	Low – less than 300 m A.S.L	I–VIII
B	Medium – 300–600 in	II–VIII
C	High – 600–1000 m	II–VIII
D	Very high – more than 914 m	IV–VIII

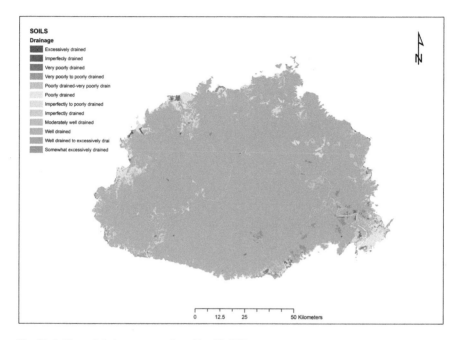

Fig. 11.6 Natural drainage categories of land in Fiji

For Jatropha, lower altitudes of 0–500 m above sea-level are preferred for plant growth within 30° limits of the equator, known as the Jatropha cultivation belt.

11.4.6 Drainage

Proper drainage provides favourable growing conditions for plants by improving soil environment. Drainage helps in reduction of soil and nutrient run-offs and as well prevents choking of plants from stagnant water. The figure below shows the natural soil drainage of the lands on Viti Levu. The legend on the figure describes the specific drainage types with individual colour codes (Fig. 11.6).

Drainage is equally important as irrigation and is entirely dependent on the crop selection. Artificial drainage enhances the natural water flow in the soil to suit farming conditions. Jatropha plants prefer semi-arid sandy and/or loam soils with well aerated, drained soils. However, it is believed that these rigid plants are tolerant to water-logged saline or irrigated alkaline soils.

11.5 Production Potential

11.5.1 Geographic Information System (GIS) Analysis

Attributes such as slope, altitude and drainage of Viti Levu were the information relevant to this research which was digitally stored in GIS framework. The data available on rainfall was not digitally stored within the GIS database. The mapping of rainfall conditions was done manually with the use of *Microsoft paint* software. These data were mapped over the satellite map of Viti Levu. The mapping revealed assorted colour coded regions of the various data spread of each attribute. These data were then further categorised as suitable (green) and unsuitable (red) for jatropha agronomy.

This procedure helped to shortlist as unsuitable the land area which was inhospitable for farming and/or currently appropriate for agricultural purposes. The suitable categories of land were those that jatropha preferred and which did not conflict with existing agricultural land. The integration of these suitable categorical lands constituted *residual land* coded as blue. Finally, the intersected sites were mapped over Viti Levu using GIS analysis to reveal the spatial locations and the associated area of the residual land.

11.5.2 Suitable Residual Land

Data manipulation commenced with the manual tracing of lowest rainfall zones of Viti Levu. The two, low and very low rainfall zones of less than 2000 mm below is highlighted in red as Jatropha has capabilities of withstanding minimum rainfall and surviving drought conditions. Dry and rigid north-western side of Viti Levu makes it more suitable for jatropha cultivation then agriculture (Fig. 11.7).

The figure below shows the sites suited for jatropha farming (green) based on land relief requirements of 8–20° leaving land with less than 8° slope for agricultural purposes while above 20° as too hostile for any commercial farming (Fig. 11.8).

The map below coded in green meets the height above sea level falling in the range of the altitude between 50 and 600 m (50 m $<$ h $<$600 m) suited for jatropha while the red represents the upper (above 600 m) and lower extremes (<50 m) of the altitudes. Land falling above 600 m has no value to jatropha's agricultural needs while the land below 50 m has been reserved for agricultural farming, low lying and swampy fields (Fig. 11.9).

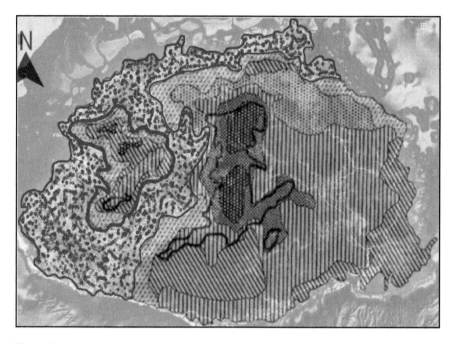

Fig. 11.7 Lowest rainfall zones of Viti Levu (<2000 mm)

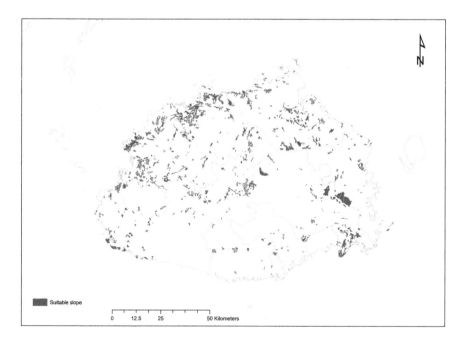

Fig. 11.8 Slopes most suited for Jatropha on Viti Levu

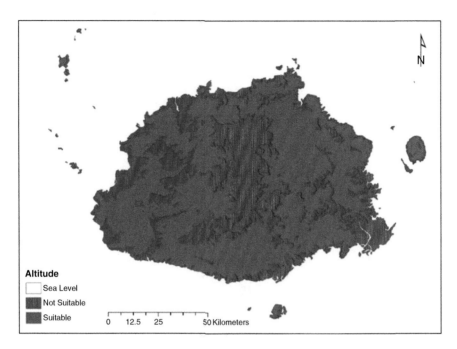

Fig. 11.9 Altitudes between 50 and 600 m on Viti Levu (Green)

Jatropha prefers semi arid and well drained natural soil. The suitable sites coded in green below are the land with well to excessively drained soils. Excessively drained soils have been selected as imposed land for jatropha to allow for soil with moisture and naturally irrigated land reserved for agricultural uses (Fig. 11.10).

Further on, all the environmental conditions deemed suitable (shown above) for jatropha agronomy were mapped over each other. Basically, the land types conditionally imposed as workable with jatropha with respect to each of the attributes above; rainfall, slope, altitude and drainage; were overlapped to find a common intersection. The intersection of this mapping revealed the final sites on Viti Levu where the natural environment were ideal for jatropha propagation and cultivation while unsuitable for agriculture. The intersection of these factors shows the land most suitable (coded blue) for jatropha production as shown below (Fig. 11.11).

The determined sites are residual land for general agriculture but suitable enough for jatropha cultivation. The environmental conditions are consciously selected as harsh in particular to ensure that the land is not compromised for jatropha over other purposes. Hence, from the combination of most suited land types, there exist ten possible sites with respective areas (in square metres) as shown below.

This yields a total of 490,816,838.348 m²of field to work with Jatropha which is equivalent to 49,081.683835 Ha (~49,000 Ha) of residual land.

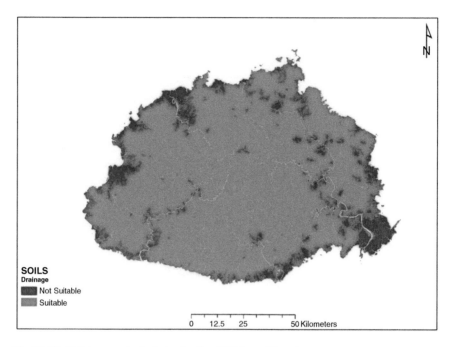

Fig. 11.10 Well to excessively drained soils of Viti Levu (Green)

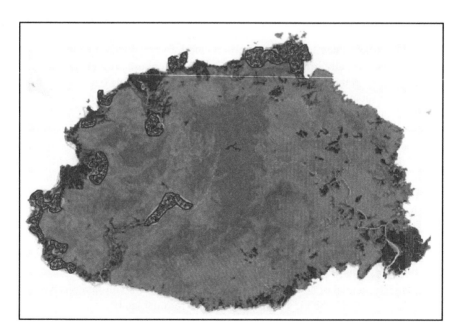

Fig. 11.11 Common intersection of all preferred land choices (blue)

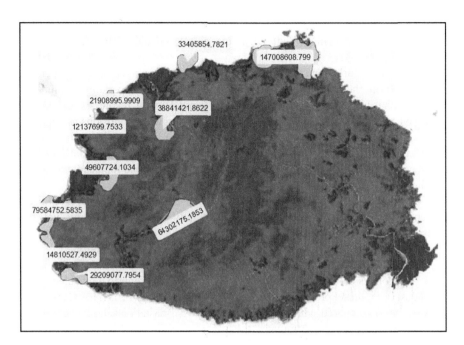

Fig. 11.12 Residual land on Viti Levu for Jatropha with respective areas (m²)

11.5.3   Capacity

From the technical data in Table 11.1 and the land availability as determined above in Fig. 11.12, the total biodiesel production capacity of Viti Levu is approximately 73,500 tonnes. There are about 2000 plants per hectare, totalling to 98 million plants on 49,000 Ha. The seed yield per hectare is approximately 5 tonnes at full maturity which sums up to 245,000 tonnes of seeds from 49,000 Ha of farm. With an average oil recovery rate of 30%, this yields 73,500 tonnes of jatropha crude oil from the total plantation. This is equivalent to 78,191 m³ in volume given a density of 940 kgm⁻³. The crude oil to biodiesel ratio is approximately 1:1; hence the total volume of biodiesel production stands at 78,191,000 litres for the 49,000 Ha of residual land.

11.6   Jatropha Economics

11.6.1   Stability Status

Jatropha *curcas Linnaeus* was initially seen as a "wonder crop" for the biofuels industry, both for SVO and biodiesel production (Bryant and Romijn 2013). It came up with the promises of a crop that would not threaten food security, providing desirable seed and oil yields on marginal and rugged waste-land, whilst demanding little irrigation or management (Brittaine and Lutaladio 2010; Robinson and Beckerlegge 2008).

In contrast, (Burley and Griffiths 2009) and the (German Technical Cooperation (GTZ) 2009)in their studies urge restraint over such optimistic claims arguing that the Jatropha crop performance is critically dependent on resources and labour-intensive processes. Similarly, (Kant and Wu 2011) argue that Jatropha as a biofuel-crop is a failure; comparing poor performance to other fuel crops like sunflower. As well as citing significant varying seed and oil yields and poor financial returns as evidence that Jatropha *curcas* should be ruled as unfavourable replacements of fossil diesel and for future biofuel production.

On the other hand, (N.A.C., Sale and Dewes 2009) believe jatropha still has potential to be utilised as an effective and financially viable method of biofuel (biodiesel) production, in spite of approximations that the cost of Jatropha biofuel production can be up to 2.5 times that of comparable fossil-fuel production.

Assumed yields of 500 tonnes of Jatropha seeds per year; (the '500 t scenario') and 1000 tonnes of Jatropha seeds per year (the '1000 t scenario') revealed higher total cost of biodiesel production i.e. US$ 2.52/L and US$ 2.11/L (N.A.C., Sale and Dewes 2009) compared with the sale prices of fossil-based diesel. However, when the by-product revenue accumulates, the total cost of biodiesel production drops to US$1.62/L for '1000 t' scenario and US$ 2.03/L for 'biodiesel 500 t' scenario (N.A.C., Sale and Dewes 2009). Whilst further inclusion of the revenue from the soap by-product further subsides the apparent total cost of biodiesel to US$ 1.04/L and US$ 1.45/L; a significant reduction from the original total cost of production. The inclusion of the actual revenue generated from by-products suggests that a financially viable Jatropha-based biodiesel production process is possible and (N.A.C., Sale and Dewes 2009) is supported by (Shumba et al. 2011) who state that selling value added products (as is the current case in Zambia and Zimbabwe) from Jatropha, such as soap, can be economically attractive.

It is seen that mixed results and conclusions are obtained from a review of the literature. This necessitates a thorough economic analysis of Jatropha biodiesel production. This is detailed hereunder.

11.6.2   Total Capital Investment (TCI)

Any development of a jatropha biofuel industry will have to be heavily financed. As the industry will be a new startup, it will require a large capital investment. The acquisition of specialised equipment and land, hiring of experts, labour and farm machinery and working capital are some of the large funding needed for such a biofuel industry to commence operation.

The estimated total capital cost of this project is determined on the sum of costs involved in purchasing of assets and the working capital involved over 30 years of jatropha's productive life. The requisition of residual land for farming of Jatropha

Table 11.6 Total capital investment

Assets	Particulars	Cost (FJD)/ha
(a) Fixed assets		
Decorticator	Assembly of 1.5 tonne capacity	10,000
Mechanical press		7000
Biodiesel reactor		60,000
Nursery with equipment		8000
Warehouse		15,000
Storage tanks (×2)	$5000 each	10,000
(b) Variable assets		
Leased land (1 Ha)	$500/year	15,000
License and registration	$150/year	4500
Documentation	$150/year	4500
(c) Liquidated assets		
Working capital	0.25 of total assets	33,500
Total		~ F$ 170,000

has not been included in these estimates. The estimated investment in setting up jatropha biodiesel venture is tabulated below with costs per hectare as the baseline for calculations (Table 11.6).

Thus, the total capital cost of this project is estimated to be approximately F$170,000.00.

11.6.3   Total Variable Cost (TVC)

The total variable cost of this project is determined on the expendables such as land preparations, nursing and nurturing, harvesting, extraction, reaction and other miscellaneous associated costs for a plantation of 49,000 ha with costs per hectare as the baseline for calculations.

For the purpose of this section, various assumptions and factors have been taken into consideration. Fertiliser requirements are based on optimum quantity as per the prescribed treatment shown in Table 11.1 and are assumed in similar proportions to the locally produced *Blend A* and *Blend C* fertilisers. Weedicide is worked on the average application rate stated on the packages available in the local market. The work rate of labour is assumed on the minimum number of man days required to carry out stipulated task and disbursement on the current average farm labour rates. Truck hire costs have also been determined from the rate at which farm trucks operate. For quick reference, these particulars have been outlined below in Table 11.7.

It is to be noted that labour requirements for each phase of the work varies. It is assumed that harvesting will take place randomly as fruit maturity will not be uniform and as such trucks will be hired once every 3 months when the harvest is of

Table 11.7 Total variable costs

Variables	Particulars	Cost/ha ($)	Cost for 49,000 Ha ($M)	Lifetime cost (FJD) ($M)
(a) Farm preparations				
Cleaning	7 man days @ $25/day	175	8.575	257.25
Ploughing	2 hrs @ $130/h	260	12.74	382.2
Tilling	1 hrs @ $80/h	80	3.92	117.6
Planting	7 man days @ $25/day	175	8.575	257.25
Nurturing	3 man days @ $25/day	75	3.675	110.25
Re-planting	3 man days @ $25/day	75	3.675	110.25
Fertiliser (blend C)	6x 50 kg bags @ $20/bag	120	5.88	176.4
Weedicide	2 bottles @ $10/bottle	20	0.98	29.4
Seedling delivery	$120/truck load	120	5.88	176.4
(b) Nursery				
Polybags	2000 @ $0.01 each	20	0.98	29.4
Fertiliser(blend A)	1 × 50 kg bags @ $20/bag	20	0.98	29.4
Labour	5 man days @ $25/day	125	6.125	183.75
Seeds	1.5 kg @ $0	0	0	0
(c) Harvesting				
Labour	52 man days @ $25/day	1300	63.7	1911
Truck hire	4 times @ $120/trip	480	23.52	705.6
(d) Extraction				
Labour	24 man days @ $25/day	600	29.4	882
Electricity	18.4 MJ on 200 L @ $0.34/ kWh	13.90	.6811	20.433
(e) Reaction				
Labour	12 man days @ $25/day	300	14.7	441
Electricity	18.4 MJ on 200 L @ $0.34/ kWh	13.90	0.6811	20.433
Water	3408 L on 0.67 Ha @ $1.06/m^3	5.40	0.2642	7.93
Methanol	15.77 L on 200 L @ $55/L	6938.8	340	10,200
Catalyst (NaOH)	0.41 kg on 200 L @ $85/kg	278.8	13.7	409.8
Sub-total (a-e)		11, 195.80	548.6314	16, 457.75
(f) Miscellaneous				
Maintenance	0.02 of assets	–	0.0034	0.102
Repair	0.03 of assets	–	0.0051	0.153
Contingency	0.02 of total costs	223.92	10.97	329.16
Administration	0.03 of total costs	335.87	16.46	493.73
Total (a–f)		11,755.59	576.0699	17,280.89

significant quantity. The cost of the seedlings/seeds has been assumed to be nil since it is available in selected regions of the country and its harnessing has been included in the labour expenses while planting includes pit digging, staking and alignment.

Thus, the total variable cost as tabulated above is estimated to be $11,755.59 per hectare and accumulates up to F$ 576.0699 million for the 49,000 Ha of farm.

11.6.4   Total Production Cost (TPC)

The total production cost will then be the sum of the fixed and variable costs (TCI and TVC). As detailed above and under prescribed assumptions and estimations, the overall costs add up to $ 11,755.59 per hectare and accumulates up to F$ 576.24 million for the 49,000 Ha of farm. This is inclusive of miscellaneous such as administrative expenses, repair and maintenance costs, and 5% as other contingencies in the operations.

11.6.5   The Return

The jatropha industry alongside with its oil (biofuel) production will have cosmetics, medicine, and fertilisers (manure) as its by-products. These however, are not our primary motive but will certainly contribute to our secondary revenue collection. The glycerine by-product will be of interest to the cosmetic and pharmaceutical industries, while the shells and husks from the seeds and fruits will be suitable constituents of the manure manufacturing process. As listed, the costs of seeds and fruits is set to $0 as this is primarily for oil production and will not be for revenue generation. However, once the oil has been extracted, the oil as biodiesel and the residues both will be for commercial trade (Table 11.8).

Based on current market prices, glycerol attracts about $0.50/kg while seed cakes and other residues, including shells and husks have been assumed to fetch almost $0.20/kg. Our core product, jatropha biofuel is the primary substitute for imported fuel and accordingly the price of this is laid down on the current market price of diesel. Hence, the revenue from biofuel is the estimated savings made from substituting equivalent volume of imported diesel. This figure is shown above as

Table 11.8 The lifetime return from Jatropha output

Outputs	Qty./ha (tonnes)	Qty./49,000 Ha (tonnes)	Lifetime production (M tonnes)	Price (cost/ yield) ($M)
Fruits	8.3	406,700	12.2	0
Seeds (60% of fruits)	5	245,000	7.35	0
Oil (30% of seeds)	1.5 (~1595 l)	73,500 (~78 M litres)	2.2 (2.3 G litres)	0
Biodiesel	1.5	73,500	2.2	4215.8
Glycerol (37.68 kg/200 L)	0.3	14,700	0.441	220.5
Seed cake (70% of seeds)	3.5	171,500	5.15	1030
Shells/husks(40% of fruits	3.3	161,700	4.85	970
Total				6436.3

cost/yield of biodiesel and the volume of equivalent diesel that it displaces is detailed in the next section.

From Table 11.6 the total capital investment has been approximated to be around $170,000 together with the variable costs of $ 11,755.59 per Ha from Table 11.7, the gross cost from the scratch sums up to F$ 576.24 million for establishing 49,000 Ha of jatropha biodiesel industry and costs up to F$ 17,280.89 million in its lifetime. From Table 11.8 the return has been approximated to be F$ 6436.3 million from the possible diesel substitution and revenue generated from by-products. A simple trial balance of the expenses against the revenue reveals a negative return from this study. Thus the loss incurred from establishing a jatropha industry on 49,000 Ha of residual land in Viti Levu is determined to be F$ 10,844.59 million over its 30 years commercial lifetime.

11.6.6 Sensitivity Analysis of Total Production Cost

The lifetime cost of establishing a 49,000 Ha jatropha methanol biodiesel industry from scratch is estimated to be F$ 17,280.89 million. Evidently from Table 11.8 above, this is in excess of F$ 10,844.59 million of the savings that it would bring if diesel was not to be substituted. However, in comparison with the economically viable alternatives available for the use of jatropha, the costs are significantly low and reveals a more profitable and positive industry.

The sensitivity analysis done below shows the total cost of JME production compared to the costs involved with the three cases of producing alternative jatropha based fuels such as Jatropha ethyl esters (JEE), jatropha ethanol hybrid fuel production using micro emulsion technique, and the use of straight jatropha oil (i.e. pure plant oil). In all these alternatives, methanol, a more expensive constituent of the biodiesel is either substituted by ethanol, a locally produced cheaper alternative, or eliminated in case of straight oil usage. Ethanol is available in the local market which is produced by Fiji Sugar Corporation (FSC) as a by-product and estimated at $27/liter while methanol costs $55/liter commercially.

11.6.6.1 Case A: JEE

In this case, the production cost of Jatropha Ethyl Esther (JEE) is compared to the initial Jatropha Methyl Esther (JME) production. With reference to Table 11.7 above and assuming all phases and constituents of biodiesel production to be identical with the exception of methanol being replaced by ethanol, the production cost is re-examined as below.

Part (e) of Table 11.7 is modified as follows (Table 11.9):

This modification reduces the cost of alcohol in the reaction process from F$ 10,200 million to F$ 5007.3 million. The production of biodiesel through transesterification itself is an expensive course of action and despite the entire production

Table 11.9 Cost of substituting methanol with ethanol in the reaction step of Table 11.7

(e) Reaction				
Labour	12 man days @ $25/day	300	14.7	441
Electricity	18.4 MJ on 200 L @ $0.34/kWh	13.90	0.6811	20.433
Water	3408 L on 0.67Ha @ $1.06/m³	5.40	0.2642	7.93
Ethanol	15.77 L on 200 L @ $27/L	3406.32	166.9	5007.3
Catalyst (NaOH)	0.41 kg on 200 L @ $85/kg	278.8	13.7	409.8

cost of biodiesel lessening to F$ 12,088 million there is still an investment loss of F$ 5651.7 million over the 30 year period. An alternative method is an inexpensive process of producing jatropha based hybrid fuels as pursued hereunder.

11.6.6.2   Case B: Micro Emulsion

Micro emulsion is the process of blending oil with alcohol into an optically transparent and thermodynamically stable emulsion producing hybrid fuels. It is a much simpler and inexpensive process when compared to transesterification process of producing biodiesel. In this case, the production cost of Jatropha-Ethanol (CJ1) blend is compared to the initial Jatropha Methyl Esther (JME) production. *CJ1* is the name given to the lab developed hybrid fuel of 90% oil blended with 10% ethanol.

It was experimentally determined that up to 16% of ethanol was miscible with jatropha crude oil without the aid of surfactants. However, CJ1 is chosen to maintain the calorific value of this hybrid fuel (39.25 MJ/kg) as close as to JME.

Micro emulsion technique will see elimination of catalyst, water, electricity and the biodiesel reactor costs as well as substitution of methanol by ethanol with 10% by volume of oil.

Other modifications to *Part (a)* of Table 11.6 and *Part (e)* of Table 11.7 are as follows (Table 11.10):

Thus, the total capital cost of this project now lessens to approximately F$ 110,000 in comparison to the initial estimation as per Table 11.6 of F$ 170,000 (Table 11.11).

The total reaction cost as per *Part (e)* of Table 11.7 was initially determined to be F$ 11,079.16 million which after the elimination of key transesterification constituents now reduces to F$ 6791 million. Consequently, the total estimated production cost of the hybrid fuel as an alternative to biodiesel now comes to F$ 12,992.67 million as compared to the initial F$ 17,280.89 million in its lifetime.

On the other hand, micro emulsification blending will also reduce the revenue generation as the sale of glycerol will have to be withdrawn since it is a by-product of transesterification process which no longer will be available.

Table 11.8 entails the changes below (Table 11.12):

Hence with modifications to Table 11.8 as shown, the final closing figures for revenue at F$ 6215.8 million less the final closing figures for expenses at F$

Table 11.10 Cost after reactor elimination

Assets	Particulars	Cost (FJD)
(a) Fixed assets		
Decorticator	Assembly of 1.5 tonne capacity	10,000
Mechanical press		7000
Biodiesel reactor		60,000
Nursery with equipment		8000
Warehouse		15,000
Storage tanks (×2)	$5000 each	10,000

Table 11.11 Cost micro-emulsion process substitution

(e) Reaction				
Labour	12 man days @ $25/day	300	14.7	441
Electricity	18.4 MJ on 200 L @ $0.34/kWh	13.90	0.6811	20.433
Water	3408 L on 0.67Ha @ $1.06/m^3	5.40	0.2642	7.93
Ethanol	20 L on 200 L @ $27/L	4320	211.7	6350
Catalyst (NaOH)	0.41 kg on 200 L @ $85/kg	278.8	13.7	409.8

Table 11.12 Cost after glycerol elimination

Outputs	Qty./ha (tonnes)	Qty./49,000 Ha (tonnes)	Lifetime production (M tonnes)	Price (cost/ yield) ($M)
Fruits	8.3	406,700	12.2	0
Seeds (60% of fruits)	5	245,000	7.35	0
Oil (30% of seeds)	1.5 (~1595 l)	73,500 (~78 M litres)	2.2 (2.3 G litres)	0
Biodiesel	1.5	73,500	2.2	4215.8
Glycerol (37.68 kg/200 L)	0.3	14,700	0.441	220.5
Seed cake (70% of seeds)	3.5	171,500	5.15	1030
Shells/husks (40% of fruits	3.3	161,700	4.85	970
Total				6215.8

12,992.67 million yields-F$ 6776.87 million and yet again returns a loss in the investment over the productive lifetime of jatropha.

11.6.6.3   Case C: Pure Oil

If pure plant oil, i.e. crude jatropha oil is to be utilized without any chemical reaction or blending, the entire cost of reaction as mentioned in part e of Table 11.7 will be nullified together with the cost of purchasing the biodiesel reactor. However, this will come at a cost of engine modification to accommodate new fuel and/or flushing of fuel lines with regular fossil fuel at the end of each work cycle.

The cost of establishment is now reduced to F\$ 6201.67 million after the elimination of both, the biodiesel reactor purchase and the entire reaction process. And as is mentioned in Case B above, use of straight oil will similarly lessen the revenue to \$ 6215.8 million from withdrawal of glycerol sales as none will be produced. Nevertheless, the net return here is then estimated to be a profit of F\$ 14.13 million.

11.7   Diesel Displacement

The primary objective of this research was to substitute diesel with renewable fuel and to reduce carbon emissions in the combat to mitigate climate change and for a sustainable environment. It has been estimated that Viti Levu can produce 78 million litres of biodiesel from jatropha farming on 49,000 Ha of estimated residual land and approximately 2.3 Giga litres in a lifetime of 30 years.

Jatropha biodiesel has a net calorific value of approximately 39 MJ/L while diesel sits close to 42 MJ/L. Therefore, the estimated 2.3 Giga L of biodiesel would generate 89.7×10^{15} J of energy and displace the quantity of diesel producing same measure of energy. Since diesel is of a little higher energy value, 2.3 Giga L of biodiesel will displace 2.14 Giga L of diesel in terms of equivalent energy.

The recommended retail price in Fiji, as at 20th February, 2019, for a litre of diesel was \$1.97. The fluctuations in world commodities lead to oscillations in the local oil costs too and as the trend has been, fuel prices have been escalating on yearly mean prices. Based on the current market price, jatropha biodiesel will save approximately F\$ 4215.8 million in its lifetime produced from 49,000 Ha of residual land on Viti Levu.

11.8   Net Avoided Emissions and Carbon Sequestration

Successful climate change mitigation is only possible if low carbon emitting technologies are available or if emission of these green house gases are significantly reduced. Net avoided emissions are the emissions of carbon that can be avoided by combustion of an alternative energy fuel rather than fossil diesel. For this research, renewable jatropha oil based fuels are sought as an alternative to diesel in reducing carbon emissions. The total jatropha biodiesel production potential of Viti Levu from the 49,000 Ha of residual land is approximated up to 2.3 Giga liters at 39 MJ/l energy content in its 30 year lifetime. This is sufficient to displace 2.14 Giga liters of diesel with calorific value of 42 MJ/l in terms of equivalent energy.

Using the results of the previous chapter on Pongamia biodiesel as an estimate, the Jatropha system can be assumed to show a net Carbon Dioxide (CO_2) emission of 18.32 g CO_2eq/MJ, which is five times lower in comparison with diesel production system that shows a net CO_2 emission of 98.03 g CO_2eq/MJ. This will give a net Carbon Dioxide (CO_2) emission of the Jatropha biodiesel system to be less than 20% of the diesel production system it replaces.

The equivalent energy measure of both these fuels is determined to be around 89.7×10^{15} J which if generated by biodiesel will emit 1.6433 tonnes of CO_2. In contrast, the same amount of energy to be generated by diesel will emit an equivalent of 8.793 tonnes of CO_2. The net avoided emissions are therefore the difference between the carbon emissions from combustion of both these fuel sources which is calculated to be 8.791 tonnes.

Seeds, the primary yield of jatropha farming, sequestrates 580–725 kg carbon per hectare (Wani et al. 2012) both in form of our principal product biofuels and as seed cake residues, which are added back to the soil as manure. Seeds are a carbon sink, which then recycles sequestrated carbon back into atmosphere when burned as biofuel. The carbon accumulation and addition to soil by jatropha's raw biomass is tabulated below:

With reference to Table 11.13, on average a jatropha plant will shed 1.45 kg leaves per year which is equivalent to 2900 kg per hectare and returning 800 kg Carbon yearly. The trimmings and prunings will return 150 kg Carbon per year from 410 kg organic mass. Live plants in the same way, are the biggest carbon sinks ranging from 5120 to 6100 kg of Carbon per hectare extracted from atmosphere depending on the plant densities on farms.

Calculations based on densities of 2000 plants per hectare, jatropha will accumulate a minimum of 19.607 Giga tonnes of carbon from atmosphere in its commercial lifetime. This is in addition to the 8.8 tonnes of carbon dioxide discharged from 2.14 Giga litres of diesel production corresponding to 98.03 g CO_2eq/MJ, which could be saved from emission into the atmosphere if 2.3 Giga L of jatropha biodiesel is produced from 49,000 Ha of residual land on Viti Levu.

Table 11.13 Carbon accumulation by jatropha biomass (Wani et al. 2012)

Stages	Particulars @ 5 + years	Carbon sequestration kg/ha	Carbon sequestration @ 49,000 Ha	Lifetime carbon accumulation
Leaf fall	1.45 kg/plant/year	800	39.2 M tonnes	1.18 G tonnes
Pruning	410 kg/ha	150	7.35 k tonnes	0.221 G tonnes
Seeds	1290–1610 kg/ha	580–725	28.42 M tonnes	0.853 G tonnes
Seed cakes	900–1130	395–495	19.36 M tonnes	0.581 G tonnes
Biofuels	390–480	185–230	9.07 M tonnes	0.272 G tonnes
Live plants	3.07 kg C per plant @ 1667 plants/ha	5120	250.9 M tonnes	7.53 G tonnes
	2.44 kg C per plant @ 2500 plants/ha	6100	298.9 M tonnes	8.97 G tonnes
Total				19.607 G tonnes

11.9   Conclusion

The estimated cost of establishing and producing 49,000 Ha of jatropha plantation is about F\$ 17,280.89 million over its 30 year commercial lifetime. This figure is about F\$ 10,844.59 million more than the revenues it would bring in financial terms. The production becomes economically viable if the oil is utilised straight rather than converting to biodiesel. The constituents of the transesterification process are excessively priced and avoiding methanol alone will almost compensate for the loss by F\$ 10,200 million. Other cheaper biofuel production process such as micro-emulsification may be of greater significance in off-setting the biofuel production costs making it more economically suited for such endeavour. The three cases shown as sensitivity analysis for the production cost compares well with the initial estimated cost and reveal three better alternatives to producing jatropha using methanol. The straight jatropha oil usage as source of fuel reveals a much reduced cost of establishment of F\$ 6201.67 million and revenue generation of \$F 6215.8 million. The result is an economically viable jatropha fuel industry with a net profit of F\$ 14.13 million.

Fiji's NDC is aimed at reducing CO_2 emissions by 30% and the energy sector expects to achieve a reduction of 627,000 tonnes of CO_2/year by 2030 at an estimated cost of US\$ 3.089 billion. 2.3 Giga liters of jatropha fuel produced in its 30 year lifetime is sufficient to displace 2.14 Giga liters of diesel in terms of equivalent energy determined to be around 89.7×10^{15}J. The net avoided emission in generating this energy is therefore calculated to be 8.791 tonnes. Although establishing jatropha farms investigated in this project require high investments, a positive aspect is that such a project will sequestrate a minimum of 19.607 Giga tonnes of carbon from the atmosphere in its commercial lifetime. This is in addition to the 8.8 tonnes of excess carbon that will be discharged from 2.14 Giga litres of diesel production in the absence of this fuel substitution.

In summary, though a Jatropha biodiesel project may be an expensive proposition for the island of Viti Levu, adopting other options such as pure plant oil usage or cheaper production processes or methods will greatly aid in compensating for costs. Further work is needed on domesticating and genetically modifying jatropha plants to produce more oil bearing varieties to increase yield. The determination and requisition of residual land, their tenure types, accessibility and the establishment cost customised to specific locations needs to be addressed. Larger series of GIS data with greater precision needs to be available first hand to make more detailed analysis of such requisition. Finally, locally produced cheaper constituents or alternative blends for Jatropha based fuels for a more profitable biofuel industry need to be considered.

References

Abreu, F. (2008). *Alternative by-products from Jatropha*. Rome: International Consultation on Pro-poor Jatropha Development.

Abugre, A. (2011). *Compatibility of Jatropha Curcas in an agroforestry system*. Kumasi: Department of Agroforestry, Faculty of Renewable Natura lResources, Kwame Nkrumah University of Science and Technology.

Achten, W., Verchot, L., Franken, Y., Mathijs, E., Singh, V., Aerts, R., & Muys, B. (2008). Atropha bio-diesel production and use. Biomass and Bioenergy, 32, 1063–1084.

Biswas, S., Kaushik, N., & Srikanth, G. (2006). Biodiesel: Technology and business opportunities an insight presented at the. *Biodiesel conference towards energy independence – Focus on Jatropha*. Rashtrapati Nilayam, Bolaram, India.

Brittaine, R., & Lutaladio, N. (2010). *Jatropha: A smallholder bioenergy crop*. Rome: Food and Agriculture Organization of the United Nations.

Bryant, S., & Romijn, H. (2013). *Not quite the end for Jatropha? A case study of the financial viability of biodiesel production from Jatropha in Tanzania*. The Netherlands: Eindhoven Centre for Innovation Studies (ECIS), School of Innovation Sciences, Eindhoven University of Technology.

Burley, H., & Griffiths, H. (2009). *Jatropha: Wonder crop*. Retrieved 01 20, 2013, from Friends of the Earth: http://www.foe.co.uk/resource/reports/jatropha_wonder_crop.pdf

Divakara, B. N., Upadhyaya, H. D., Wani, S. P., & Laxmipathi Gowda, C. L. (2009). Biology and genetic improvement of Jatropha curcas L.: A review. *Applied Energy, 87*(2010), 732–742.

FACT. (2007). Position Paper on Jatropha curcas L. State of the art, small. *Fuels from Agriculture in Communal Technology*.

Fiji NDC Implementation Roadmap 2017–2030. (2017). Republic of Fiji: Ministry of Economy.

Francis, G., Edinger, R., & Becker, K. (2005). A concept for simultaneous wasteland reclamation, fuel production, and socio-economic development in degraded areas in India: Need, potential and perspectives of Jatropha plantations. *Natural Resources Forum, 29*, 12–24.

German Technical Cooperation (GTZ). (2009). *Jatropha Reality Check: A field assessment of the agronomic and economic viability of Jatropha and other oilseed crops in Kenya*. Retrieved 11 16, 2016, from http://www.worldagroforestry.org/downloads/publications/PDFs/B16599.PDF

Gour, V. K. (2006). *Production practices including post-harvest management*. Hyderabad: Rashtrapati Bhawan.

Graves, L. (2015). *McDonald's UAE completes 5 million kilometres on McFuel bio-diesel*. Retrieved 11 4, 2018, from https://www.thenational.ae/business/mcdonald-s-uae-completes-5-million-kilometres-on-mcfuel-biodiesel-1.81399

Heller, J. (1996). *Physic nut, Jatropha curcas L. promoting the conservation and use of underutilized and neglected crops. 1, institute of plant genetics and crop plant research*. Garrtersleben: International Plant Genetic Resources Institute.

Henning, R. K. (2008). *Identification, selection and multiplication of high yielding Jatropha curcas L plants and economic key points for viable Jatropha oil production costs*. Rome: International Consultation on Pro-poor Jatropha Development.

Janick, J., & Robert, E. (2008). The encyclopedia of fruits and nuts: *Jatropha curcas L. Euphorbiaceae*. In *CABI publishing series*. ISBN 0851996388, 9780851996387 (p. 371).

Jongschaap, R., Corre, W., Bindraban, P., & Brandenburg, W. (2007). *Claims and facts on Jatropha curcas L*. Wageningen: Plant Research International. Retrieved 11 10, 2017, from Wageningen, The Netherlands: Plant Research International: www.factfuels.org/media_en/Claims_and_Facts_on_Jatropha_WUR?session=isgsklbna58j.

Kant, P., & Wu, S. (2011). The extraordinary collapse of Jatropha as a global biofuel. *Environmental Science & Technology, 45*, 7114–7115.

Kantar, J. (2008). Air New Zealand flies on engine with Jatropha biofuel blend. *New YorkTimes*.

Kumar, A., Yadav, S., Thawale, P., Singh, S., & Juwarka, A. A. (2008). Growth of *Jatropha curcas* on heavy metal contaminated soil amended with industrial wastes and Azotobacter: A greenhouse study. *Bioresource Technology, 99*, 2078–2082.

Kureel, R. (2006). Prospect and potentials of Jatropha curcas for biodiesel production. *Biodiesel conference towards energy independence focus on Jatropha.* Bolaram, Hydrabad: Rashtrapati Nilayam.

Land Use Planning Section. (n.d.). *A Fiji guideline for the classification of land for agriculture.* Suva: Land Resource Planning & Development, Department of Agriculture.

McKenna, P. (2009). All washed up for jatropha: The draught – Resistant dream biofuel is also a water hog. In *Technology review.* Retrieved 9 30, 2019, from https://www.technologyreview.com/s/413746/all-washed-up-for-jatropha/.

N.A.C. Sale, & Dewes, H. (2009). Opportunities and challenges for the international trade of Jatropha curcas-derived biofuel from developing countries. *African Journal of Biotechnology, 8*, 515–523.

Openshaw, K. (2000). A review of Jatropha curcas: An oil plant of unfulfilled. *Biomass and Bioenergy, 19*, 1–15.

Parawira, W. (2010). Biodiesel production from *Jatropha curcas: A review. Scientific Research and Essays, 5*, 1796–1808.

Prasad, B. (2010). *Volume 4: Agricultural resources inventory of the Fiji Islands* (p. 9). Suva: Natural Resource Inventory Report Of The Fiji Islands.

Prueksakorn, K., & Gheewala, S. H. (2008). Full chain energy analysis of biodiesel from Jatrophacurcas L. in Thailand. *Environmental Science & Technology, 43*(9), 383–388.

Robinson, S., & Beckerlegge, J. (2008). *Jatropha in Africa – Economic potential.* Retrieved 11 16, 2013, from Bio Diesel Fuels Inc.: http://www.jatropha.pro/PDF%20bestanden/Jatropha_in_Africa_Economic_Potential-2008.pdf

Secretariat I. E (Ed.). (2011). *Technology roadmap – biofuels for transport.* (I. R. Division, Producer, & International Energy Agency) Retrieved 04 28, 2019, from The International Energy Agency (IEA): https://www.iea.org/publications/freepublications/publication/Biofuels_Roadmap_WEB.pdf

Shumba, E., Roberntz, P., Mawire, B., Moyo, N., Sibanda, M., & Masuka, M. (2011). *Community level production and utilization of Jatropha feedstock in Malawi.* Zambia: World Wildlife Fund (WWF), Zimbabwe.

Trading Economics. (2019, 03 20). *Fiji – forest area (% of land area).* Retrieved 03 20, 2019, from Trading Economics: https://tradingeconomics.com/fiji/forest-area-percent-of-land-area-wb-data.html

Tuilau, T. (2012). *Urbanization and land use change in urban.* Fiji: Yonsei University.

Wahl, N., Jamnadass, R., Baur, H., Munster, C., & Iiyama, M. (2009). *Economic viability of Jatropha curcas L. plantations in northern Tanzania.* Nairobi: World Agroforestry Centre.

Wani, S., Chander, G., Sahrawat, K., Rao, C., Raghvendra, G., Susanna, P., & Pavani, M. (2012). Carbon sequestration and land rehabilitation through Jatropha curcas (L.) plantation in degraded lands. *Agriculture, Ecosystems and Environment, 161*, 112–120.

Chapter 12
Potential for Biobutanol Production in Fiji from Sugarcane and Timber Industry Residues: Contribution to Avoided Emissions

Shaleshni Devi Prasad

Abstract Biobutanol provides an alternative fuel for petrol (i.e. Spark-Ignition) engines and has far better fuel properties than ethanol which is traditionally used in fuel blends. This biofuel can be produced from (second generation) ligno-cellulosic feedstock such as forestry and agricultural residues. There is an abundance of such agro-industrial residues available in Fiji. Biofuels produced from such residues can contribute towards the emission reduction targets as determined by Fiji's Nationally Determined Contributions (NDC) Implementation Roadmap. Unlike the first generation biofuels, the production of biobutanol from lignocellulosic materials requires the additional steps of pretreatment and enzymatic hydrolysis followed by fermentation using the *Clostridium bacteria*, leading to the production of acetone, butanol and ethanol. This chapter first describes the methodology for the production of bio-butanol from sugarcane bagasse and hog fuel and reports on the yields that can be obtained. It then assesses the potential of producing butanol from sugarcane and timber industry residues and by-products. It is found that 115,203 tonnes of butanol can be produced from available residues like sugarcane bagasse, trash (cane tops and leaves), molasses and hog fuel which have the potential to avoid approximately 259 kt of CO_2 emissions per year, which is approximately 41% of Fiji's total emissions reduction target of 627 kt of CO_2 per year.

Keywords Cellulose · Hemicellulose · Extractives · Hydrolysis · Pretreatment · Fermentation · Carbon emissions reduction

S. D. Prasad (✉)
School of Engineering and Physics, The University of the South Pacific, Suva, Fiji

© Springer Nature Switzerland AG 2020
A. Singh (ed.), *Translating the Paris Agreement into Action in the Pacific*,
Advances in Global Change Research 68,
https://doi.org/10.1007/978-3-030-30211-5_12

12.1 Introduction- Background to Fiji's Biofuel Industry

Fuel reserves are depleting continuously and becoming a major concern since access to fossil fuels will be difficult in a few decades (Kumar and Gayen 2011). This has led to continuous rise in the fuel prices and the scarcity of the fossil fuel reserves has driven researchers towards alternative fuel sources. In addition to depletion of fossil fuel reserves, its usage leads to serious environmental concerns such as global warming and climate change (Kumar and Gayen 2011). For this reason there is an urgent need to shift from fossil fuels to renewable biofuels.

Biofuels are produced from biomass from plants, animals or microorganisms which can be in any of solid, liquid or gaseous forms (Chang 2010). Recently there has been a significant increase in the amount of research done in second generation biofuels from agricultural residues. Using lignocellulosic materials is an excellent way to produce economical biofuels and at the same time avoid the "food versus fuel" debate. Previously the research focus was primarily on ethanol production. More recently however, the research focus has shifted towards bio-butanol production. Butanol has many chemical uses such as in plastic, food and flavour industries. It is better than ethanol because of its superior qualities such as it has 30% more energy than ethanol (Qureshi and Ezeji 2008), lower volatility, less ignition problems in winter, high intersolubility, less corrosive, low vapour pressure and high flash point, thus potential to replace gasoline.

Fiji currently relies highly on imported fossil fuels. According to the Fiji Bureau of Statistics the mineral product import by Standard International Trade Classification (SITC) for Fiji in 2017 amounted to approximately $824 million (Fiji Bureau of Statistics 2017). Diesel, motor spirit and aviation fuel are the main imported fuels. Other imported fuels include liquid petroleum gas (LPG), kerosene and heavy fuel oil (HFO). Fiji's reduction in fuel imports can be achieved if more renewable sources of energy are used by the transport sector. The Fiji Government encourages development of local energy sources to curb the petroleum imports. Fuel blends B5 (5% vegetable oil ester and 95% diesel) and E10 (10% anhydrous ethanol and 90% petrol) have been approved by the cabinet (Fiji National Energy Policy 2013–2020, 2013). Use of coconut oil (CNO) biofuel has also been promoted. However, recent studies have raised concerns about the economic viability of using CNO as a replacement fuel and also the food versus fuel issue (FDoE 2013). It has been suggested that in order to move away from total dependence on imported fuels, diesel fuel in Compression Ignition (CI) engines could be replaced by biodiesel and petroleum in Spark Ignition (SI) engines with ethanol blends (Singh 2012).

Fiji adopted its Nationally Determined Contributions (NDC) Implementation Roadmap 2017–2030 in November 2017 as its contribution towards the Paris Agreement reached at the UNFCCC's COP21 2 years earlier (Ministry of Economy 2017). This roadmap aims to implement mitigation actions in Fiji with a target for carbon dioxide (CO_2) emission reduction of 30% by 2030. It is divided into the three sub-sectors of electricity generation and transmission, demand-side energy efficiency and transportation (Ministry of Economy 2017). The total estimated annual mitigation target through actions in the roadmap to be achieved is 627 ktCO$_2$/year by 2030. Mitigation actions related to transportation include vehicle replacement programmes which has a mitigation target of 95 ktCO$_2$/year, use of biodiesel (37 ktCO$_2$/year), maintenance of sea vessels and utilising fuel-efficient outboard motors (5 ktCO$_2$/year). It also includes fuel use from increased sustainable biomass/waste to energy (WTE) generation at the sugar mills.

It is of interest to see if biobutanol can make significant contributions towards the aims of Fiji's NDC Roadmap. The primary objective of this study is to consider the production of biobutanol from industry-relevant biomass such as sugarcane and timber industry residues, and to assess the extent of the contribution that such an energy source can make towards avoiding emissions.

12.2 Biobutanol as an Alternative Fuel for Transportation and Power Generation

The age of fossil fuels is coming to an end since fossil fuel reserves are depleting rapidly, which has drawn a lot of interest in biofuels as an alternative. Other alternatives to biofuels are solar, hydropower, geothermal and wind which are replacing the carbon intensive fossil fuels (Nesbit, 2015). The International Energy Agency (IEA) foresees a rapid increase in biofuel demand, in particular for second generation biofuels, in an energy sector that aims on stabilizing atmospheric carbon dioxide concentration below 450 parts per million (ppm) (Robert 2014).

First generation biofuel production like bioethanol has increased rapidly over the past years. Millions of gallons of ethanol are produced per year and blended with gasoline. In 2017 biofuel production reached 143 billion litres which is equivalent to 3.5 EJ (exajoule). Approximately 65% of the total biofuel production in terms of energy was ethanol (IRENA 2018) which is equivalent to 105 billion litres. During this year US produced 60 billion litres of ethanol and used more than 90% of it as fuel with a record average blend rate of 10.08% (IRENA 2018) while the remaining was exported to 60 other countries. US, Brazil, China, Canada and Thailand are still the largest producers of ethanol, despite its numerous disadvantages. Ethanol is mainly produced from food crops like corn and sugarcane sources, which has led to food versus fuel competition. The second generation fuels that are produced from lignocellulosic materials (LCM) such as agricultural and forest residues (Kumar et al. 2014) do not suffer from the short-comings of first generation biofuels such as corn ethanol.

At the 245th National Meeting and Exposition of the American Chemical Society, Duncan Wass explained that even though ethanol has become a leading biofuel, it has lower energy content than gasoline, has corrosive effect on car engines and cannot be easily used in amounts higher than 10–15% (Bernstein and Woods 2013). Thus butanol is an ideal fuel that can be produced from agricultural cellulosic feedstock. The production of this second generation biofuel is considered promising and economically feasible. An added advantage of this fuel is that it can be used to replace gasoline without engine modification (Cheng et al. 2012). Liquid fuel distribution system already exists in cars therefore, biobutanol is a high potential fuel. It also has higher energy content than ethanol, can be blended with gasoline or used directly in cars without engine modifications, has low corrosiveness, so can be easily transported through pipelines, it's less explosive, has high heat of vaporization, has low contamination with water (Al-Shorgani 2015) and a high market value. The A:B:E ratio is 3:6:1 which indicates high butanol concentration in ABE fermentation (Green 2011).

12.2.1 Advantages of Using Butanol as a Fuel

Currently, ethanol is the leading fuel blended with gasoline due to its less carbon monoxide and hydrocarbon emissions, however, its drawbacks include low heating value which leads to high fuel consumption or more fill ups and is highly hygroscopic which makes it corrosive (Lapuerta et al. 2017). Butanol has the potential to overcome the disadvantages due to low-carbon alcohols as some of its parameters are comparable to gasoline as depicted in Table 12.1. The advantages of using butanol as an alternative fuel are as follows:

- It is a four carbon alcohol thus has 30% more energy than ethanol. This implies less fuel consumption and better mileage.
- Has lower volatility therefore, has less tendency towards cavitation and vapour lock problems.
- Less ignition problems since butanol has lower heat of vaporisation than ethanol and methanol thus, reduces ignition problems during winter.
- Has high intersolubility due to high carbon (long chain) makes it fairly non-polar content as a result is very soluble with diesel and gasoline.

Table 12.1 Comparison of basic parameters of ethanol and n-butanol and gasoline

Parameter	Ethanol	Butanol	Gasoline
Chemical formula	C_2H_5OH	C_4H_9OH	
Density at 15 °C (kg dm^{-3})	0.79	0.81	~0.73
Kinematic viscosity at 20 °C (mm^2 s^{-1})	1.54	3.64	0.4–0.8
Boiling point (°C)	78	118	30–215
Calorific value (MJ kg^{-1})	26.8	32.5	42.9
Heat of vaporisation (MJ kg^{-1})	0.92	0.43	0.36
Vapour pressure (kPa)	19.3	18.6	60–90
Oxygen content (% by weight)	34.7	21.6	‹2.7
Research octane number (RON)	106–130	94	95
Motor octane number (MON)	89–103	80–81	85

Mařík et al. (2014)

- Kinematic viscosity is higher due to longer carbon chain. Viscosity is much higher than gasoline and similar to diesel thus it will not cause harm to pumps designed for diesel engines.
- Safer to use at high temperatures since it has low vapour pressure and high flash point.
- Easy to distribute because it is less corrosive.

Nevertheless, several draw backs are still associated with butanol production which includes high substrate cost due to using edible biomass, low final butanol concentration (<20 g/L), high cost of butanol recovery (recovery is energy intensive), low volumetric butanol productivity (<0.5 g L^{-1} h^{-1}) (Zheng et al. 2015).

Research Octane Number (RON) is known for describing the anti-knock quality as high RON fuels have high anti-knock quality thus higher thermal efficiency (Nakata et al. 2007). Motor Octane Number (MON) and RON constitute to the main characteristics of gasoline as they provide a sensitive indication of fuels (RON-MON) (Kalghatgi 2001). The average of the RON and MON gives the Anti-Knock Index (AKI). Ethanol, butanol and gasoline have the AKI values as 97.5, 87 and 90 respectively. Since the AKI values of the alcohols are high, that is, similar to gasoline (ordinary European Petrol) both alcohols create very little particulates.

12.2.2 Pretreatment

Pretreatment is a very important process to lignocellulosic materials carried out before the fermentation process. Lignocellulosic or herbaceous materials comprise of five major components: extractives, cellulose, hemicellulose, lignin and ash (Williams et al. 2017). Since most of the processes are cost intensive, it requires proper treatment at industrial scale so that inhibitors like furfural are minimized (Thulluri et al. 2013). It loosens the crystalline structure and lignin in the biomass thus increasing the surface area to saccharification enzymes. Different pretreatment methods like physical (mechanical communition), chemical (dilute acid, ammonia, etc.), physio- chemical (steam explosion), and biological (white rot fungi) can be used for different biomass. Pretreatment method for bagasse include acid, alkali, steam explosion, alkali dewaxing, biological treatment, wet oxidation, organic solvent and liquid hot water pretreatment (Khuong et al. 2014). Alkali, dilute acid and steam explosion pretreatment are the most common methods used.

Pretreatment improves the yield of the formation of fermentable sugars while avoiding the degradation and loss of carbohydrates and formation of inhibitors (Garcia 2010).

12.2.3   Pretreatment Methods

12.2.3.1   Physical (Mechanical Comminution)

Comminution of lignocellulosic material through a combination of chopping, grinding and milling reduces the crystallinity of cellulose (Kumar et al. 2009). Chopping reduces the particle size to 10–30 mm which is further reduced to 0.2–2 mm by milling or grinding (Sun and Cheng 2002). Ball milling or vibratory ball mill and rotary drums are effective physical pretreatment methods. Maceration, sonication and high pressure homogeniser are examples of physical pretreatments suitable for lignocellulosic biomass and municipal solid wastes (MSW) (Das et al. 2015). Mervate et al. 2013 investigated the effect of three different physical pretreatment methods on bagasse. In this study, it was discovered that the milling pretreatment produced the lowest total reducing sugars (TRS), followed by milling and autoclaving then gamma (γ) irradiation which produced the highest amount of TRS (Abo-State et al. 2013). γ- irradiation cleavages the β-1,4-glycosidic bonds giving a larger surface area and lower crystallinity (Takacs et al. 2000).

12.2.3.2   Alkali Pretreatment

Alkali pretreatment is done by using bases of sodium, calcium, potassium and ammonium hydroxides which breakdown ester bonds (hemicellulose), remove lignin, partially decrystallises cellulose, partially solubilizes hemicellulose (Kapoor et al. 2015). Calcium hydroxide is considered the most effective and cheapest of all bases while sodium hydroxide is the most studied (Kumar et al. 2009). Alkali pretreatment uses lower temperatures and pressure compared to other methods (Mosier et al. 2005) but takes longer. Compared to acid pretreatment, alkali pretreatment causes less sugar degradation.

12.2.3.3   Dilute Acid Pretreatment

Dilute acid pretreatment is done mainly by using dilute sulphuric acid to breakdown the complex lignocellulosic structures by solubilizing hemicellulose (Mosier et al. 2005). In the current research dilute acid pretreatment with dilute sulphuric acid was used for pretreatment of bagasse and hog fuel.

12.2.3.4   Steam Explosion

Steam explosion involves both physical and chemical techniques to breakdown hemicelluloses. The factors affecting steam explosion are temperature, residence time, moisture content and chip size (Kumar et al. 2009). This method requires a temperature of 160–260 °C which is equivalent to a pressure of 0.69–4.83 MPa for a few minutes then exposed to atmospheric pressure, which causes cellulose and hemicellulose degradation.

12.2.3.5 Autohydrolysis

Autohydrolysis is another cost effective and environmentally friendly method of pretreatment from which oligomers can be obtained. It allows selective hydrolysis of hemicellulose at high temperature using water as the reagent.

12.2.3.6 Liquid Hot Water Pretreatment (LHW)

LHW and steam explosion are the two common methods used for sugarcane bagasse. These methods do not involve chemical addition. It uses high pressure to maintain water in liquid state at high temperatures. A xylose recovery of 73.3% was obtained under a pressure of 6.0 MPa (Yu et al. 2013). LHW uses water only as a reagent and is capable of partially solublising hemicellulose and disrupting lignin and cellulose structures (Mosier et al. 2005).

12.2.4 Enzymatic Hydrolysis

Enzymatic hydrolysis is the most effective way to optimise butanol production from lignocellulosic biomass. Hydrolysis breaks the β-(1, 4)-glycosidic bonds in cellulose and hemicellulose. Hemicellulose is broken down mostly during the pretreatment while cellulose is broken down by cellulase which mainly consists of endoglucanase, exoglucanase and β-glucosidase. These enzymes are found mainly in fungal species *Trichoderma reesei* (Steffien et al. 2014), *Aspergillus niger* (Bravo et al. 2000), *Penicillium veruculosum* (Steffien et al. 2014), *Sporotrichum thermophile* (Margaritis and Creese 1981). Endoglucanase releases oligosaccharides by working from inside the cellulose chains in the amorphous region to convert it to cellobiose. Beta glucosidase converts cellobiose to glucose.

Hydrolysis of lignocellulosic material is a very important process in the production of biofuels produced through fermentation process because most bacteria or yeast cannot use cellulose directly. Due to the rigid structure of cellulose it is difficult for cellulose to hydrolyse directly, therefore, the complex structures are broken down by pretreatment processes.

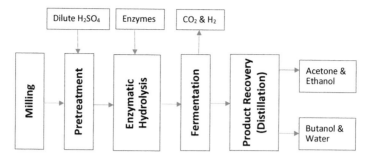

Fig. 12.1 Schematic diagram for the processes involved for butanol production

12.2.5 Fermentation

Butanol is produced by solvent producing Clostridium species with acetone, butanol and ethanol as the main products (Guan et al. 2018) from sugars (pentoses,
hexoses, starch and even cellulosic materials) under anaerobic fermentation. During
the early stages (acidogenesis phase) of fermentation acetic acid and butyric acids
are produced. The acids produced are utilised as carbon source during the solvento-
genesis phase to produce solvents acetone, butanol and ethanol. Production yield is
greatly affected due to inhibition of the microbial cells by the products. Various
methods of solvent recovery techniques are employed to overcome this problem
such as liquid-liquid extraction, pervaporation, perstraction and others (Fig. 12.1).

12.3 Production and Yield of Bio-Butanol

12.3.1 Methodology

12.3.1.1 Compositional Analysis of Raw Sugarcane Bagasse and Hog Fuel

Compositional analysis of SCB and hog fuel was done to identify the composition
of extractives, hemicellulose, lignin and ash content. Extractives were determined
by gravimetric method using soxhlet extraction. Two step extraction was done by
removing water soluble materials followed by ethanol soluble materials as described
in the National Renewable Energy Laboratory (NREL) procedure (Sluiter et al.
2008). The hemicellulose, lignin and ash content were determined using the method
used by (Adeeyo et al. 2015) which is modified NREL methods for determination
of hemicellulose, lignin and ash content.

12.3.1.2 Pretreatment of the Feedstock

The pretreatment methods for the raw ligno-cellulosic material chosen for this study was the acid pretreatment method. A 10 g ground biomass was placed in a 250 ml conical flask and treated with 100 ml, 1% v/v concentrated H_2SO_4, at a temperature of 121 °C, for 60 minutes in an autoclave (Jonglertjunya et al. 2014). The pretreated slurry was neutralised with 0.5 mol L^{-1} sodium hydroxide (NaOH) solution. A 10 mL volume of slurry was filtered using vacuum filtration which was analysed for total reducing sugars (mixture of pentoses and hexoses like glucose, xylose, mannose, arabinose) using dinitrosalicylic acid (DNS) method. The rest of the neutralised slurry was subjected to enzymatic hydrolysis.

12.3.1.3 Enzymatic Hydrolysis of Feedstock After Pretreatment

The hydrolysate from pretreatment contained low levels of sugar (needed for the fermentation stage). Enzymatic hydrolysis was done so that the sugar content in the hydrolysate increases. Before enzymatic hydrolysis an optimisation curve of 0–50 µl/ml endoglucanase and 0–5 mg/ml β-glucosidase was done to ensure the optimum concentration of enzyme is used for hydrolysis. About 6 ml/l of cellulase (β-glucosidase and endoglucanase with enzyme activity ≥2 units/mg solid and ≥2 units/mg protein respectively), obtained from Sigma Aldrich, Australia, was placed in neutralised pretreated biomass slurry in a 125 mL Erlenmeyer flasks, sealed with aluminium foil and incubated at 50 °C for 72 h with agitation. After 72 h of enzymatic hydrolysis the samples were heated on a boiling water bath to denature the enzymes. The slurry was centrifuged at 3000 g to separate the supernatant from the hydrolysate obtained after enzymatic hydrolysis. The supernatant (liquid that is separated after the slurry is centrifuged) was filtered using a 0.45 µm filter disc to remove any sediments and the hydrolysate (i.e. the filtrate) obtained was analysed for sugars and sterilised for fermentation.

12.3.1.4 Fermentation

Clostridium beijerinkii was used for fermentation of the hydrolysate (containing sugars) obtained from enzymatic hydrolysis to produce acetone, butanol and ethanol.

Inoculum Preparation

For inoculum (active microbial culture) preparation the dried spores of *Clostridium beijerinkii* were rejuvenated in sterilized reinforced clostridium media (RCM) broth and incubated in an anaerobic chamber. For inoculum development 100 mL freshly prepared RCM was added to 200 mL Duran bottle and approximately 6 ml of the actively growing culture was transferred into it (Nanda et al. 2014). The RCM contained peptone (10gL^{-1}), beef extract (10 gL^{-1}), yeast extract (3 gL^{-1}), sodium chloride (NaCl) (5 gL^{-1}), dextrose (5 gL^{-1}), sodium acetate (3 gL^{-1}), starch (1 gL^{-1}), L- cysteine HCl (0.5 gL^{-1}) and 0.025% resazurin (4 mL/L). The inoculum was incubated for 16–18 h at 35 °C before transferring it into the fermentation media.

Fermentation Control

The control experiment was used to compare the results of fermentation of the samples. Glucose was used in the control instead of the filtrate (hydrolysate) from enzymatic hydrolysis. Butanol production from glucose with 45 gL^{-1} was used as a control which is similar to the method used by Nanda et al. 2014. One hundred milliliters of glucose solution was transferred into the Duran bottles and 2.5 mL of 40 gL^{-1} pre-sterilised yeast extract solution was added. A 6 mL volume of inoculum with optical density (OD$_{600}$) approximately 1.5 was added to the fermentation media. The fermentation media was placed in an incubator at 35 °C with Gas Pak to maintain an anaerobic condition for 96 h.

Fermentation of Sample

Butanol fermentation of the hydrolysate was done similarly. During fermentation 2 mL of the sample was removed every 12 h filtered with 0.45 μm filter disc to remove any sediments and analysed for sugars, acids and alcohols.

12.3.2 Analysis

12.3.2.1 Sugar Analysis

Reducing sugars contain a free aldehyde or ketone which act as the reducing agent while non-reducing sugars do not. All monosaccharides are examples of reducing sugars like glucose, fructose and xylose. Other examples of reducing sugars are maltose and lactose. Sucrose is an example of a non-reducing sugar. Reducing sugars are required for fermentation.

DNS method was used for analysis of total reducing sugars for hydrolysis and fermentation products. A glucose standard curve (curve prepared from multiple samples of standard glucose solutions for quantitative research which can be interpolated to identify concentration of unknown samples) was obtained using glucose concentrations of 2–10 mgL^{-1} in distilled water. Different concentrations of standards were transferred to test tubes with a 500 μL volume and 500 μL of DNS solution was added to it. The mixture was boiled for 15 min then cooled and 4 mL of distilled water was added to it. The mixture was homogenised and absorbance was measured at 540 nm using Perkin Elmer UV- Visual spectrophotometer, with UV Express software version 4.0.0. The relation between glucose concentration and absorbance was used to determine the sugar concentration in the sample using the method above (Narkprasom and Wongputtisin 2013).

12.3.2.2   Acid Analysis

The acid analysis was done to identify the presence of acetic and butyric acids. In the process of fermentation by *Clostridium beijerinkii,* the metabolic pathway of Clostridia during fermentation involves two paths; acidogenic and solventogenic. In the acidogenic phase acetic acid and butyric acids are produced, while in the solventogenic phase the acids are used to produce acetone, butanol and ethanol.

The acid analysis was done using a Waters High Performance Liquid Chromatography (HPLC). A Biorad HPX- 87H column equipped with a guard column (300 × 78 mm) was used. Phosphoric acid (0.1%) at a flow rate of 0.8 ml/ min was used as mobile phase. The column temperature was kept at 25 °C. Dual absorbance UV detector was used for detecting the acetic acid and butyric acid signals at 210 nm wavelength. The injection volume was 20 μL. Peaks were identified by comparing the retention times of sample peaks with the pure standard solution peaks. The concentrations were determined using the standard curves of the pure standard solutions.

12.3.2.3   Acetone, Butanol and Ethanol Analysis

Gas chromatography is used to separate volatile liquids that are injected into the instrument which turns into gas. The gaseous compound is carried by the carrier gas (nitrogen used in this study) and the different components are separated in the column. The retention times of standards are used to identify the acetone, butanol and ethanol.

The volatiles (acetone, butanol and ethanol) were determined by using the Perkin Elmer instruments Clarus 500 Gas Chromatograph equipped with a flame ionised detector (FID). An AT-1 15 m × 0.53 mm × 5.0 μm capillary column was used from *Alltech* with serial No. 8543. The oven was set to 60 °C while the injector and detector temperature was 250 °C. The FID temperature was 250 °C. The carrier gas used was nitrogen at a flow rate of 5 mL/min. The flow rate was 150 mL/min with a split ratio of 1:30. Propan-1-ol of concentration 1 gL^{-1} was used as an internal standard.

12.3.2.4   Statistical Analysis

All experiments were done in duplicates and glucose analysis was done in triplicates while solvent analysis were done in duplicates. Results presented in this research are mean values ± standard deviation. One-way Analysis of Variance (ANOVA) with Post Hoc – Tukey's test was done to determine significant difference between the samples with level of significance set at 5%. International Business Machines (IBM) Statistical Package for the Social Sciences (SPSS) Statistics 23 software package was used for statistical analysis of data.

12.4 Results and Discussion

12.4.1 Compositional Analysis of Raw Lignocellulosic Feedstocks

Compositional analysis was performed to quantify the lignocellulosic material, extractives and ash. Lignocellulosic matter mostly consists of cellulose, lignin and hemicellulose. These are the complex carbohydrate structures which breakdown during pretreatment and enzymatic hydrolysis and contribute towards the reducing sugar content. The exhaustive extraction process by water and ethanol removed the extractives. Table 12.2 shows the compositions of SCB and hog fuel.

The yield of total extractives in bagasse and hog fuel were 8.88% and 14.41% respectively. Hog fuel contained higher amount of extractives. Since hog fuel consists of extractive rich bark therefore, the extractives content is higher.

The extractives content of hog fuel is similar to the amount of extractives reported by (Burkhardt et al. 2013) using water extraction only. The extractive content in bagasse was similar to extractive content determined by (Krishnan et al. 2010; Masarin et al. 2011). Extractives content in literature ranged from 1.5% to 9% in bagasse.

The cellulose content was 20% and 4.05% in bagasse and hog fuel respectively. This was lower than the cellulose content of untreated bagasse reported in literature which ranged from 16% to 52%. The hemicellulose content in bagasse was also higher than in hog fuel. This could be due to the fact that fast growing plants like sugarcane have abundant hemicellulose content which assist in conducting and concentrating tissue for mineral solutions (Nanda et al. 2014).

Hog fuel on the other hand, showed a higher lignin content than bagasse. Sugarcane is herbaceous plant, therefore, the lignin content is much lower in bagasse than in hog fuel which is from woody biomass. The lignin content of 21.8% in bagasse is comparable to 22% lignin content reported by (Monteiro & Seleghim Jr, 2014) and 20.88% recorded by (Jonglertjunya et al., 2012).

Table 12.2 Compositional analysis of raw lignocelluloses of sugarcane bagasse and hog fuel (% w/w)

	Sugarcane bagasse	Hog fuel
Extractives	8.88 ± 0.62	14.41 ± 0.02[a]
Cellulose	20.00 ± 2.70	4.05 ± 1.86
Hemicellulose	43.6 ± 2.40	30.5 ± 2.5
Lignin	21.18 ± 0.85	43.04 ± 0.68
Ash	6.35 ± 0.25	8.00 ± 1.33

[a]Values are means ± standard deviations of two separate experiments

12.4.2   Dilute Acid Pretreatment

Dilute acid pretreatment is done using dilute sulphuric acid to breakdown the complex lignocellulosic structures by solubilizing hemicellulose. This step is necessary prior to enzymatic hydrolysis as it leads to high reaction rates and improves cellulose hydrolysis significantly since it removes hemicellulose and exposes the cellulose fibres for enzymatic hydrolysis (Nanda et al. 2014). Table 12.3 shows the reducing sugar concentrations of various samples after pretreatment.

Acid treatment breaks down the hemicellulose structure, therefore, higher reducing sugar content was obtained from sugarcane bagasse (SCB) which correlates with its higher hemicellulose content than hog fuel. SCB was most susceptible to dilute acid pretreatment under the conditions used in this research. When the hemicellulose is removed by acid pretreatment the cellulose becomes more exposed to enzymatic hydrolysis.

12.4.3   Enzymatic Hydrolysis

The neutralised pretreated slurry was further subjected to enzymatic hydrolysis to escalate the reducing sugar content in the slurry. The reducing sugar content of various samples is shown in Table 12.4.

Enzymatic hydrolysis of the samples was done at 50 °C for 72 h. After 72 h the hydrolysate was filtered using a 0.45 µm filtering disc and the filtrate was analysed for reducing sugar content.

Endoglucanase hydrolyse β-(1,4)-glycosidic links in cellulose to release cellobiose which consists of two glucose molecules and β-glucosidase hydrolyse cellobi-

Table 12.3 Reducing sugar content after dilute acid pretreatment

Sample	Reducing sugar content after dilute acid pretreatment (gL^{-1})
Bagasse	11.26 ± 0.85
Hog fuel	8.60 ± 0.14
Control bagasse	0.60 ± 0.00
Control hog fuel	0.57 ± 0.00

Table 12.4 Reducing sugar content after enzymatic hydrolysis

Sample	Reducing sugar content after enzymatic hydrolysis (gL^{-1})
Bagasse	56.52 ± 1.41[a]
Hog fuel	44.40 ± 0.49
Control bagasse	22.13 ± 0.03
Control hog fuel	23.13 ± 0.02

[a]Values are means ± standard deviations of two separate experiments

ose to release two molecules of glucose (Nanda et al. 2014).The reducing sugar content increased greatly after enzymatic hydrolysis of the substrates. The sugar content in bagasse increased from 11.26 to 56.52 gL^{-1} while in hog fuel it increased from 8.60 to 44.40 gL^{-1}. A greater increase in sugar content was observed in bagasse which indicates that it is more susceptible to dilute acid pretreatment and enzymatic hydrolysis. A study done by Monteiro and Seleghim Jr. using 100 gL^{-1} initial concentration dry bagasse gave a yield of 63 gL^{-1} glucose concentration (Monteiro and Seleghim 2014).

Statistical analysis performed using one way ANOVA with Tukey's Test showed that there was significant difference in the concentration of glucose from bagasse and hog fuel since $p < 0.001$.

12.4.4 *Fermentation of Substrates*

The hydrolysate obtained from enzymatic hydrolysis was used for fermentation by *Clostridium beijerinkii*. Glucose solution was also fermented as a control substrate. Figures 12.2, 12.3 and 12.4 show the concentrations of sugars and solvents during the fermentation process while Table 12.5 shows the fermentation results.

Fermentation by Clostridium species produces organic acids (acetic acid and butyric acid) and solvents (acetone, butanol and ethanol). Figures 12.2 and 12.3 and represent the trend in the production of acids and solvents from bagasse and hog fuel respectively during fermentation. The acids and solvent levels were recorded for 96 h.

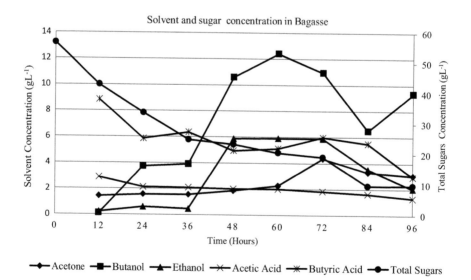

Fig. 12.2 Concentrations of total sugars and solvents during fermentation of bagasse

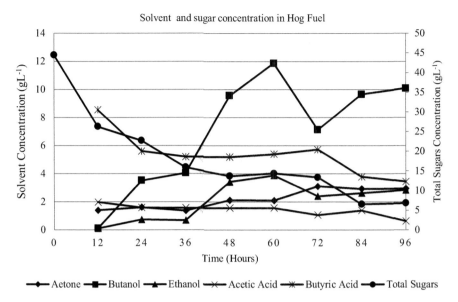

Fig. 12.3 Concentrations of total sugars and solvents during fermentation of Hog Fuel

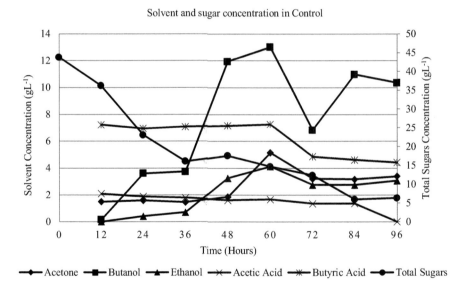

Fig. 12.4 Concentrations of glucose and solvents during fermentation of control

The maximum concentrations of butanol recorded were 12.39 gL^{-1} and 11.86 gL^{-1} for bagasse and hog fuel respectively at 60 h of fermentation. The initial sugar concentration for bagasse and hog fuel were approximately 57 gL^{-1} and 44.4 gL^{-1} respectively. Higher concentration of ethanol (5.90 gL^{-1}) was exhibited in bagasse

Table 12.5 Fermentation results

Parameter	Bagasse	Hog fuel
Total reducing sugar (gL^{-1})	56.52 ± 1.41	44.40 ± 0.49
Acetone (gL^{-1})	4.35 ± 1.20	3.11 ± 0.32
Ethanol (gL^{-1})	5.90 ± 1.30	3.88 ± 1.60
Butanol (gL^{-1})	12.39 ± 1.26	11.86 ± 4.62
Total ABE (gL^{-1})	21.16 ± 3.44	17.83 ± 5.21
Sugar yield (gg^{-1}) substrate	0.57	0.44
Sugar consume rate $(gL^{-1} h^{-1})$	0.59	0.46
ABE productivity $(gL^{-1} h^{-1})$	0.22	0.19
Butanol productivity $(gL^{-1} h^{-1})$	0.13	0.12
ABE yield $(gg^{-1})*$sugar	0.37	0.41
Butanol yield (gg^{-1}) sugar	0.23	0.27
Butanol yield (gg^{-1}) substrate	0.12	0.12

*ABE yields were calculated based on ABE produced (g)/sugar consumed (g)

at 60 h whereas higher concentration of acetone $(4.35 \ gL^{-1})$ was obtained from bagasse at 72 h. The control showed maximum concentration of $13.02 \ gL^{-1}$ butanol at 60 h with initial glucose concentration of $45 \ gL^{-1}$ which is comparable to the results obtained from the two substrates. This result agrees with the study by (Al-Shorgani et al. 2015) which recorded $12.30 \ gL^{-1}$ butanol under anaerobic condition using $50 \ gL^{-1}$ glucose. Similar results of butanol production using *C. beijerinkii* with corn stover also obtained in another study, which produced $13.2 \ gL^{-1}$ butanol at 60 h, giving an ABE productivity of $0.25 \ gL^{-1} \ h^{-1}$ ($0.22 \ gL^{-1} \ h^{-1}$ and $0.19 \ gL^{-1} \ h^{-1}$ for bagasse and hog fuel respectively in this study) and ABE yield of $0.38 \ gg^{-1}$ ($0.37 \ gg^{-1}$ in this study) (Qureshi et al. 2008). Although the fermentation was for 96 h butanol production stopped at 60 h in all samples. Glucose consumption rate was $0.59 \ gL^{-1} \ h^{-1}$ and $0.46 \ gL^{-1} \ h^{-1}$ while ABE yield recorded was $0.37 \ gg^{-1}$ and $0.41 \ gg^{-1}$ for bagasse and hog fuel respectively. Li et al. 2014 investigated cassava with and without pervaporation coupled process and obtained $0.33 \ gg^{-1}$ and $0.36 \ gg^{-1}$ ABE yield respectively in batch fermentation. However, similar result to this study was obtained in continuous fermentation ($0.38 \ gg^{-1}$ ABE yield).

High concentrations of butyric acids were recorded during the lag phase in all the samples and control with the maximum at 12 h. *Clostridium beijerinkii* is one of the *Clostridia* genera that produces butyric acid and acetate as the main product (Huang et al. 2018). Butyric acid concentrations approximately $8 \ gL^{-1}$ were quite high in both fermentation broth which resulted in higher concentration of butanol. Higher butyric acid concentrations enhance the butanol production (Al-Shorgani et al. 2016). In addition, ethanol concentrations were higher than acetone concentrations in all cases.

12.4.5   Availability of Agro-Industrial Feedstock in Fiji

According to the FSC Annual Report 2018, the production of sugarcane fluctuated from 2008 to 2017 as shown in Fig. 12.5. However, in 2017 there was an increase in the production from 1.39 million tonnes from 36,795 ha to 1.63 million tonnes from 38,040 hectares compared to 2016. Fibre content in the cane was 12% and the average bagasse produced over the 5 year period was approximately 398,880 tonnes. The average molasses production for the same period was approximately 68,000 tonnes.

Sugarcane bagasse, molasses and sugarcane leaves and stalks are the main residues of the sugarcane industry. Approximately 140 kg bagasse and 140 kg trash on dry weight basis is produced per ton of sugarcane (Pereira et al. 2015). Trash consists of dry leaves, green leaves and tops. Based on this approximation the total amount of bagasse and trash produced per year was calculated in Table 12.6.

Advantages of using sugarcane bagasse as a substrate for butanol production:

- Does not require a separate harvest
- High carbon content
- Readily available and cheap
- Already ground in the process of sugar production
- Advantages of using molasses as a substrate for butanol production:
- It is one of the cheapest carbon source
- Does not require pretreatment
- Easier to handle since it is in liquid form
- Molasses mash is easy to sterilise

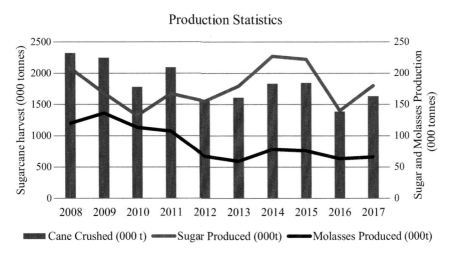

Fig. 12.5 Production statistics of sugar industry based on data from FSC Annual Report, 2018

Table 12.6 Data on sugarcane production

Year	Tonnes of sugar cane produced[a] (millions)	Tonnes of bagasse (dry weight basis)	Tonnes of trash (dry weight basis)	Tonnes of molasses[a] (000)	Area of sugarcane farming[a] (hectares)
2017	1.63	391,200	228,200	66	38,040
2016	1.39	333,600	194,600	63	36,795
2015	1.85	444,000	259,000	76	41,304
2014	1.83	439,200	256,200	78	38,427
2013	1.61	386,400	225,400	59	38,248
Average	1.66	398,880	232,680	68	38,562

[a]Source: FSC Annual Reports 2014–2018

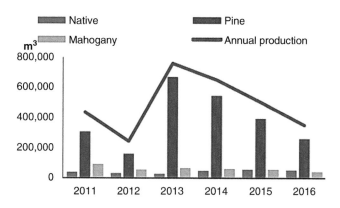

Fig. 12.6 Log production chart for 6 years. (Adapted from: Fiji's Forestry Sector- Developments in recent years Reserve Bank of Fiji 2018)

Sugarcane tops and straws were more susceptible to enzymatic hydrolysis than bagasse thus released 80% and 40% higher glucose respectively (Pereira et al. 2015). It has been also reported that straws were more susceptible to enzymatic hydrolysis than bagasse since it produced 18% higher glucose than bagasse (Moutta et al. 2014). Thus, sugarcane trash that is left behind in the fields is a promising lignocellulosic feedstock for butanol production.

The Tropik Wood Industries Fiji Limited (TWIL) and Fiji Hardwood Corporation Limited (FHCL) are the two key players of Fiji's forestry sector. Fiji's timber production exports are woodchips from pine, mahogany and other products from wood. Annual wood production generally declined from 2013 to 2016 as shown in Fig. 12.6. Pine showed the highest average commodity of 338, 575 m³ from 2011 to 2016, followed by mahogany and native forests as shown in Table 12.7.

Table 12.7 Average timber commodity and approximate coverage

Timber commodity	Average for 6 years – (m³) (2011–2016)	Approximate coverage (ha) as per global canopy programme, 2017
Pine	338,575	76,171
Mahogany	60,163	54,000
Native forests	40,961	526,453

12.5   Estimation of Annual Butanol Production Capacity in Fiji Using Agro-Industrial Residues

12.5.1   Butanol from Bagasse as per Values Obtained from This Study

Based on data from FSC's Annual Report on average 398,880 tonnes of bagasse (dry weight basis) is produced from the total sugarcane production. In this study 12.39 gL^{-1} butanol was produced from a hydrolysate concentration of 100 gL^{-1}. Using this data 59 million litres (equivalent to 47,865 tonnes) of butanol can be produced per year from bagasse. It can be estimated that utilising bagasse alone could provide 1556 TJ of energy per year from butanol production. The amount of energy produced from butanol (calorific value = 32.5 MJ kg^{-1}) is higher than ethanol (calorific value = 26.8 MJ kg^{-1}) due to its higher calorific value.

12.5.2   Butanol from Molasses as per Values from Literature

Based on the following assumptions from data reported by (Artış 2008):

Yield of butanol from backlash molasses per gram substrate reported was 0.367 gg^{-1} with total sugar concentration. In Fiji on average 68,000 tonnes of molasses is produced per annum. Assuming the yield reported by Artış 2008 is true for molasses in Fiji, it can be estimated that 24,956 tonnes of butanol can be prepared annually. This is equivalent to 811 TJ of energy.

12.5.3   Butanol from Sugarcane Trash

Sugarcane trash (leaves and cane tops) is the most unexploited agro-industrial residue (Pereira et al. 2015) which can be successfully utilised for biofuel generation due to the cellulose content in it. According to Pereira et al. 2015 the average cellulose, hemicellulose and lignin content in the four varieties of sugarcane tops studied were 35.2%, 37.2% and 8.4% respectively. Approximately 66% of the trash from the sugarcane fields can be removed without affecting the soil fertility (Franco et al. 2011). In Brazil approximately 50% of the residues were transported to the biorefinery plants per harvest for biogas and *n*-butanol production (Mariano et al. 2013).

In present study, literature values were adopted to estimate the butanol production from the tops and straws. It has been reported that straws had 18% higher sugar content than bagasse (Moutta et al. 2014). The butanol yield (gg^{-1}) in this research from bagasse and hog fuel were similar (0.23 and 0.27 respectively). Assuming similar yield from straws available in Fiji, 66.69 gL^{-1} total sugar can be obtained from enzymatic hydrolysis and considering the butanol yield per gram sugar 16.67 gL^{-1} butanol can be produced.

Franco et al. 2013 reported that on average dry weight basis the trash composition is 54% leaves (straws) and 46% tops. Taking into consideration that 66% of trash is removed so that soil quality is not compromised, the total mass of trash would be approximately 153,569 tonnes, which comprises of 82,927 tonnes of straws and 70,642 tonnes of tops dry weight basis that could be removed per year. The butanol yield per gram substrate (straws) would be 0.15 gg^{-1}, thus, it can be estimated that 12,439 tonnes of butanol can be produced from straws with an energy content 404 TJ.

Menandro et al. 2017 investigated the glucose content in straws and tops and had similar glucose content from each substrate (55.9% and 57% respectively). Pereira et al. 2015 also reported similar contents of glucose in the two substrates (66.9% and 51.8% respectively). Assuming same glucose and butanol yield from sugarcane tops as straws, it can be estimated that 10,596 tonnes of butanol can be produced from tops with equivalent energy content of 344 TJ.

12.5.3.1 Assumptions

1. Straw contains 18% more sugar than bagasse.
2. Butanol yield (gg^{-1}) sugar as 0.23 gg^{-1}
3. Percentage trash removal- 66%
4. Straw and tops composition in trash is 54% and 46% respectively.

12.5.3.2 Sample Calculations

$$\text{Average trash removal} = \frac{66}{100} \times 232,680 \text{ tonnes} = 153,569 \text{ tonnes}$$

$$\text{Straw composition} = \frac{54}{100} \times 153,569 \text{ tonnes} = 82,927 \text{ tonnes}$$

$$\text{Tops composition} = \frac{46}{100} \times 153,569 \text{ tonnes} = 70,642 \text{ tonnes}$$

$$\text{Total Reducing Sugar in bagasse} = 56.52 \times \frac{118}{100} = 66.69 \, gL^{-1}$$

$$\therefore \text{Butanol production from straw} = \frac{\text{Butanol Production}\left(gL^{-1}\right)}{\text{Total Sugar Concentration}\left(gL^{-1}\right)}$$

$$0.23 = \frac{x}{66.69}$$

$$= 15.34 \, gL^{-1}$$

$$\text{Butanol yield per gram straw} = \frac{\text{Butanol Production}\left(gL^{-1}\right)}{\text{Substrate Concentration}\left(gL^{-1}\right)}$$

$$= \frac{15.34\left(gL^{-1}\right)}{100\left(gL^{-1}\right)}$$

$$= 0.15 \, gg^{-1}$$

$$\Rightarrow \text{Mass of butanol per kg straw} = 0.15 \, kg \, per \, kg \, straw$$

$$\text{Estimated mass of butanol per year} = \text{Mass of butanol per kg straw} \times \text{Mass of straw}$$

$$= 0.15 \times 82,927,000$$

$$= 12,439,000 \, kg$$

$$\text{Energy Content} = \text{Calorific value of butanol} \times \text{Mass of butanol}$$

$$= 32.4 \, MJ \, kg^{-1} \times 12,439,000 \, kg$$

$$= 404 \, TJ$$

12.5.4   Butanol from Hog Fuel

The yield of butanol per gram substrate (hog fuel) from this study was 0.12 gg^{-1}. Tropik Woods Industries Fiji limited (TWIL) processes pine. According to a report in the Fiji's Forestry Sector – Developments in recent years (2017), around 338,575 m^3 (equivalent to (161,226 tonnes) annually. Assuming approximately 14% (Fiji Department of Energy 2013) biomass residue is generated during processing. Using the yield from this study it can be estimated that 19,347 tonnes of butanol can be produced using hog fuel equivalent to 629 TJ of energy.

12.5.5   Contribution to Emission Reduction

Despite being a minor contributor of greenhouse gases (GHGs), it is significantly important for Fiji to continue with its mitigation measures of GHGs. The total GHG emission for Fiji according to the World Bank data for 2012 was 2258 kilo tonnes (kt) of CO_2 (https://tradingeconomics.com/fiji/total-greenhouse-gas-emissions-kt-of-co2-equivalent-wb-data.html). The estimated total amount of butanol produced annually from this research would be 115,203 tonnes which is equivalent to 3744 TJ of energy as shown in Table 12.8 based on calorific value of butanol (45.5 MJ/kg). The mass of gasoline replaced by butanol production would be 82,286 tonnes. Using the IPCC emission factor for gasoline as 69.3 tonnes CO_2/TJ for gasoline, butanol production from the substrates considered in this study has the potential to reduce approximately 259 kt of CO_2 per year which is equivalent to 11% of the total CO_2 emitted.

The targets for Fiji's NDC Implementation Roadmap are a total reduction of 627 $ktCO_2$-eq/year by 2030, of which 37 kt CO_2-eq/year is assigned to imported biodiesel (Ministry of Economy 2017). Biobutanol from the substrates considered under this research, thus, has the potential to contribute towards 41% of the national CO_2 reduction target of Fiji's Nationally Determined Contributions (NDC) towards GHG emission reductions through its commitments to the Paris agreement. This far exceeds the 137 kt of CO_2 per year target set for the transport sector.

According to the World Data Atlas the CO_2 emission for Fiji is 1.89 tonnes per capita, thus neat butanol has the potential to reduce CO_2 emission by 0.25 tonnes per capita. CO_2 reduction through biobutanol production from agro-industrial residues will fulfil one of the obligations of the United Nations Framework Convention on Climate Change (UNFCCC), as well as this mitigation measure will enable Fiji towards technology transfer.

In order to use butanol as an alternative fuel in a more practical way for transportation in SI engines, it is important to enhance its properties as far as pollutant emissions are concerned (Elfasakhany and Mahrous 2016). Ternary blends and dual blends of butanol have shown promising results in terms of reduced emissions.

Table 12.8 Summary of estimated butanol production from available residues

Feedstock	Estimated butanol production (Tonnes)	Equivalent energy (TJ)
Bagasse	47,865	1556
Trash	23,035	748
Molasses	24,956	811
Hog fuel	19,347	629
Total	115,203	3744

Elfasakhany and Mahrous 2016 used butanol-methanol-gasoline fuel blends in SI engines. Higher concentrations of methanol and butanol blended with gasoline showed improved engine efficiency and reduced emission concentrations compared to lower concentrations of methanol and butanol blended with gasoline and neat gasoline and single alcohol blended with gasoline. Gu et al. 2012 also conducted an experiment on SI engines with fuel blends of n-butanol with gasoline (Bu10, Bu30, Bu40 and pure butanol). In this study less than 40% n-butanol blends emitted less hydrocarbon and carbon monoxide (CO) at different loads compared to that of gasoline (Gu et al. 2012). However, it was observed that pure n-butanol gave higher HC and CO emission than gasoline.

12.6   Summary and Conclusion

Biobutanol can be produced in Fiji from the available agro-forestry residues which can contribute significantly towards meeting Fiji's emission reduction targets as per its NDC Roadmap. A wide range of agro-forestry residues are available which can be utilised for biobutanol production.

Sugarcane bagasse (SCB) and hog fuel in this study have shown high sugar content which makes them suitable feedstocks for butanol production. Using *Clostridium beijerinkii* for fermentation, approximately 47,865 tonnes and 19,347 tonnes of butanol can be produced from SCB and hog fuel respectively. In addition, based on literature values, 24,956 tonnes of butanol can be produced from molasses while 23,035 tonnes can be produced from sugarcane trash. Therefore, the overall estimated biobutanol production from agro-industrial residues in Fiji would be 115,203 tonnes which is equivalent to 3744 TJ of energy.

Thus, biobutanol production from the substrates considered in this study has the potential to avoid 259 kt of CO_2 emissions per year. This is approximately 41% of the total CO_2 emissions reduction target of Fiji's NDC Implementation Roadmap for GHG emissions reductions as part of the country's commitments to the Paris agreement. This is far in excess of the 137 kt of CO_2 per year target for the transport sector alone.

Finally, it is of interest to note that about 30% Fiji's total merchandise is fuel import (Pacific Regional Data Repository Sustainable Energy for All 2017). Thus, production of biobutanol in Fiji as elaborated by this research indicates the possible local manufacture of a fuel product that can substantially reduce Fiji's total fuel imports. Apart from strengthening the country's economy generally, this new fuel manufacture possibility will reduce market risks for the sugar and timber industries, while at the same time offering social and environmental benefits by converting waste to energy.

References

Abo-State, M. A., Ragab, A. M., EL-Gendy, N. S., Farahat, L. A., & Madian, H. R. (2013). Effect of different pretreatments on egyptian sugar-cane bagasse saccharification and bioethanol production. *Egyptian Journal of Petroleum, 22*(1), 161–167.

Al-Shorgani, N. K. N. (2015). Biobutanol production by a new aerotolerant strain of Clostridium acetobutylicum YM1 under aerobic conditions. *Fuel*, 855–863.

Al-Shorgani, N. K. N., Kalil, M. S., Yusoff, W. M. W., & Hamid, A. A. (2015). Biobutanol production by a new aerotolerant strain of Clostridium acetobutylicum YM1 under aerobic conditions. *Fuel, 158*, 855–863.

Al-Shorgani, N. K. N., Shukor, H., Abdeshahian, P., Kalil, M. S., Yusoff, W. M. W., & Hamid, A. A. (2016). Enhanced butanol production by optimization of medium parameters using Clostridium acetobutylicum YM1. *Saudi Journal of Biological Sciences.*

Artış, Ü. (2008). Enhanced butanol production by mutant strains of Clostridium acetobutylicum in molasses medium. *Türk Biyokimya Dergisi (Turkish Journal of Biochemistry–Turk J Biochem), 33*(1), 25–30.

Adeeyo, O. A., Oresegun, O. M., & Oladimeji, T. E. (2015). Compositional analysis of lignocellulosic materials: Evaluation of an economically viable method suitable for woody and non-woody biomass. *American Journal of Engineering Research (AJER), 4*(4), 14–19.

Bravo, V., Paez, M. P., Aoulad, M., & Reyes, A. (2000). The influence of temperature upon the hydrolysis of cellobiose by β-1,4-glucosidases from Aspergillus Niger. *Enzyme and Microbial Technology, 26*(8), 614–620. https://doi.org/10.1016/S0141-0229(00)00136-8.

Burkhardt, S., Kumar, L., Chandra, R., & Saddler, J. (2013). How effective are traditional methods of compositional analysis in providing an accurate material balance for a range of softwood derived residues? *Biotechnology for Biofuels, 6*(1), 90. https://doi.org/10.1186/1754-6834-6-90.

Chang, W. L. (2010). *Acetone-butanol-ethanol fermentation by engineered Clostridium beijerinckii and Clostridium tyrobutyricum* (Doctoral dissertation). The Ohio State University.

Cheng, C.-L., Che, P.-Y., Chen, B.-Y., Lee, W.-J., Lin, C.-Y., & Chang, J.-S. (2012). Biobutanol production from agricultural waste by an acclimated mixed bacterial microflora. *Applied Energy, 100*, 3–9.

Das, A., Mondal, C., & Roy, S. (2015). Pretreatment Methods of Ligno-Cellulosic Biomass: A Review. *Journal of Engineering Science & Technology Review, 8*(5).

Elfasakhany, A., & Mahrous, A.-F. (2016). Performance and emissions assessment of n-butanol–methanol–gasoline blends as a fuel in spark-ignition engines. *Alexandria Engineering Journal, 55*(3), 3015–3024.

Fiji Department of Energy. (2013). Fiji National Energy Policy 2013–2020. Retrieved from: http://www.fijiroads.org/fiji-national-energy-policy/

Fiji Sugar Corporation Ltd. (2014–2018). FSC Annual Reports. Retrieved from: http://www.fsc.com.fj/annualreport.html

Franco, H., Magalhães, P., Cavalett, O., Cardoso, T., Braunbeck, O., Bonomi, A., & Trivelin, P. (2011). How much trash to removal from sugarcane field to produce bioenergy? *Proceedings Brazilian BioEnergy Science and Technology; Campos do Jordão.*

Franco, H. C. J., Pimenta, M. T. B., Carvalho, J. L. N., Magalhães, P. S. G., Rossell, C. E. V., Braunbeck, O. A.,. & Rossi Neto, J. (2013). Assessment of sugarcane trash for agronomic and energy purposes in Brazil. Scientia Agricola, 70(5), 305–312.

Garcia, V. (2010). Challenges in biobutanol production: How to improve the efficiency. *Renewable and Sustainable Energy Reviews, 15*(2), 964–980.

Green, E. M. (2011). Fermentative production of butanol – the industrial perspective. *Current Opinion in Biotechnology, 22*(3), 337–343. https://doi.org/10.1016/j.copbio.2011.02.004.

Gu, X., Huang, Z., Cai, J., Gong, J., Wu, X., & Lee, C.-f. (2012). Emission characteristics of a spark-ignition engine fuelled with gasoline-n-butanol blends in combination with EGR. *Fuel, 93*, 611–617.

Guan, W., Xu, G., Duan, J., & Shi, S. (2018). Acetone–butanol–ethanol production from fermentation of hot-water-extracted hemicellulose hydrolysate of pulping woods. *Industrial & Engineering Chemistry Research, 57*(2), 775–783.

Huang, J., Tang, W., Zhu, S., & Du, M. (2018). Biosynthesis of butyric acid by Clostridium tyrobutyricum. *Preparative Biochemistry and Biotechnology, 48*(5), 427–434.

IRENA, I. (2018). REN21 (2018). *Renewable Energy Policies in a Time of Transition, 4*, 62–64.

Jonglertjunya, W., Pranrawang, N., Phookongka, N., Sridangtip, T., Sawedrungreang, W., & Krongtaew, C. (2012). Utilization of sugarcane bagasses for lactic acid production by acid hydrolysis and fermentation using Lactobacillus sp. *World Academy of Science, Engineering and Technology, 66*, 173–178.

Jonglertjunya, W., Makkhanon, W., Siwanta, T., & Prayoonyong, P. (2014). Dilute acid hydrolysis of sugarcane bagasse for butanol fermentation. *Chiang Mai Journal of Science, 41*(1), 60–70.

Kalghatgi, G. T. (2001). Fuel anti-knock quality-Part II. Vehicle studies-how relevant is Motor Octane Number (MON) in modern engines? *Shell Global Solutions* (pp. 1–11), Chester, UK, (0148–7191).

Kapoor, M., Raj, T., Vijayaraj, M., Chopra, A., Gupta, R. P., Tuli, D. K., & Kumar, R. (2015). Structural features of dilute acid, steam exploded, and alkali pretreated mustard stalk and their impact on enzymatic hydrolysis. *Carbohydrate Polymers, 124*, 265–273. https://doi.org/10.1016/j.carbpol.2015.02.044.

Khuong, L. D., Kondo, R., De Leon, R., Kim Anh, T., Shimizu, K., & Kamei, I. (2014). Bioethanol production from alkaline-pretreated sugarcane bagasse by consolidated bioprocessing using Phlebia sp. MG-60. International Biodeterioration & Biodegradation, 88, 62–68. doi:https://doi.org/10.1016/j.ibiod.2013.12.008.

Krishnan, C., Sousa, L. d. C., Jin, M., Chang, L., Dale, B. E., & Balan, V. (2010). Alkali-based AFEX pretreatment for the conversion of sugarcane bagasse and cane leaf residues to ethanol. *Biotechnology and Bioengineering, 107*(3), 441–450.

Kumar, M., & Gayen, K. (2011). Developments in biobutanol production: New insights. *Applied Energy, 88*(6), 1999–2012.

Kumar, P., Barrett, D. M., Delwiche, M. J., & Stroeve, P. (2009). Methods for pretreatment of lignocellulosic biomass for efficient hydrolysis and biofuel production. *Industrial & Engineering Chemistry Research, 48*(8), 3713–3729.

Kumar, R., Satlewal, A., Sharma, S., Kagdiyal, V., Gupta, R. P., Tuli, D. K., & Malhotra, R. K. (2014). Investigating Jatropha prunings as a feedstock for producing fermentable sugars and chemical treatment for process optimization. *Journal of Renewable and Sustainable Energy, 6*(3), 1–11.

Lapuerta, M., Ballesteros, R., & Barba, J. (2017). Strategies to introduce n-butanol in gasoline blends. *Sustainability, 9*(4), 589.

Li, J., Chen, X., Qi, B., Luo, J., Zhang, Y., Su, Y., & Wan, Y. (2014). Efficient production of acetone–butanol–ethanol (ABE) from cassava by a fermentation–pervaporation coupled process. *Bioresource Technology, 169*, 251–257.

Margaritis, A., & Creese, E. (1981). Thermal stability characteristics of cellulase enzymes produced by Sporotrichum thermophile. *Biotechnology Letters, 3*(9), 471–476. https://doi.org/10.1007/BF00147556.

Mariano, A. P., Dias, M. O., Junqueira, T. L., Cunha, M. P., Bonomi, A., & Maciel Filho, R. (2013). Utilization of pentoses from sugarcane biomass: Techno-economics of biogas vs. butanol production. *Bioresource Technology, 142*, 390–399.

Mařík, J., Pexa, M., Kotek, M., & Hönig, V. (2014). Comparison of the effect of gasoline-ethanol E85-butanol on the performance and emission characteristics of the engine Saab 9-5 2.3 l turbo. *Agronomy Research, 12*(2), 359–366.

Masarin, F., Gurpilhares, D. B., Baffa, D. C., Barbosa, M. H., Carvalho, W., Ferraz, A., & Milagres, A. M. (2011). Chemical composition and enzymatic digestibility of sugarcane clones selected for varied lignin content. *Biotechnology for Biofuels, 4*(1), 55.

Menandro, L. M. S., Cantarella, H., Franco, H. C. J., Kölln, O. T., Pimenta, M. T. B., Sanches, G. M., et al. (2017). Comprehensive assessment of sugarcane straw: Implications for biomass and bioenergy production. *Biofuels, Bioproducts and Biorefining, 11*(3), 488–504.

Ministry of Economy. (2017). *Fiji NDC Implementation Roadmap 2017–2030- Setting a pathway for emissions reduction target under the Paris Agreement*. Global Green Growth Institute. Republic of Fiji Islands.

Monteiro, P., & Seleghim Jr, P. (2014). Enzymatic hydrolysis of sugarcane bagasse in rotating drum reactor. *Florianopolis/SC*.

Mosier, N., Wyman, C., Dale, B., Elander, R., Lee, Y., Holtzapple, M., & Ladisch, M. (2005). Features of promising technologies for pretreatment of lignocellulosic biomass. *Bioresource Technology, 96*(6), 673–686.

Moutta, R. D. O., Ferreira-Leitão, V. S., & Bon, E. P. D. S. (2014). Enzymatic hydrolysis of sugarcane bagasse and straw mixtures pretreated with diluted acid. *Biocatalysis and Biotransformation, 32*(1), 93–100.

Nakata, K., Uchida, D., Ota, A., Utsumi, S., & Kawatake, K. (2007). *The Impact of RON on SI Engine Thermal Efficiency*. https://doi.org/10.4271/2007-01-2007

Nanda, S., Dalai, A. K., & Kozinski, J. A. (2014). Butanol and ethanol production from lignocellulosic feedstock: Biomass pretreatment and bioconversion. *Energy Science & Engineering, 2*(3), 138–148.

Narkprasom, N., & Wongputtisin, R. A. P. (2013). Optimization of reducing sugar production from acid hydrolysis of sugarcane bagasse by Box Behnken Design. *Journal of Medical and Bioengineering, 2*(4).

Pacific Regional Data Repository Sustainable Energy for All. (2017). *Fiji's NDC energy sector implementation roadmap (2017–2030)*. Retrieved from: http://prdrse4all.spc.int/sites/default/files/1._fijis_energy_sector_outlook_ndc_roadmap.pdf.

Pereira, S. C., Maehara, L., Machado, C. M. M., & Farinas, C. S. (2015). 2G ethanol from the whole sugarcane lignocellulosic biomass. *Biotechnology for Biofuels, 8*(1), 44.

Qureshi, N., & Ezeji, T. C. (2008). Butanol, 'a superior biofuel' production from agricultural residues (renewable biomass): Recent progress in technology. *Biofuels, Bioproducts and Biorefining: Innovation for a sustainable economy, 2*(4), 319–330.

Qureshi, N., Ezeji, T. C., Ebener, J., Dien, B. S., Cotta, M. A., & Blaschek, H. P. (2008). Butanol production by Clostridium beijerinckii. Part I: Use of acid and enzyme hydrolyzed corn fiber. *Bioresource Technology, 99*(13), 5915–5922.

Reserve Bank of Fiji. (2018). *Fiji's forestry sector- developments in recent years*. Fiji. Retrieved from: https://www.rbf.gov.fj/getattachment/c6c9ef57-9eaf-428d-9f65-8af3bed6cd1c/Fiji-s-Forestry-Sector-Developments-23-September.

Robert, N. (2014). Techno-economics of carbon preserving butanol production using a combined fermentative and catalytic approach. *Bioresource Technology, 161*, 263–269.

Singh, A. (2012). Biofuels and Fiji's roadmap to energy self-sufficiency. *Biofuels, 3*(3), 269–284.

Sluiter, A., Ruiz, R., Scarlata, C., Sluiter, J., & Templeton, D. (2008). Determination of extractives in biomass. *Laboratory analytical procedure (LAP)*, TP-510-42619.

Steffien, D., Aubel, I., & Bertau, M. (2014). Enzymatic hydrolysis of pre-treated lignocellulose with Penicillium verruculosum cellulases. *Journal of Molecular Catalysis B: Enzymatic, 103*, 29–35. https://doi.org/10.1016/j.molcatb.2013.11.004.

Sun, Y., & Cheng, J. (2002). Hydrolysis of lignocellulosic materials for ethanol production: A review. *Bioresource Technology, 83*(1), 1–11.

Takacs, E., Wojnarovits, L., Földváry, C., Hargittai, P., Borsa, J., & Sajo, I. (2000). Effect of combined gamma-irradiation and alkali treatment on cotton–cellulose. *Radiation Physics and Chemistry, 57*(3–6), 399–403.

Thulluri, C., Goluguri, B. R., Konakalla, R., Reddy Shetty, P., & Addepally, U. (2013). The effect of assorted pretreatments on cellulose of selected vegetable waste and enzymatic hydrolysis. *Biomass and Bioenergy, 49*, 205–213. https://doi.org/10.1016/j.biombioe.2012.12.022.

Williams, C. L., Emerson, R. M., & Tumuluru, J. S. (2017). Biomass compositional analysis for conversion to renewable fuels and chemicals. In *Biomass volume estimation and valorization for energy* (pp. 251–270). Intech Open Science Open Mind. https://doi.org/10.5772/65777.

Yu, Q., Zhuang, X., Lv, S., He, M., Zhang, Y., Yuan, Z.,... Tan, X. (2013). Liquid hot water pretreatment of sugarcane bagasse and its comparison with chemical pretreatment methods for the sugar recovery and structural changes. Bioresource Technology, 129, 592–598.

Zheng, J., Tashiro, Y., Wang, Q., & Sonomoto, K. (2015). Recent advances to improve fermentative butanol production: Genetic engineering and fermentation technology. *Journal of Bioscience and Bioengineering, 119*(1), 1–9.

Part III
Outcomes

Chapter 13
Summary of Outcomes and Implications for the Fiji NDC Implementation Roadmap

Anirudh Singh

Abstract The main objective of this study was to investigate the range of possible RE projects that could contribute significantly to Fiji's NDC Implementation Roadmap and to assess the extent of emissions savings they could provide. This final chapter summarises the results obtained in Part II of this work, and interprets them in the light of the aims of the Roadmap. The chapter ends by making specific recommendations that were informed by these observations and interpretations.

Keywords Fiji NDC implementation roadmap · Renewable energy · Avoided emissions · Climate change mitigation · Biofuels · CNO · WTE power generation

13.1 Introduction

This work set out to demonstrate how renewable energy (RE) can play a role in the successful implementation of the Paris Agreement by providing one of the two main tools for mitigating climate change.

The full range of RE options available for consideration include hydro, solar, wind, biomass and geothermal and ocean energy to lesser extent. The order of importance of these resources depends on both their efficacy as energy sources as well as their availability in a specific country or regional setting. How these factors play a role in the present consideration is dealt with in detail in Chap. 2.

To be an effective contributor to Fiji's Nationally Determined Contribution (NDC) Implementation Roadmap, the RE project considered must be new. This is simply because the emissions reductions of established RE projects (via displacement of fossil fuel-based energy) have already been accounted for, and therefore add nothing to the additional reductions sought under the NDC Roadmap.

A. Singh (✉)
School of Science and Technology, The University of Fiji, Lautoka, Fiji
e-mail: anirudhs@unifiji.ac.fj

© Springer Nature Switzerland AG 2020

A. Singh (ed.), *Translating the Paris Agreement into Action in the Pacific*,
Advances in Global Change Research 68,
https://doi.org/10.1007/978-3-030-30211-5_13

The RE projects undertaken for this study were those deemed to be the most efficacious for bringing about reductions in GHG emissions for Fiji. They ranged from solar, wind and hydro to biomass and biofuels. An important chapter was devoted to the investigation of new wind energy potential, an option that was not considered in the Roadmap. Several biofuel options were considered, ranging from the first generation feedstock provided by coconut oil (CNO) to three examples of second generation biodiesel.

The results are revealing, and should have a significant bearing on both the future of renewable energy utilization as well as the emissions reduction ambitions of the country. In this final part of this work, we appraise the findings of this exercise and enumerate the main outcomes. We begin by highlighting the main results of Part II of this book.

13.2   Summary of Outcomes

13.2.1   Hydropower

Hydro is arguably the most important RE resource due to its dispatchable nature, which makes it the ideal energy source to provide the baseload requirements for grid power. Fiji's current installed hydro capacity of some 160 MW provides more than half of the country's grid electricity requirements at any one time.

Resource assessment using modern techniques of GIS and hydrological modelling elaborated in Chap. 5 reveal that the potential for new hydro is far greater than that identified earlier by the Fiji Department of Energy (FDoE) and Energy Fiji Limited (EFL – Fiji's main electricity utility) using traditional techniques. At least 40% of the land area of the main island of Viti Levu is suitable for hydro development. Many of these regions have so far not been investigated.

13.2.2   Solid Biomass and WTE for Power Generation

Fiji's solid biomass resources consists of fuelwood, hog-fuel and sugarcane bagasse as well as some 100,000 tonnes of municipal solid waste (MSW) per annum which is generated mostly on the main island of Viti Levu. Bagasse and hog-fuel are currently used for power generation by Fiji's two IPPs (Fiji Sugar Corporation (FSC) and Tropik Woods) while the newly-established Nabou Green Energy Ltd. (NGEL) is currently un-operational.

EFL has considered plans to establish a Waste-to-Energy (WTE) project at Fiji's largest landfill site at Naboro (near the capital city Suva). While this project may be feasible at the chosen site for the purpose of power generation, investigations reported in Chap. 6 of this study of the power generation potential of the much smaller Vunato dump in Lautoka reveals that WTE may not be an important player in emissions savings as postulated in the NDC Roadmap.

The study reveals that 20,000 tonnes of MSW can typically generate 258 kW of power and 0.55 kT of avoided CO_2 eq per year from biogas produced from the MSW, and 825 kW of power and 2.12kT of avoided CO_2eq per year from the incineration of the MSW. The avoided emissions amount to only 0.25% of the NDC target in the case of biogas-generated power and 1.0% for the case of incineration. Making the reasonable assumption of a national MSW production rate of 100,000 tonnes per year, these figures indicate that the maximum avoided emission via MSW power generation in Fiji are 1.3% and 5.0% of the NDC target for biogas and incineration-generated power respectively. These assessments indicate clearly that Fiji falls very short of generating enough MSW to meet the contribution to its NDC target for avoided emissions through MSW power generation.

13.2.3   Viability of Onshore Wind Farms

Wind energy has been a noticeable omission in the Fiji NDC Roadmap for consideration as a viable emissions reduction option. This is probably because of the dismal performance of the Butoni wind farm in Sigatoka. The rated capacity factor of this onshore wind farm is 12%, but this has seldom, if ever, been realised over its entire history of existence.

Contrary to popular belief however, there are sites on the island of Viti Levu where economically viable onshore wind farms can be established. An assessment (reported in Chap. 7) based on Fiji Meteorological Office wind resource monitoring shows that Rakiraki in Northern Viti Levu can produce 22,603 MWh/year at a cost of $0.053/kWh and a total investment of only $20.38 million from a wind farm with the same rated capacity (10 MW) as the unproductive wind farm at Butoni in Southern Viti Levu. Such a wind farm will produce an emissions savings of 15.37kT CO_2eq per annum. The viability of such a project is further enhanced by a relatively large capacity factor of 0.267 and a payback period of only 3.4 years.

13.2.4   Solar Energy for Power Generation

With an average insolation somewhere in the vicinity of 5.4 sun-hours/day, Fiji has a good potential for solar PV generation. However, this power generation opportunity remained virtually un-utilized till about 2010. There has, however, been a surge of interest lately, bringing up the total roof-top solar to the current 3.6 MW installed capacity.

Chapter 8 uses two separate methods (including the much-acclaimed LEAP tool) to assess the country's actual solar PV potential. It is estimated that a total of 100 MW of newly-installed solar PV is possible. This will bring about an emissions savings of 151 kT CO_2 eq/year. This is 210% of Fiji's NDC target of 72 kT CO_2 eq/year.

13.2.5 Coconut Oil as a Source of Transportation Fuel

Fiji is capable of producing enough coconut oil (CNO)-based biodiesel to provide the entire diesel transportation needs of the country. However, its actual copra production has been declining relentlessly from a high of ~ 33,000 tonnes in 1977. The government has set up a coconut oil industry task force to reverse this decline. Based on a scenario that the government can increase the copra yield to 3.25 tonnes/ha on the land area of 15,000 ha occupied by the current plantations, Chap. 9 reveals that it is possible to replace the entire 450 million litres of estimated diesel fuel import by B5. The estimated emissions savings is then evaluated using the LCA procedure to amount to 47.5 kT CO_2 eq/year. This is comparable to the NDC targeted savings of 37 kT CO_2 eq/year by the use of imported B5 fuel for diesel transportation.

13.2.6 Second Generation Feedstocks for Biodiesel Production

The prospects of producing second generation biofuels for transportation have also been considered in this study. Chapter 11 considers the possibility of the non-edible Jatropha-based biodiesel production on available marginal land in Viti Levu and finds it to be economically non-viable. However as Chap. 10 shows, the production of the inedible pongamia-based biofuel in Vanua Levu will produce significant emissions savings and should be considered seriously.

The Life Cycle Impact Assessment (LCIA) of the Global Warming Potential (GWP) of pongamia biodiesel produced on the 59,000 ha of available marginal land in Vanua Levu shows emissions of 18.3 g CO_2 eq/MJ. This is much smaller than the 98 g CO_2 eq/MJ for petroleum diesel.

An existing pongamia farm of 154 ha in Vanua Levu can produce 1,223,145 l of pongamia biodiesel (PME). When used in diesel fuel blend B5, this will yield an emissions savings of 3204 tonnes of CO_2 eq/year. For future reference, it is also worthy to note that some 1800 ha of pongamia plantation in Vanua Levu is sufficient to satisfy the entire B5 target requirements of 37 kT/year stipulated in Fiji's NDC Roadmap.

13.2.7 Bio-Butanol as a Fuel Additive for Petrol Engine Fuels

Bio-butanol provides an alternative fuel for petrol (i.e. Spark-Ignition or SI) engines and has fuel properties that are better than those of bio-ethanol. It can be produced from (second generation) lignocellulosic feedstock such as forestry, agricultural and industry residues in addition to first generation feedstocks such as molasses.

The total potential from all such feedstock for Fiji is assessed to be 115,203 tonnes of bio-butanol per annum. This will provide a total of 3744 TJ of energy for the transportation sector, producing a possible emissions savings of 259.5 kT CO_2 eq/year. This value far exceeds the 37 kT/year targeted emissions savings from imported B5 in the NDC Roadmap.

Note however that B5 is a fuel substitute for diesel engines. No mention has been made in the NDC Roadmap to emissions savings through petrol engine fuel substitution, a fuel category to which bio-butanol as well as bio-ethanol belongs.

The above estimate for butanol production includes the use of sugarcane trash and molasses in addition to bagasse and hog fuel. It must be noted that

- There will be competition from the power generation sector for bagasse and hog fuel supply
- Molasses can also be used for ethanol production.

A more realistic assessment of the actual potential for bio-butanol production is only obtained after considering these factors. However, it is informative to note that Fiji can produce a total of 67,212 tonnes of bio-butanol from all the bagasse and hog fuel produced in the country per annum. This will provide 2185 TJ of energy for the transportation sector, yielding an emissions savings of 151.42 kT CO_2 eq/year. This is 71% of the NDC Roadmap target of 212 kT/year from sustainable biomass plantation and Waste-to-Energy (WTE).

13.3   A Critique of the Present Energy Strategies and Implications on the Roadmap

The last section summarised the results of specific RE projects which could contribute meaningfully to emissions reductions. Before coming to a final conclusion about the efficacy of such emissions reductions projects, several observations about the present energy strategies and their likely impacts on the NCD Roadmap ought to be noted.

(a) The importance of resource assessment as a pre-requisite to the development of any RE project cannot be ignored. This is particularly so in the case of biomass resources for thermal power generation. This requirement must be addressed before any further attempts are made to develop new thermal power plants in Fiji.

   The case of the faltering Nabou Green Energy Limited (NGEL) Power Plant in Western Viti Levu provides a pertinent example of the importance of such assessments. While the circumstances surrounding its current closure are complex, the basic problem stems from the failure to ascertain a viable supply of biomass feedstock. It was evident from the outset that this power producer did not have the benefit of any thorough quantitative assessment of the required resources. The situation was compounded by the failure to take into account the possibility of competition by other independent power producers (IPPs).

The situation points to the need for a nation-wide assessment of the available biomass resources in Fiji for power generation purposes. Because of the often geographically-scattered nature of this resource, it also becomes imperative to carry out a full Life Cycle Impact Assessment of the Global Warming Potential of this resource before it is adopted as a legitimate contributor to any emissions reduction strategy.

(b) Indications are that waste-to-Energy (WTE) is not a viable contender as an RE project for Fiji's NDC Roadmap. As inferred from the Vunato rubbish dump study, Fiji's total production of 100,000 tonnes of MSW per annum can contribute at most 5% of the targeted savings of 212 kT CO_2/year. As sustainable biomass plantation was also a component of this savings mechanism, the larger share of the burden must be borne by this latter resource.

(c) Solar energy, in the form of Solar PV Power, offers much more opportunity for power production and consequent emissions reductions than previously envisaged. As Chap. 8 showed, Fiji is in a position to produce an additional 100 MW of Solar PV power by 2030. This would produce twice as much of the 72 kT CO_2/year savings targeted by the NDC Roadmap. This is a clear revelation of the untapped opportunity that lies unexplored.

(d) Wind power is a viable option for Fiji. The case study of Rakiraki proves that economically viable wind farm sites do exist in Fiji, and more specifically that high capacity factor on-shore wind farms are possible on Viti Levu.

(e) Fiji's NDC Roadmap has considered diesel engine fuel substitutes for emissions reductions, but totally ignored substitute fuels for spark-ignition engines. The country is producing copious amounts of molasses that can be used to produce first generation bio-ethanol. Moreover, there is also significant potential for second-generation bio-butanol in Fiji. The latter can produce an emissions savings of 151.42 kT CO_2 eq/year. This is 71% of the NDC Roadmap target set for sustainable biomass and WTE.

(f) The government's plan to revive the ailing coconut oil industry may be justified for producing the oil as a traditional source of food and other uses. However, pongamia oil produced in Vanua Levu provides a more viable feedstock for biodiesel production. It has been demonstrated that pongamia oil produced from just 1800 ha of marginal land in Vanua Levu is sufficient to produce the entire quantity of biodiesel needed to meet Fiji's corresponding NDC target.

13.4 Recommendations

(a) Pongamia is a better alternative than coconut oil (CNO) as a feedstock for biodiesel production. The determining requirement is that of economic viability. Unpublished results of the Vanua Levu pongamia study show that the production of pongamia biodiesel (PME) in Vanua Levu is economically viable. Therefore, Fiji should consider this alternative starting material for the indigenous production of biodiesel.

(b) Fuel alternatives for petrol (i.e. spark-ignition) engines should be an integral part of Fiji's NDC Roadmap. The first consideration should be given to first generation ethanol production from molasses, a by-product of the sugar industry. The possibility of producing bio-butanol, an advanced biofuel that can be produced from the abundantly available bagasse and hog fuel feedstocks, should be considered.

(c) Wind energy is a viable option for Fiji. Greater efforts should be devoted to wind energy monitoring projects to identify the locations around the country where wind farms with high capacity factors can be established and that ensure economic viability.

(d) Fiji should carry out a quantitative assessment of the biomass resources of the country, and conduct full Life Cycle Analyses of net avoided emissions before implementing projects to ensure that they bring about actual savings in emissions.

9783030302139